Advances in Soft Computing

T0137902

Springer

London
Berlin
Heidelberg
New York
Hong Kong
Milan
Paris
Tokyo

Jose Manuel Benítez, Oscar Cordón,
Frank Hoffmann, Rajkumar Roy (Eds.)

Advances in
Soft Computing

Engineering Design and Manufacturing

With 125 Figures

Springer

Jose Manuel Benítez
Oscar Cordón

Department of Computer Science and Artificial Intelligence,
University of Granada, C/Daniel Saucedo Aranda, s/n, E-18071, Granada, Spain

Frank Hoffmann
Royal Institute of Technology, Dalagatan 64, S-11324, Stockholm, Sweden

Rajkumar Roy
Department of Enterprise Integration, School of Industrial and Manufacturing
Science, Cranfield University, Cranfield, Bedford MK43 0AL, UK

British Library Cataloguing in Publication Data
Online World Conference on Soft Computing in Industrial
Applications (7th)
    Advances in soft computing : engineering design and
    manufacturing
    1.Soft computing – Congresses 2.Application software –
    Development – Congresses 3.Engineering design – Data
    processing – Congresses 4.Intelligent control systems –
    Congresses
    I.Title II.Benitez, Jose Manuel
    006.3
    ISBN 978-1-84996-905-5

Library of Congress Cataloging-in-Publication Data
A catalog record for this book is available from the Library of Congress

Apart from any fair dealing for the purposes of research or private study, or criticism or review,
as permitted under the Copyright, Designs and Patents Act 1988, this publication may only be
reproduced, stored or transmitted, in any form or by any means, with the prior permission in
writing of the publishers, or in the case of reprographic reproduction in accordance with the
terms of licences issued by the Copyright Licensing Agency. Enquiries concerning reproduction
outside those terms should be sent to the publishers.

Springer-Verlag London Berlin Heidelberg
a member of BertelsmannSpringer Science+Business Media GmbH
http://www.springer.co.uk

© Springer-Verlag London Limited 2010

The use of registered names, trademarks, etc. in this publication does not imply, even in the
absence of a specific statement, that such names are exempt from the relevant laws and
regulations and therefore free for general use.

The publisher makes no representation, express or implied, with regard to the accuracy of the
information contained in this book and cannot accept any legal responsibility or liability for any
errors or omissions that may be made.

Preface

Soft Computing is a computing paradigm representing a group of methodologies, which aim to exploit tolerance for imprecision, uncertainty, and partial truth to achieve tractability, robustness, and low solution cost. The group includes fuzzy computing, neuro-computing, evolutionary computing, probabilistic computing and computing based on chaos theory. In recent years, Soft Computing has become popular to solve real life complex problems. Many industries are using Soft Computing as a regular business or operational tool within the organization. The computing paradigm is now utilized even in consumer products, such as washing machines and camcorders.

World Conference on Soft Computing (WSC) series of conferences, organized by the World Federation on Soft Computing (WFSC), promotes application of the paradigm to solve complex industrial problems. WSC7 focuses on industrial applications of Soft Computing. We are very proud to present you this book from the WSC7 conference papers. The area of application of soft computing is very wide. The book presents advances in fuzzy, neural, and evolutionary computing techniques and its application in real life problems. The book deals with a wide range of applications, such as intelligent control, clustering, fault diagnosis and face recognition to weather analysis. The book also presents new frontiers for soft computing.

This truly international conference includes 36 papers from 18 countries and 4 continents (America, Europe, Asia and Australia). The conference is sponsored by IEEE Systems, Man and Cybernetics Society, IEEE Industrial Electronics Society (IES), EUSFLAT and University of Granada. WSC7 is hosted on the Cyberspace by University of Granada. This book is divided into four main sections: neural networks and neuro-fuzzy systems, fuzzy systems, evolutionary computing and miscellaneous applications. Each section presents latest applications of a soft computing technique and the challenges for the application. The book also presents future directions in Applied Soft Computing.

Cyberspace,
18.03.2003

Jose Manuel Benítez
Oscar Cordón
Frank Hoffmann
Rajkumar Roy

Organization

The 7th Online World Conference on Soft Computing in Industrial Applications was organized by the World Federation of Soft Computing, Department of Computer Science and Artificial Intelligence, University of Granada.

Organizing Commitee

General Chair:	Jose M. Benítez, University of Granada, Spain
	Oscar Cordón , University of Granada, Spain
International Co-Chair:	Takeshi Furuhashi, Nagoya University, Japan
	Mario Köppen, Fraunhofer IPK Berlin, Germany
	Rudolf Kruse, University of Magdeburg, Germany
	Rajkumar Roy, Cranfield University, United Kingdom
Scientific Chair:	Juan L. Castro, University of Granada, Spain
Technical Program Chair:	Francisco Herrera, University of Granada, Spain
Web Chair:	Luis Castillo, University of Granada, Spain
	J.M. Fernandez, University of Granada, Spain
Publicity and Online Tutorial Chair:	Luis Martinez, University of Jaen, Spain
Publication Chair:	Frank Hoffmann, Royal Institute of Technology, Stockholm, Sweden
Computer Chair:	Miguel Garcia-Silvente, University of Granada, Spain
Proceedings Chair:	Jorge Casillas, University of Granada, Spain
Online Chair:	Jose Raimundo, University of Campinas, Brazil
Best Paper Award:	Larry Hall, University of South Florida, USA
Sponsorship Chairman:	Seppo J. Ovaska, Helsinki University of Technology, Finland

International Scientific Board

Janos Abonyi, Hungary
Ajith Abraham, Australia
R. Babuska, The Netherlands
Senen Barro, Spain
Juan C. Cubero, Spain
Katrin Franke, Germany
Fernando Gomide, Brazil
Antonio Gonzalez, Spain
Enrique Herrera, Spain
Hisao Ishibuchi, Japan
Maria Jose del Jesus, Spain
Frank Klawonn, Germany
Bart Kosko, USA
L. Kuncheva, UK

Luis Magdalena, Spain
Cristophe Marsala, France
L. Martinez, Spain
J.J. Merelo, Spain
Nikhil R. Pal, India
S. Pal, India
Javier Ruiz-del-Solar, Chile
A. Saffioti, Sweden
Luciano Sanchez, Spain
Vicenc Torra, Spain
G. de Tre, Belgium
Igor Zwir, Argentina

Sponsoring Institutions

IEEE Systems, Man, and Cybernetics Society
IEEE Industrial Electronics Society (IES)
European Society for Fuzzy Logic and Technology (EUSFLAT)
University of Granada

Contents

Fault Diagnosis of Air-conditioning System Using CMAC Neural Network Approach

Chin-Pao Hung and Mang-Hui Wang

Department of Electrical Engineering
National Chin-Yi Institute of Technology
Taichung, Taiwan 411, R.O.C.

Summary. In this paper, a CMAC neural network application on fault diagnosis of a large air-conditioning system is proposed. This novel fault diagnosis system contains an input layer, binary coding layer, and fired up memory addresses coding unit. Firstly, we construct the configuration of diagnosis system depending on the fault patterns. Secondly, the known fault patterns were used to train the neural network. Finally, the diagnosis system can be used to diagnose the fault type of air-conditioning system. By using the characteristic of self-learning, association and generalization, like the cerebellum of human being, the proposed CMAC neural network fault diagnosis scheme enables a powerful, straightforward, and efficient fault diagnosis. Furthermore, the following merits are obtained: 1) High accuracy. 2) High noise rejection ability. 3) Suit to multi-faults detection. 4) Memory size is reduced by new excited addresses coding technique.

Index Terms: fault diagnosis, air-conditioning system, neural network, CMAC.

1 Introduction

The fault diagnosis of a large air-conditioning system, however, is not an easy task. Traditional fault diagnosis schemes employed a look-up table, which recorded the mapping relation of the fault reasons and symptoms and supplied by equipment manufacturer[1]. A large system has a large amount of symptom detection points. The number of symptom code combinations is larger than the number of fault reasons (types), and "no match" may be indicated in the fault diagnosis. When the "no match" conditions exist, the fault diagnosis will become time-consuming and depend strong on the maintenance man's expertise. Long time failure of air-conditioning system may cause a break in suitable air supply and loss of profits. Therefore, it is of great importance to detect incipient failures in air-conditioning system as soon as possible, so that the people can maintain them quickly and improve the reliability of air-conditioning system.

In the past decade, various intelligent fault diagnosis techniques have been proposed. For example, the fault diagnosis of power transformer using the expert systems [2], neural network (NN) [3, 4]and fuzzy logic approaches [5, 6, 7]. Our recent work also developed a CMAC-based fault diagnosis scheme [8]. However, these researches consider fewer detection points and the fault symptoms are analog signals. When the fault signals are large amount binary value, the method described above can not diagnose them directly.

In this paper, a novel digital CMAC (Cerebellar Model Articulation controller) neural network methodology is presented for the fault diagnosis of air-conditioning system, especially for system with large amount symptom detection points. By using the fault type look-up table as the training data, the characteristic of association and generalization make the CMAC fault diagnosis scheme enables a powerful, straightforward, and efficient fault diagnosis. With application of this scheme to the training data and simulated data that adding noise or uncertainty, the diagnoses demonstrate the new scheme with high accuracy and high noise rejection abilities. Moreover, the results also proved the ability of multiple incipient faults detection.

The remainder of this paper is organized as follows. Section 2 is a brief description about the digital CMAC neural network. Section 3 is the configuration of proposed diagnosis system. It contains the training rule, learning performance evaluation, and the diagnosis algorithm. Section 4 shows the diagnosis results of training patterns and simulated data. Finally, the conclusion is stated in section 5.

2 Brief description of digital CMAC neural network

Figure 1 schematically depicts the digital CMAC networks [9,11], which like the models of the human memory to perform a reflexive processing. The CMAC, in a table look-up fashion, produced a vector output in response to a state vector input. The mapping processes include the segmentation, excited addresses coding, and summation of the excited memory weights. The characteristic of the mapping processes is that similar inputs activate similar memory addresses; restated, if the input states are close in input space, then their corresponding sets of association cells overlap. For example, if x_1 and x_2 are similar (close), then x_1 activates the memory addresses a_1, a_2, a_3, a_4, and x_2 should activate the memory addresses a_2, a_3, a_4, a_5 or a_3, a_4, a_5, a_6. The inputs are said to be highly similar if two inputs activate the same memory addresses. Lower similarity outputs would activate fewer same memory addresses. Therefore, we can use the known training patterns to train the CMAC network. The CMAC will distribute the specific fault type feature on fired memory addresses. When the input signal is same as the training data, it will activate the same memory addresses and output exactly fault type. Non-training data input the CMAC will activate different memory addresses depending on the similarity degree with the training data. Therefore, the summation of fired memory weights diverges from the original output of training data fired. Assuming the output is trained to equal one to denote a specific fault type, that is the node output one confirms what fault type is. Then input the diagnosed data to the diagnosis system, the node output can express the possibility of the specific fault type.

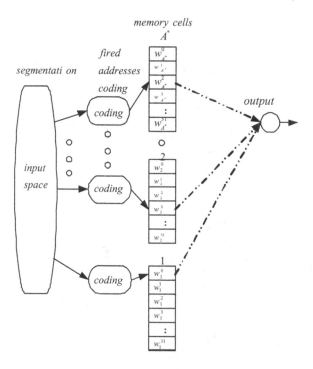

Fig. 1. Functional schematic of digital CMAC

3 The configuration of proposed fault diagnosis system

Extending from the Fig. 1 and using Table 1 as training patterns, Figure 2 is the proposed fault diagnosis configuration of air-conditioning system. We obtained Table 1 from Industrial Technology Research Institute-Energy & Resources Laboratories (in Taiwan). Some books refer to similar fault table also, such as P.C. Koelet's book[12]. Table 1 shows the air-conditioning system has 44 fault types and 40 symptom detection points. Therefore we construct the network with 40 symptom detection points, 44 parallel memory layers, and 44 output nodes. The operation of the proposed diagnosis system illustrated as follows.

3.1 Segmentation and fired addresses coding

As shown in Table 1, the binary code of first fault type is

$$0010100100110011000100000000000000000000$$

Take five digits as a segment (group), then from LSB to MSB the excited memory addresses are coded as $a_1 = a_2 = a_3 = a_4 = 00000B = 0$, $a_5 = 1001B = 17$, $a_6 = 11001B = 25$, $a_7 = 00100B = 4$, $a_8 = 00101B = 5$. Assuming all the initial memory

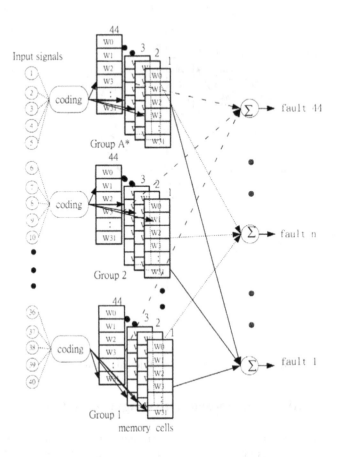

Fig. 2. The configuration of CMAC NN fault diagnosis system of air-conditioning system

weights are zero, then the summation of the fired up memory addresses w_1^0, w_2^0, w_3^0, w_4^0, w_5^{17}, w_6^{25}, w_7^4, w_8^5 are zero. The output of CMAC can be expressed as

$$y = \sum_{i=1}^{A^*} w_i^{ai}, \quad A^* : \text{the number of fired up memory cells} \tag{1}$$

Assuming the i layer ($i = 1, \cdots, 44$) outputs one denotes the fault type i is confirmed, then one can be thought as the teacher and the supervised leaning algorithm can be described as[8, 11]

$$w_{i(new)}^{ai} \leftarrow w_{i(new)}^{ai} + \beta \frac{y_d - y}{A^*}, i = 1, 2, ..., A^* \tag{2}$$

where $w^{ai}_{i(new)}$ are the new weight values after the weights tuning, $w^{ai}_{i(new)}$ are the old weight values before weight tuning, and ai denotes the fired memory addresses, β the learning gain, y_d the desired output. Using the well known learning algorithm, the convergence can be guaranteed[10].

3.2 Noise rejection

The noise rejection ability can be illustrated as follows. Assuming the first fault type coding has following error.

Original coding: 001010010011001100010000000000000000000000

Error coding: 001010010011001100010000000000000000000011

Then the fired up memory addresses coding $(a_1, a_2, a_3, a_4, a_5, a_6, a_7, a_8)$ changed from $(0,0,0,0,17,25,4,5)$ to $(3,0,0,0,17,25,4,5)$, only the a_1 is wrong. Since the fault features are stored on eight different addresses, the output will preserve 87% features at least and the noise rejection ability can be obtained.

3.3 Learning performance evaluation

Assuming the i-th (i=1,...,44) layer outputs one denotes the system has fault type i, and the number of training pattern for fault type i is d. Let the performance index be

$$E = \sum_{i=1}^{d}(y_i - 1)^2 \qquad (3)$$

when $E < \varepsilon$ the training process will stop. (ε is a small positive constant).

3.4 Wrong diagnosis learning

If the diagnosis output is wrong, the following modified learning rule is used to update the memory weights on line.

$$w^{ai}_{i(new)} \leftarrow w^{ai}_{i(new)} + \alpha\frac{\eta - y_{err}}{A^*}, i = 1, 2, ..., A^* \qquad (4)$$

where η is the threshold value, y_{err} the error output value, and α learning gain. Since the nodes outputs denote the possibility of fault type, the learning law merely modifies the memory weights to lead the error output nodes smaller or higher than threshold value.

3.5 Diagnosis algorithm

Based on the configuration of Fig. 2, the diagnosis algorithms are summarized as follows.

Off-line mode

Step 1 Build the configuration of CMAC fault diagnosis system. It includes 40 binary input signals, 44 parallel memory layers and 44 output nodes.

Step 2 Input the training patterns, through segmentation, fired memory addresses coding, and summation of fired memory addresses weights to produce the node output.

Step 3 Calculating the difference of actual output and the desired output ($y_d = 1$) and using Eq. (2) to update the weight values.

Step 4 Does the training finished? Yes, next step. Otherwise, go to **step 2**.

Step 5 Training performance evaluation. If , the training is finished. Save the memory weights. Otherwise, go to **step 2**.

On-line mode

When the training is finished, the diagnosis system can be used to the diagnosis of air-conditioning system. The diagnosis steps are described as follows.

Step 6 Load the up to date memory weights from the saved file.

Step 7 Input the diagnosed data.

Step 8 Segmentation, fired memory address coding, and summation of the fired memory weights using Eq. (1).

Step 9 Does the diagnosis correct? Yes, go to **step 10**. Otherwise, go to **step 11**.

Step 10 Does the next data to be diagnosed? Yes, go to **step 7**. Otherwise, go to **step 12**.

Step 11 Update the fired memory weights using Eq. (4).

Step 12 Save the up to date memory weights to file.

4 Some test results

Using the training patterns of Table 1 to train the CMAC neural network, the weights distribution plot of group 1 memory layers is shown in Fig. 3. Input the training patterns to the CMAC again, Table 2 are the nodes output (partial results). As shown in Table 2, output value 1 denotes the fault type is confirmed. Input the simulated non-training patterns of Table 3 (that modify from Table 1, include noise and uncertainty), the node outputs are shown in Table 4. Of course, no outputs are exact equal to 1, but the outputs still denote the most possible fault reason. The maintenance man can obtain useful information from the table. For example, the first row indicates the most possible fault type is type 1 (87.5%) and the probability of fault type 11 is 62.5%. The third raw denotes the most possible fault types are 3 and 4 (75%), the multiple faults detection ability is proven. Assuming Table 3 is the new added patterns, the memory weights can be retraining. After the retraining processes, the up to date weights distribution plot is shown in Fig. 4. Input the patterns of Table 1 and Table 3 to the CMAC again, the nodes outputs are shown in Table 5 and Table

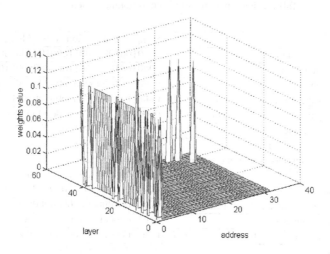

Fig. 3. Memory weights distribution plot after training

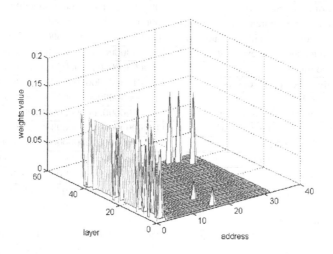

Fig. 4. Memory weights distribution plot after training

Table 1. Fault patterns of air-conditioning system

Fault reason	Symptoms detection coding(40digits)
1: shortage of refrigerant charge	0010100100110011000100000000000000000000
2: expansion value jammed	0000100100001011000000000001000000000001
3: expansion value closed fully	0000000100000001000000000000001000000000
4:membrane leakage of expansion value	0000000100001010000000000000001000000000
5: the hole of expansion value too small	0000100100001010000000000000000000000000
6: electromagnetic value cannot open	0000100100111011010000000001000000000001
7: filter valve blocked of liquid pipe	0000100000110011010000000001000000000001
8: output valve of container does not open fully	0000100000110011001000000000000000000000
9: liquid pipe blocked	0000100000000110101010000001000000000001
10 cooling water of condenser not enough	0101000000000000000000000000000000000001
11 cooling water of condenser too much	0010100100000010000000000000000000000000
12 air exist in the condenser	0101000000000000000000000000000000000000
13 water or air not enough on the evap. and cond.	0000000000000000000000000000000000000001
14 refrigerant too much	0100000001000000001000000000000000000000
15 expansion value stick on the open position	0000000001000100000000000000000000000000
16 expansion value with wrong adjustment	0000100101000110000000000000000000000000
17 pressure drop of steam pipe too large	0000000000000010000000000000000000000000
18 the capacity of evaporators too large	0001000100000000000000001000000000000000
19 load distribution of evaporator is not uniform	0000100001000010100000000000000000000000
20 the distribution of refrigerant is not uniform	0000100001000001000000000000100000000000
21 refrigerate oil gather in the evaporator	0000000000000100000010000000000000000000
22 refrigerant too much	0000000000000000010100000000000000000000
23 air return value of compressor blocked	0000100000000000001000010000000000000001
24 actuator overload	0000000000000000000000000010000000000000
25 low pressure control value too high	1000000000000000000000001000000000000001
26 high pressure control value too low	1000000000000000000001000000000000000001
27 compressor value cracked	0011000000000000000011000000100000000000
28 leakage of low pressure value	0000000100000000000000000000000000000000
29 compressor bearing can not rotate	0000000000000000000000001000010000000000
30 compressor movable element can not action	0000000000000000000000001000110000000000
31 compressor oil pump damage	0000010000000000000101000010100000000000
32 compressor oil pipe blocked	0000001000000000000001100000100000000000
33 compressor capacity controller damage	1000000000000000000001110000000000000000
34 compressor over reduce load	0001000100000010000010000000000000000000
35 compressor piston or link damage	0011000000000000001000000000000000000000
36 power failure	0000000000000000000000001000001000000000
37 voltage supply too low	0000000000000000000000011000100111100000
38 fuse burn down	0000000000000000000000001000001000000000
39 short circuit of control board	0000000000000000000000001000000000000000
40 switch contact element burn out or dusty	0000000000000000000000001000000001110000
41 wire terminal loose	0000000000000000000000000000000000011010
42 motor coil burn out	0000000000000000000000001000010011000000
43 temperature control switch failure	0000000000000000000000010001000000000000
44 instrument damage	0111111000000000000000000000001100000000

4. It is clear the '1' output confirms what fault type is, and other outputs indicate the possibility of other fault type.

Note: Symptoms detection points contain the pressure normal, high pressure too high, high pressure too low, low pressure too high, low pressure too low, oil pressure too low, oil pressure too high, backflow tube temperature too high, backflow tube near the compressor temperature too high, backflow tube with cold sweat, refrigerant window with air bubble..., breaker fault, the switch junction with spark, the amount of refrigerant.

Table 2. Diagnosis output of training pattern (partial results

No.	1	2	3	4	5	6	7	8	9	10	11
1	1	0.375	0.5	0.5	0.625	0.375	0.375	0.625	0.25	0.375	0.75
2	0.375	1	0.5	0.5	0.625	0.75	0.625	0.375	0.625	0.375	0.375
3	0.5	0.5	1	0.875	0.5	0.375	0.25	0.375	0.25	0.375	0.5
4	0.5	0.5	0.875	1	0.5	0.375	0.25	0.375	0.25	0.25	0.5
5	0.625	0.625	0.5	0.5	1	0.5	0.375	0.625	0.375	0.5	0.75
6	0.375	0.75	0.375	0.375	0.5	1	0.75	0.375	0.625	0.375	0.375
7	0.375	0.625	0.25	0.25	0.375	0.75	1	0.625	0.75	0.5	0.25
8	0.625	0.375	0.375	0.375	0.625	0.375	0.625	1	0.5	0.5	0.5
9	0.25	0.625	0.25	0.25	0.375	0.625	0.75	0.5	1	0.375	0.375
10	0.375	0.375	0.375	0.25	0.25	0.375	0.5	0.5	0.5	1	0.5
11	0.75	0.375	0.5	0.5	0.75	0.375	0.25	0.5	0.375	0.5	1

Table 3. Simulated non-training patterns (include uncertainties or noise)

Fault reason	Symptoms detection coding(40digits)
1: shortage of refrigerant charge	0010100100110011000100000000000000000001
2: expansion value jammed	0000100100001011000000000010000000000010
3: expansion value closed fully	0000000100001001000000000000001000000000
4:membrane leakage of expansion value	0000000100000101000000111000010000000000
5: the hole of expansion value too small	0000100100011010000000000100000000000000
6: electromagnetic value cannot open	1000100100111011100000000010000000000011
7: filter valve blocked of liquid pipe	0000100000110011100000000010010000000001
8: output valve of container does not open fully	1000100000110011001000010000000000000001
9 liquid pipe blocked	0000100001000110101010000010000000000001
10 cooling water of condenser not enough	1101000000000000000000000000000000000000
11 cooling water of condenser too much	0010100100000010000000011000000000001111

Table 4. Diagnosis output of non-training data Table 3(partial results)

No.	1	2	3	4	5	6	7	8	9	10	11
1	0.875	0.5	0.375	0.375	0.5	0.5	0.5	0.5	0.375	0.5	0.625
2	0.375	0.875	0.5	0.5	0.625	0.625	0.5	0.375	0.5	0.25	0.375
3	0.375	0.5	0.75	0.75	0.375	0.375	0.25	0.25	0.25	0.25	0.375
4	0.375	0.375	0.75	0.875	0.375	0.25	0.125	0.25	0.125	0.125	0.375
5	0.5	0.5	0.5	0.5	0.75	0.5	0.375	0.5	0.375	0.375	0.625
6	0.375	0.5	0.375	0.375	0.375	0.625	0.375	0.5	0.375	0.25	0.375
7	0.375	0.5	0.25	0.25	0.375	0.5	0.75	0.625	0.625	0.5	0.25
8	0.375	0.25	0.125	0.125	0.25	0.25	0.5	0.625	0.375	0.5	0.25
9	0.125	0.375	0.25	0.25	0.125	0.375	0.375	0.125	0.625	0.25	0.25
10	0.5	0.25	0.5	0.375	0.625	0.25	0.375	0.625	0.375	0.75	0.625
11	0.5	0.25	0.25	0.25	0.5	0.25	0.125	0.25	0.25	0.375	0.75

Table 5. Diagnosis output of Table 1 data after retraining (partial results)

No.	1	2	3	4	5	6	7	8	9	10	11
1	1.109	0.421	0.593	0.546	0.75	0.515	0.468	0.765	0.296	0.5	0.875
2	0.437	1.109	0.625	0.546	0.75	0.937	0.75	0.462	0.765	0.437	0.437
3	0.546	0.562	1.187	0.968	0.625	0.515	0.312	0.421	0.343	0.5	0.562
4	0.546	0.562	1.062	1.109	0.625	0.515	0.312	0.421	0.343	0.343	0.562
5	0.687	0.703	0.593	0.546	1.187	0.640	0.468	0.718	0.421	0.656	0.875
6	0.437	0.828	0.468	0.406	0.625	1.234	0.875	0.468	0.765	0.437	0.25
7	0.437	0.687	0.312	0.265	0.468	0.890	1.187	0.812	0.890	0.593	0.281

Table 6. Diagnosis output of non-training data after retraining (partial results)

No.	1	2	3	4	5	6	7	8	9	10	11
1	1	0.546	0.468	0.406	0.593	0.640	0.625	0.687	0.468	0.593	0.75
2	0.421	1	0.625	0.546	0.75	0.812	0.593	0.421	0.593	0.312	0.437
3	0.421	0.562	1	0.828	0.468	0.515	0.312	0.296	0.343	0.312	0.437
4	0.406	0.421	0.906	1	0.468	0.343	0.156	0.296	0.218	0.187	0.437
5	0.546	0.562	0.593	0.546	1	0.640	0.468	0.546	0.421	0.5	0.718
6	0.421	0.562	0.468	0.406	0.468	1	0.468	0.343	0.462	0.312	0.437
7	0.437	0.546	0.312	0.265	0.468	0.640	1	0.812	0.718	0.593	0.281

5 Conclusion

This work presents a novel digital CMAC neural network fault diagnosis scheme of air-conditioning system. Using limited training patterns to train the CMAC neural network, like the brain of human being, each fault type feature is distributed and memorized on an assigned memory layer. When a diagnosed data input the CMAC, the diagnosis system will output the possibility of all fault types. It provides useful information to system fault diagnosis and maintenance. As the accumulation of training patterns and learning, the diagnosis system will become a more powerful and accurate diagnosis tool. The simulation results demonstrate the objectives of high diagnosis accuracy, multiple faults detection, suit to non-training data, and alleviate the dependency to expert's expertise are obtained.

References

1. Chung Hsin Electric & Machinery Mfg. Corp. Ltd., " The catalog of Water Cooled Chiller Units",1992.
2. C. E. Lin, J. M. Ling, and C. L. Huang, "An expert system for transformer fault diagnosis using dissolved gas analysis", IEEE Trans. on Power Delivery, Vol. 8, No. 1, pp. 231-238, Jan. 1993.
3. Y. Zhang, X. Ding, Y. Liu and P. J. Griffin, " An artificial neural network approach to transformer fault diagnosis", IEEE Trans. on PWRD, Vol. 11, No. 4, pp. 1836-1841, Oct. 1996
4. Z. Wang, Y. Liu, and P. J. Griffin, "A combined ANN and expert system tool for transformer fault diagnosis", IEEE Trans. on Power Delivery, Vol. 13, No. 4, pp. 1224-1229, Oct. 1998.
5. G. Zhang, K. Yasuoka, and S. Ishii," Application of fuzzy equivalent matrix for fault diagnosis of oil-immersed insulation", Proc. of 13th Int. Conf. On Dielectric Liquids (ICDL'99), Nara, Japan, pp. 400-403, July 20-25, 1999.
6. Q. Su, C. Mi, L. L. Lai, and P. Austin, " A Fuzzy Dissolved Gas Analysis Method for the Diagnosis of Multiple Incipient Faults in a Transformer," IEEE Trans. on Power Systems, Vol. 15, No. 2, pp. 593-598, May 2000.
7. J. J. Dukarm, "Transformer oil diagnosis using fuzzy logic and neural networks", 1993 Canadian Conference on Electrical and Computer Engineering, Vol. 1, pp. 329-332, 1993.
8. W.S.Lin,C.P.Hung, Mang-Hui Wang, "CMAC_based Fault Diagnosis of Power Transformers", Proc. IJCNN Conf., 2002,#1266
9. J. S. Albus, "A new approach to manipulator control: the cerebeller model articulation controller (CMAC)1 , Trans. ASME J. Dynam., Syst., Meas., and Contr., vol. 97, pp.220-227, 1975
10. Y. F. Wong, A. Sideris, "Learning convergence in the cerebellar model articulation controller", IEEE Trans. on Neural Network, vol. 3,no. 1, pp. 115-121, 1992.
11. D. A. Handeiman, S. H. Lane, and J. J. Gelfand, "Integrating neural networks and knowledge-based systems for intelligent robotic control";AIEEE Control System Magazine, pp. 77-86, 1990.
12. P.C. Koelet, "Industrial Refrigeration Principle, Design and Application", Macmillan, 1991.

Prototypes Stability Analysis in the Design of Radial Basis Function Neural Networks

Lino Ramirez and Witold Pedrycz

Dept. of Electrical and Computer Engineering,
University of Alberta, Edmonton Canada, T6G 2V4
ramirez@ee.ualberta.ca, pedrycz@ee.ualberta.ca

Summary. This paper introduces a notion of prototypes stability. Prototypes obtained by applying a clustering algorithm to different random portions of a data set should be stable, meaning that they should not differ significantly. If they are stable, prototypes could be used in the design of different learning architectures. In this work, we study the use of prototypes stability analysis in the design of radial basis function (RBF) neural networks, especially their layer composed of RBF's. The usefulness of the proposed method is discussed and illustrated with the aid of numeric studies including two well-known data sets available on the web.

Key words: clustering, radial basis functions, classification, regression

1 Introduction

Partitional clustering techniques attempt to find partitions in a sample data set by minimizing the within-cluster dispersion or by maximizing the between-cluster dispersion. So, the problem of partitional clustering can be formulated in the following way: given n patterns in a p-dimensional space, determine a partition $\mathbf{U}$ of the patterns into c clusters such that the patterns in a cluster are more similar to each other than to patterns of different clusters. One of the most important considerations before applying a partitional clustering technique is the selection of the number of clusters (c) that best represent the data structure. Once the structure is determined, the results could be used for developing models of the data. The objective of this study is to employ a prototypes stability analysis to determine the appropriate number of clusters for the application of partitional clustering techniques. In this work, we deal with fuzzy c-means [1]. The prototypes obtained with the fuzzy c-means are afterwards used as the modal values of the receptive fields of an RBF neural network [2].

The material is arranged as follows. In Section 2, we summarize the underlying concepts of RBF neural networks to help identify some emerging problems. In the sequel, Section 3, we present the algorithm of prototypes stability analysis to determine the best number of centres to be used in modelling. Experimental studies involving two widely used data sets are reported in Section 4. Finally, a summary of the conclusions from the present work is presented in Section 5.

2 Radial Basis Function Neural Networks: A Brief Introduction

The RBF network [2] is a feed-forward neural network with two layers. The first layer consists of neurons normally equipped with Gaussian-like functions, even though other types of basis functions might be used, centred at the points specified by the weights associated with this layer. The second layer performs the linear combination of the outputs obtained from the first layer. The decision boundary defined by the RBF network with Gaussian basis functions is given by

$$f(\mathbf{x}) = \sum_{i=1}^{Ns} w_i \exp(-\gamma \| \mathbf{x} - \mathbf{c_i} \|) + b \qquad (1)$$

where Ns is the number of centres; $\mathbf{x}$ is an input pattern; w_i are the weights for the second layer; γ is the spread; b is the bias; and $\mathbf{c_i}$ is the ith centre vector used as a reference. Note that this decision boundary is continuous. To perform binary classification, the final output is given by

$$g(\mathbf{x}) = \mathrm{sgn}(f(\mathbf{x})) \qquad (2)$$

For the training of the RBF neural network, one may consider the two layers separately, leading to a two-stage training procedure. In the first stage, the centres $\mathbf{c_i}$ and the spread γ are determined. In the second stage, the basis functions are kept fixed and the weights of the second layer are found. The most commonly used approaches to find the centres are the application of clustering methods (such as the c-means and its fuzzy counterpart the fuzzy c-means [1]) and the application of the orthogonal least squares (OLS) approach [3]. In the clustering methods, the input data set is partitioned into several groups in such a way that the similarity among members of a group is larger than the similarity among members of different groups. In this case, the centres represent the prototypes of each one of the resulting groups. In the OLS approach, centres (training points) are added until some preset performance is reached or all the training points have been used.

3 Prototype Stability Analysis

One of the most important considerations before applying a partitional clustering technique is the selection of the number of clusters (c). An error in the choice of the number of clusters may prevent for a correct detection of clustering structure [4]. This can be illustrated by considering the data set depicted in Fig. 1. If we apply clustering to ten, randomly selected, different portions of the data set, we may have the following results: for $c=5$, the resulting prototypes for each cluster are very close to each other (see Fig. 2); while for $c=4$, some of the prototypes are split into two visually different clusters (see Fig. 3). From Figs. 2 and 3, we conclude that if the number of clusters is appropriate, any prominent data structure ought to survive even if clustering is applied only to a random portion of the data. This idea will be used in the development of a method to determine the appropriate number of clusters for clustering a data set. This method is explained below.

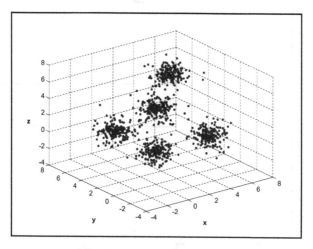

Fig. 1. Synthetic data. Five clusters with 200 elements each. The patterns for each cluster are distributed according to a normal density function with unit covariance matrices.

To do the prototypes stability (or variability) analysis, the original data set is randomly re-sampled to produce n_p non-overlapping subsets. Clustering is then applied to each subset. The resulting prototypes are sorted and the average scattering (or stability index) per dimension is calculated by

$$\tau_{cj} = \frac{\frac{1}{c}\sum_{i=1}^{c}\sigma_{ij}}{r_j}, 1 \leq j \leq p, c_{min} \leq c \leq c_{max} \tag{3}$$

where p is the number of dimensions of the training data; c_{min} and c_{max} are the minimum and maximum number of clusters that are tested; σ_{ij} is the standard deviation of the jth dimension for the prototypes of the ith cluster; and r_j is the range per dimension of the training data ($\mathbf{X}$) given by

$$\mathbf{r} = \max(\mathbf{X}) - \min(\mathbf{X}), \mathbf{r}\varepsilon\Re^p \tag{4}$$

The average scattering indicates the average of the variations within clusters for a number of clusters set to c. A small value of the average scattering is caused by a

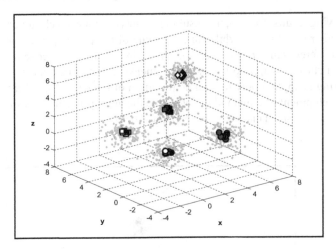

Fig. 2. Prototypes generated by the fuzzy c-means with $c=5$.

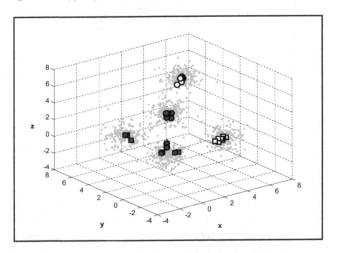

Fig. 3. Prototypes generated by the fuzzy c-means with $c=4$.

partition where the prototypes in each cluster are close to each other. As the scattering within the clusters increases, the prototypes are getting farther away. Bearing this in mind, a number of clusters which minimizes the scattering index can be considered as an appropriate value to be used for clustering the data.

The algorithm for applying prototypes stability analysis to determine the appropriate number of clusters is given by the following steps.

Table 1. Prototypes stability analysis algorithm.

Given	The data set $\mathbf{X}=\mathbf{x}_1,\mathbf{x}_2,\ldots,\mathbf{x}_n \subset \mathfrak{R}^p$, a set of n feature vectors in a p-dimensional space.
Defined	* c_{min}, the minimum number of clusters. * c_{max}, the maximum number of clusters. * n_p, the number of portions in which the data set will be randomly divided. * Any parameter needed for the clustering algorithm at hand.
Initialization	Randomly divide the data set in n_p non-overlapping subsets.
Processing	* Calculate the standard deviation per dimension of the n_p prototypes in each one of the clusters. * Calculate the stability coefficient by using $$\tau_{cj} = \frac{\frac{1}{c}\sum_{i=1}^{c}\sigma_{ij}}{r_j}, 1 \leq j \leq p, c_{min} \leq c \leq c_{max}$$ with r_j being the range per dimension. * Determine the appropriate number of clusters (per dimension) by $nc_j = \arg[\min_{c_{min},c_{min}+1,\ldots,c_{max}}(\tau_{cj})], 1 \leq j \leq p$ * Determine the number of clusters to be used in modelling by calculating the mean of the number of clusters per dimension $$nc_{average} = \frac{1}{p}\sum_{j=1}^{p} nc_j$$ and rounding $nc_{average}$ to the nearest integer.
Result	Number of clusters to be used in modelling.

4 Experimental Results

In this section, we study the applicability of prototypes stability analysis in the design of RBF neural networks. This study relies on two widely available data sets, a Wisconsin breast cancer database and a Boston housing data. Both data sets are available online at UCI Repository of Machine Learning at http://www.ics.uci.edu/ mlearn/MLRepository.html. The experiments compare the performance of an RBF neural network with the centres found using fuzzy clustering (RBF1) (the number of centres was determined beforehand by prototypes stability analysis) and an RBF neural network with the centres found using the orthogonal least squares approach (RBF2). For both approaches the spread γ was set to one. For the RBF2, the performance criterion was set to reduce the mean square error during training up to a value G=1. The mean square error (MSE) is given by

$$MSE = \frac{1}{n}\sum_{i=1}^{n}(t(\mathbf{x}) - f(\mathbf{x}))^2 \tag{5}$$

where n is the number of patterns; $\mathbf{x}$ is an input pattern; $t(\mathbf{x})$ is the expected output; and $f(\mathbf{x})$ is the RBF neural network output.

4.1 Wisconsin Breast Cancer Classification

Dr. William H. Goldberg created the Wisconsin breast cancer database as part of a study that aimed to diagnose cancer via linear programming [5]. The data set consists of 683 records, of which 444 were confirmed to correspond to benign tumours and 239 were labelled as malign tumours. Each record has the following attributes (used to determine whether or not a tumour is benign): clump thickness, uniformity of cell type, uniformity of cell shape, marginal adhesion, single epithelial cell size, bare nuclei, bland chromatin, normal nucleoli, and mitoses. All the variables attain integer values between one and ten.

Methodology

In the experiments, 60% of the data set (i.e. 409 records) was used as a training (learning) set and the remaining 40% (274 records) was used as a testing set. To obtain reliable results, a rotation method was employed, repeating the experiments ten times. Subsequently, the training data for each one of the ten experiments were normalized to attain a zero mean and one standard deviation. The values for the mean and the standard deviation were then used to normalize the testing data for each experiment.

Comparative Analysis

The experiments were performed independently over the ten randomly re-sampled groups. The results averaged over the ten iterations are summarized in Table 2. The results include the average of the percentage of classification error in training and testing and the standard deviation of the percentage of classification error in training and testing. From these results, it can be seen that, in testing, the RBF1 showed lower classification error than the RBF2 (5.57% of classification error for the RBF1 against 6.85% of classification error for the RBF2).

Table 2. Summary of results (percentage of classification error in the form *mean $\pm$ standard deviation*) for the Wisconsin breast cancer data.

Results	RBF1	RBF2
Training	5.80±0.82	7.14±0.88
Testing	5.57±1.27	6.85±1.83

4.2 Boston Housing Price Estimation

The Boston housing data set first appeared in [6]. It is concerned with house prices in the suburbs of Boston. This data set has 13 continuous attributes and one binary-valued attribute. See Table 3 for a description of all the attributes.

Table 3. Features for the Boston housing data set.

Feature	Information
CRIM	Per capita crime rate by town
ZN	Proportion of residential land zoned for lots over 25,000 sq. ft.
INDUS	Proportion of non-retail business acres per town
CHAS	Charles River dummy variable (=1 if tract bounds river; 0 otherwise)
NOX	Nitric oxides concentration (parts per 10 million)
RM	Average number of rooms per dwelling
AGE	Proportion of owner-occupied units built prior to 1940
DIS	Weighted distances to five Boston employment centres
RAD	Index of accessibility to radial highways
TAX	Full-value property - tax rate per $10,0000
PTRATIO	Pupil-teacher ratio by town
B	$1000(Bk - 0.63)^2$ where Bk is the proportion of blacks by town
LSTAT	% Lower status of the population
MEDV	Median value of owner-occupied homes in $1000's

Methodology

In the experiments, 60% of the data set (i.e. 304 records) was used as a training (learning) set and the remaining 40% (202 records) was used as a testing set. To obtain reliable results, a rotation method was employed, repeating the experiments ten times. Subsequently, the training data for each one of the ten experiments were normalized to attain a zero mean and one standard deviation. The values for the mean and the standard deviation were then used to normalize the testing data for each experiment.

Comparative Analysis

The experiments were performed independently over the ten randomly re-sampled groups. The results averaged over the ten iterations are summarized in Table 4. The results include the average of the mean square error in training and testing and the standard deviation of the mean square error in training and testing. From these results, it can be seen that, in testing, the RBF1 showed a better performance than the RBF2 (a mean square error of 69.08 for the RBF1 against a mean square error of 84.92 for the RBF2).

Table 4. Summary of results (mean square error in the form *mean ± standard deviation*) for the Boston housing data.

Results	RBF1	RBF2
Training	67.77±5.63	0.98±0.02
Testing	69.08±9.48	84.92±18.53

4.3 Discussion

From the results shown in Tables 2 and 4, it can be seen that the RBF1 outperformed the RBF2 in both pattern recognition problems. Moreover, in both problems, the RBF1 showed similar results in training and testing. This may indicate that certain structure in the data set was found. In the case of the Boston housing data, it could be seen that the RBF2 produced overly optimistic results. The mean square error on the training set was lower than the mean square error on the training set for the RBF1 yet it increased significantly for the testing set. This may suggest that for the Boston housing data, the RBF2 over fitted the training data set.

5 Conclusions

In this paper, we have introduced a notion of prototypes stability analysis whose intent is to determine the appropriate number of clusters or prototypes to be used in modelling. In this particular work, the proposed method was used to determine the number of centres for the design of the receptive fields of an RBF neural network. The centres were then found using the fuzzy c-means clustering method. Compared to the results obtained with an RBF neural network with the centres found using the orthogonal least squares method, the proposed method achieved superior performance in the numerical experiments. Therefore, we conclude that for an application point of view, the proposed method could be seen as a promising technique for determining the number of clusters or prototypes to be used in modelling.

References

1. Bezdek J C (1981): Pattern Recognition with Fuzzy Objective Functions. Plenum Press, New York
2. Bishop C M (1995): Neural Networks for Pattern Recognition. Claredon Press, Oxford
3. Chen S, Cowan C F N, Grant P M (1991): Orthogonal Least Squares Learning Algorithm for Radial Basis Function Networks, IEEE Trans. on Neural Networks 2(2): 302–309
4. Dumitrescu D, Lazzerini B, Jain L C (2000): Fuzzy Sets and their Application to Clustering and Training. CRC Press, New York
5. Mangasarian O L, Woldberg W H (1990): Cancer Diagnosis Via Linear Programming, SIAM News 23(5): 1–18
6. Harrison D, Rubinfeld D L (1978): Hedonic Prices and the Demand for Clean Air, Journal of Environmental Economics and Management 5: 81–102

Canadian Weather Analysis Using Connectionist Learning Paradigms

Imran Maqsood[1], Muhammad Riaz Khan[2], and Ajith Abraham[3]

[1] Environmental Systems Engineering Program, Faculty of Engineering, University of Regina, Regina, Saskatchewan S4S 0A2, Canada maqsoodi@uregina.ca
[2] AMEC Technologies – Transtech Interactive Training Inc., 400-111 Dunsmuir Street, Vancouver, BC V6B 5W3, Canada riaz.khan@amec.com
[3] Department of Computer Science, Oklahoma State University, 700 N Greenwood Avenue, Tulsa, USA ajith.abraham@ieee.org

Summary. In this paper, we present a comparative study of different neural network models for forecasting the weather of Vancouver, British Columbia, Canada. For developing the models, we used one year's data comprising of daily maximum and minimum temperature and wind-speed. We used a Multi-Layered Perceptron (MLP) and an Elman Recurrent Neural Network (ERNN) trained using the one-step-secant and Levenberg-Marquardt algorithms. To ensure the effectiveness of neurocomputing techniques, we also tested the different connectionist models using a different training and test data set. Our goal is to develop an accurate and reliable predictive model for weather analysis. Experimental results obtained have shown Radial Basis Function Network (RBFN) produced the most accurate forecast model compared to ERNN and MLP.

Key words. Weather forecasting, multi-layered perceptron, radial basis function network, Elman recurrent neural network, one-step-secant algorithm, Levenberg-Marquardt algorithm.

1 Introduction

Weather forecasts provide critical information about future weather. Weather prediction could be one day/one week or a few months ahead [1][15][19]. The accuracy of weather forecasts however, falls significantly beyond a week. Weather forecasting remains a complex business, due to its chaotic and unpredictable nature [10][12]. It is known that persons with little or no formal training can develop considerable forecasting skill [8]. For example, farmers often are quite capable of making their own short-term forecasts of those meteorological factors that directly influence their livelihood, and a similar statement can be made about pilots, fishermen, mountain climbers, etc. Accurate weather forecast models are important to third world countries, where the entire agriculture depends upon weather [19]. Several artificial intelligence techniques have been used in the past for modeling chaotic behavior of weather [4][5][7][10][11][12][14][16][19]. One aim of this research is to develop an accurate weather forecast model so as to minimize the impact of extreme summer and winter weather. To improve the learning capability of the connectionist paradigms, we used second order error information using the one-step-secant and the Levenberg-Marquardt approaches. The Radial Basis Function (RBF) network is a popular alternative to the Multi-Layered Perceptron (MLP), which, although is not as well suited to larger applications, can offer advantages over the MLP in some applications [3][6][13][18]. We also used a radial basis function network, which is also a well-established technique for function approximation.

2 Multi-Layered Perceptron (MLP) Networks

Typical MLP network is arranged in layers of neurons (nodes), where every neuron in a layer computes the sum of its inputs and passes this sum through a nonlinear function (an activation function) as its output. Each neuron has only one output, but this output is multiplied by a weighting factor if it is to be used as an input to another neuron (in a next higher layer). For backpropagation learning [9], the activation functions must be differentiable and saturating at both extremes. If the transfer functions were chosen to be linear, then the network would become identical to a linear filter.

3 Elman Recurrent Neural Networks (ERNN)

ERNN are a subclass of recurrent networks [2][6]. The Elman network has context units, which store delayed hidden layer values and present these as additional inputs to the network. The Elman network can learn sequences that cannot be learned with other recurrent neural network e.g. with Jordan recurrent neural network (which is a similar architecture with a context layer fed by the output layer) since networks with only output memory cannot recall inputs that are not reflected in the output. Usually, both hidden and output units have nonlinear activation functions. The network is thus a discrete dynamical system. We investigated one-step-secant and the Levenberg Marquardt algorithms to train ERNN.

4 Radial Basis Function Network (RBFN)

The RBF network has a linear dependence on the output layer weights, and the nonlinearity is introduced only by the cost function for training, which helps to address the problem of local minima. Additionally, this network is inherently well suited for weather prediction, because it naturally uses unsupervised learning to cluster the input data [17][18]. There are two basic methods to train an RBFN in the context of neural networks. One is to jointly optimize all parameters of the network similarly to the training of the MLP. This method usually results in good quality of approximation but also has some drawbacks such as a large amount of computation and a large number of adjustable parameters. Another method is to divide the learning of an RBFN into two steps. The first step is to select all the centers μ in terms of an unsupervised clustering algorithm such as the K-means algorithm proposed by Linde et al. (denoted as the LBG algorithm) [20], and choose the radii σ by the k-nearest neighbor rule. The second step is to update the weights B of the output layer, while keeping the μ and σ fixed. The two-step algorithm has fast convergence rate and small computational burden.

We used a two-step learning algorithm to speed up the learning process of the RBFN. We adopted a self-organized learning algorithm for selection of the centers and radii of the RBF in the hidden layer, and a stochastic gradient descent of the contrast function for updating the weights in the output layer. For the selection of the centers of the hidden units, we used the standard k-means clustering algorithm [18]. This algorithm classifies an input vector x by assigning it the label most frequently represented among the k-nearest neighbor samples. Specifically, it places the centers of RBF neurons in only those regions of the input space, where significant data are present. Once the centers and radii are established, we can make use of the minimization of the contrast function to update the weights of the RBFN.

5 Experimentation Setup and Results

In the normal case, architecture of the connectionist models is determined after a time-consuming trial-and-error procedure. To circumvent this disadvantage, we use a more systematic way of finding good architectures. A sequential network construction [20] is employed to select an appropriate number of hidden neurons for each of the connectionist model considered. First, a network with a small number of hidden neurons is trained. Then a new neuron is added with randomly initialized weights and the network is retrained with changes limited to these new weights. Next all weights in the network are retrained. This procedure was repeated until the number of hidden neurons reaches a preset limit and then substantially reduces the training time in comparison with time needed for training of new networks from scratch. More importantly, it creates a nested set of networks having a monotonously decreasing training error and provides some continuity in the model space, which makes a prediction risk minimum more easily noticeable.

Due to sudden variation in weather parameters, the model becomes obsolete and inaccurate. Thus, model performance and accuracy should be evaluated continuously [20]. Sometimes, periodic update of parameters or change of model structure is also required. We used the weather data from 01 September 2000 to 31 August 2001 for analyzing the connectionist models. For MLP and ERNN, we used the dataset from (11-20) January 2001 for testing and remaining data for training the networkss. We also used the dataset (01-15) April 2001 for testing and the remaining for training of RBFN, MLP and ERNN networks. We used this method to ensure that there is no bias on the training and test datasets. We used a Pentium-III, 1GHz processor with 256 MB RAM and all the experiments were simulated using MATLAB. The following steps were taken before starting the training process:

The error level was set to a relatively small value (10^{-4}).

The hidden neurons were varied (10-80) and the optimal number for each network were then decided as mentioned previously by changing the network design and running the training process several times until a good performance was obtained.

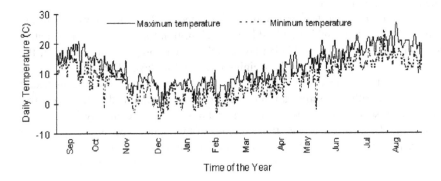

Fig. 1. Actual daily min. and max. temperatures for the year 2001 in B.C, Canada

When the network faces local minima (false wells), new ones to escape from such false wells replace the whole set of network weights and thresholds. Actually, a random number generator was used to assign the initial values of weights and thresholds with a small bias as a difference between each weight connecting two neurons together since similar weights for different connections may lead to a network that will never learn.

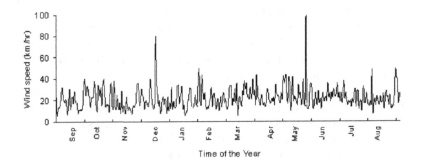

Fig. 2. Actual daily wind-speed for the year 2001 in B.C., Canada

5.1 Weather Parameters

Our initial analysis of the data has shown that the most important weather parameters are the minimum and maximum temperature variables. These variables also represent a strong correlation with other weather-parameters. Temperature, in general, can be measured to a higher degree of accuracy relative to any of the other weather variables. Anyhow, forecasting temperature requires the consideration of many factors; day or night, clear or cloudy skies, windy or calm, or will there be any precipitation? An error in judgment on even one of these factors may cause forecasted temperature to be off by as much as 20 degrees. Historical temperature data recorded by a weather station at the prominent meteorological center of the Vancouver, BC is used for the analysis. The maximum and minimum temperatures for period of one year (2000-2001) are plotted in Figure 1.

Space heating and cooling are the human response to how hot and how cold it 'feels'. Temperature in this case is only one of the contributors to such human response. Factors such as wind-speed in the winter and humidity in the summer have to be accounted for when describing how cold or hot it 'feels'. For that purpose, the weather
department developed a measure of how cold the air feels in the winter or how hot the air feels in the summer. These two new measurements are called the *wind-chill* and the *heat index*, respectively. Figure 2 presents the wind-speed (in km/hr) recorded in Vancouver, British Colombia for the year 2001.

5.2 Discussions and Test Results

Table 1 summarizes the architecture of the different network models used in this study. The training convergence of MLP and ERNN are illustrated in Figures 3 (a) and (b), respectively. RBFN took approximately 2 seconds to construct the network.

Table 1. Comparison of training of connectionist models

Network model	Number of hidden neurons	Number of hidden layers	Activation function used in hidden layer	Activation function used in output layer
MLP	45	1	Log-sigmoid	Pure linear
ERNN	45	1	Tan-sigmoid	Pure linear
RBFN	180	2	Gaussian	Pure linear

The optimal network is the one that should have the lowest error on test set and reasonable learning time. All the obtained results were compared and evaluated by the Maximum Absolute Percentage Error (MAPE), Root Mean Squared Error (RMSE), and Mean Absolute Deviation (MAD). Test results (August 2001) for the actual versus predicted minimum/maximum temperature and wind speed using MLP and ERNN with OSS and LM approaches are plotted in Figures 4, 5 and 6 respectively. Empirical results are depicted in Tables 2 through 4.

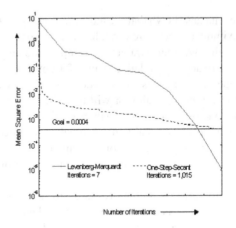

Fig. 3. Convergence of the LM and OSS training for MLP (left) and ERNN (right)

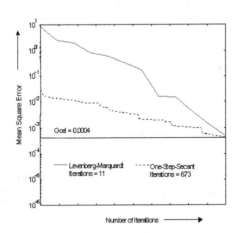

Fig. 4. Desired and predicted min. temp. using OSS and LM for MLP and ERNN

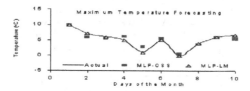

Fig. 5. Desired and predicted max. temp. using OSS and LM for MLP and ERNN

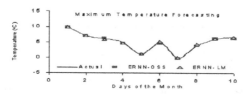

Fig. 6. Desired and predicted wind-speed using OSS and LM for MLP and ERNN

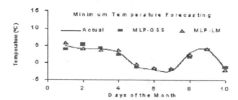

Fig. 7. Desired and 15-day ahead forecast values for **(a)** max. temperature, **(b)** min. temperature, **(c)** max. wind-speed using MLP, ERNN and RBFN

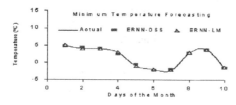

Fig. 8.

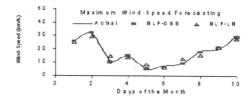

Fig. 9.

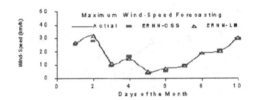

Fig. 10.

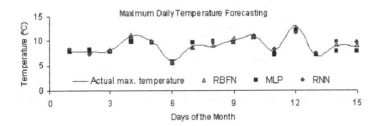

Fig. 11.

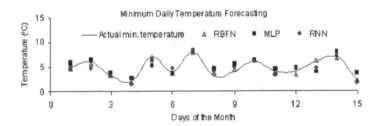

Fig. 12.

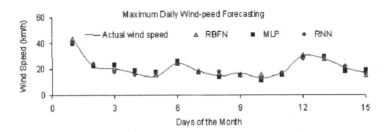

Fig. 13.

Test results (April 2001) for the actual versus predicted minimum/maximum temperature and wind-speed using MLP (OSS) and ERNN (OSS) and RBFN are illustrated in Figures 7 (a), (b) and (c). Empirical results are depicted in Table 5. Various forecasting error measures between the actual and forecasted weather parameters are defined, however the most commonly adopted by weather forecasters are used here as listed below:

$$MAD = \frac{\sum_{i=1}^{N} \left| P_{actual,\,i} - P_{predicted,\,i} \right|}{N} \tag{1}$$

$$MAPE = \frac{\sum_{i=1}^{N} \frac{\left| P_{actual,\,i} - P_{predicted,\,i} \right|}{P_{actual,\,i}}}{N} \times 100 \tag{2}$$

where P_{actual} and $P_{predicted}$ are the actual and forecasted weather parameter, respectively, and N is the number of days in the data set.

Table 2. Performance of MLP and ERNN for peak temperature forecast

Maximum temperature	MLP Network		ERNN	
	OSS	LM	OSS	LM
MAPE	0.0170	0.0087	0.0165	0.0048
RMSE	0.0200	0.0099	0.0199	0.0067
MAD	0.8175	0.4217	0.7944	0.2445
Correlation coefficient	0.96474	0.9998	0.9457	0.9826
Training time (minutes)	0.4	30	1.8	30
Epochs	850	7	1135	10

Table 3. Performance of MLP and ERNN for minimum temperature forecast

Minimum temperature	MLP Network		ERNN	
	OSS	LM	OSS	LM
MAPE	0.0221	0.0202	0.0182	0.0030
RMSE	0.0199	0.0199	0.0199	0.0031
MAD	0.7651	0.8411	0.7231	0.1213
Correlation coefficient	0.9657	0.9940	0.9826	0.9998
Training time (minutes)	0.3	1	0.3	7
Epochs	1015	7	673	11

Table 4. Performance of MLP and ERNN for wind-speed forecast

Wind-speed	MLP Network		ERNN	
	OSS	LM	OSS	LM
MAPE	0.0896	0.0770	0.0873	0.0333
RMSE	0.1989	0.0162	0.0199	0.0074
MAD	0.8297	0.6754	0.7618	0.3126
Correlation coefficient	0.9714	0.9974	0.9886	0.9995
Training time (minutes)	0.3	1	0.5	8
Epochs	851	8	1208	12

Table 5. Performance of MLP / ERNN and RBFN

	Evaluation parameters	Max. temp.	Min. temp.	Wind speed
RBFN	MAP	3.821	3.622	4.135
	MAD	0.420	1.220	0.880
	Correlation coefficient	0.987	0.947	0.978
MLP	MAP	6.782	6.048	6.298
	MAD	1.851	1.898	1.291
	Correlation coefficient	0.943	0.978	0.972
ERNN	MAP	5.802	5.518	5.658
	MAD	0.920	0.464	0.613
	Correlation coefficient	0.946	0.965	0.979

6 Conclusions

In this paper, we compared the performance of MLP network, ERNN and RBFN. Compared to the MLP neural network, the ERNN could efficiently capture the dynamic behavior of the weather, resulting in a more compact and natural internal representation of the temporal information contained in the weather profile. ERNN took more training time but it is dependent on the training data size and the number of network parameters. It can be inferred that ERNN could yield more accurate results, if good data selection strategies, training paradigms, and network input and output representations are determined properly. LM approach appears to be the best learning algorithm for mapping the different chaotic relationships. Due to the calculation of Jacobian matrix at each epoch, LM approach requires more memory and is computationally complex while compared to OSS algorithm. On the other hand, RBFN gave the overall best results in terms of accuracy and fastest training time. Empirical results clearly demonstrate that radial basis function networks are much faster and more reliable for the weather forecasting problem considered. The proposed RBFN network can also overcome several limitations of the MLP and ERNN networks such as highly nonlinear weight update and slow-convergence rate. Since the RBFN has natural unsupervised learning characteristics and modular network structure, these properties make it a more effective candidate for fast and robust weather forecasting.

7 Acknowledgements

The authors are grateful to the staff of the meteorological department, Vancouver, B.C., Canada for the useful discussions and providing the weather data used in this research work.

References

1. Maqsood I., Khan M.R. and Abraham A., Neuro-computing based Canadian weather analysis, In Proc .of the 2nd International Workshop on Intelligent Systems Design and Applications, Dynamic Publishers Inc., USA, pp. 39-44, 2002.
2. Elman J. L., Distributed representations, simple recurrent networks and grammatical structure, *Machine Learning*, Vol. 7, No. 2/3, pp. 195-226, 1991.
3. Moody J. and Utans J., Architecture selection strategies for neural networks: Application to corporate bond rating prediction, *Neural Networks in the Capital Markets*, J. Wiley and Sons, 1994.
4. Cholewo J. T. and Zurada M. J., Neural network tools for stellar light prediction, *IEEE Aerospace Conference*, Vol. 3, pp. 514-422, USA, 1997.
5. Neelakantan T.R. and Pundarikanthan N.V., Neural network-based simulation-optimization model for reservoir operation, *Journal of Water Resources Planning and Management*, Vol. 126, No. 2, pp. 57-64, 2000.
6. Elman J. L., Finding structure in time, *Cognitive Science*, Vol. 14, pp. 179-211, 1990.
7. Kugblenu S., Taguchi S. and Okuzawa T., Prediction of the geomagnetic storm associated D_{st} index using an artificial neural network algorithm, *Earth Planets Space*, Vol. 51, pp. 307-313, Japan, 1999.
8. Khan M. R. and Ondrusek C., Short-term load forecasting with multilayer perceptron and recurrent neural network, *Journal of Electrical Engineering*, Vol. 53, pp. 17-23, Slovak Republic, 2002.
9. Kuligowski R. J., Barros A. P., Localized precipitation forecasts from a numerical weather prediction model using artificial neural networks, *Weather and Forecasting,* Vol. 13, No. 4, pp.1194, 1998.
10. Moro Sancho Q. I., Alonso L. and Vivaracho C. E., Application of neural networks to weather forecasting with local data, *Applied Informatics*, Vol. 68, 1994.
11. Aussem A., Murtagh F. and Sarazin M., Dynamical recurrent neural networks and pattern recognition methods for time series prediction, Application to Seeing and Temperature Forecasting in the Context of ESO's VLT Astronomical Weather Station, Vistas in Astronomy, Vol. 38, No. 3, pp. 357, 1994.
12. Allen G. and Le Marshall J. F., An evaluation of neural networks and discriminant analysis methods for application in operational rain forecasting, Australian Meteorological Magazine, Vol. 43, No. 1, pp.17-28, 1994.
13. Doswell C. A., Short range forecasting: Mesoscale Meteorology and Forecasting, Chapter 29, American Meteor. Society, pp. 689-719, 1986.
14. Murphy A. H. and et al., Probabilistic severe weather forecasting at NSSFC: An experiment and some preliminary results, 17th conference on Severe local storms, American Meteor. Society, pp. 74-78, 1993.
15. Tan Y., Wang J. and Zurada J. M., Nonlinear blind source separation using a radial basis function network,IEEE Transactions on Neural Networks, Vol. 12, No. 1, pp. 124-134, 2001.

16. Chen, S., MacLaughlin, S., and Mulgrew, B., Complex-valued radial basis function network, Part I: Network architecture and learning algorithm, Signal Processing, Vol. 35, pp. 19-31, 1994.

17. Abraham A., Philip S. and Joseph B., Will We Have a Wet Summer? Long-term Rain Forecasting Using Soft Computing Models, Modeling and Simulation 2001, In Proceedings of the 15th European Simulation Multi Conference, Czech Republic, pp. 1044-1048, 2001.

18. Linde, Y., Buzo, A. and Gray, R., An algorithm for vector quantizer design, IEEE Transactions on Communications, Vol. 28, pp. 84-95, 1980.

Application of Neural Networks and Expert Systems in a hierarchical approach to the intelligent greenhouse control

J.V. Capella, A. Bonastre, and R. Ors

Dept. of Computer Engineering, Polytechnical University of Valencia, E-46022 Valencia, Spain {jcapella,bonastre,rors}@disca.upv.es

Summary. A novelty methodology based on the hierarchical combination of neural networks and expert systems is proposed in a centralized approach for the intelligent greenhouse control. The knowledge-based system is in charge of carrying out the determination of PH land value, composition, carbonic anhydride artificial atmosphere, external and internal temperature, wind and humidity measurements. From the results obtained, and by means of a neural network developed and trained for this application, the land quality is evaluated. On the other hand, the expert system, apart from supervising the system function and implementing fault tolerance mechanisms, performs the opportune actions in function of the results obtained by the neural network and others variables directly controlled by the expert system, in order to maintain the optimum microclimate and land composition. Satisfactory results have been obtained in the application of this approach to different quality lands and climatic conditions.

Key words: Intelligent control systems, hierarchical architecture, expert systems, neural networks.

1 Introduction

At the present time, the supervision and control systems for agricultural applications are in a world scale development stage. As it has become necessary to control more environmental variables for the cultivations it has been necessary the advance in the greenhouses technology, leaning in the results of biological investigations in diverse areas, as for example the vegetable physiology, that provide the appropriate parameters for these variables. All this growing complexity has made necessary that new methodologies for the respective control systems implementation are developed. Following this line, the application of artificial intelligence (AI) techniques to the implementation of control systems offers several advantages [1].

The use of expert systems for the implementation of control systems offers some interesting advantages, not only offering certain intelligence level to the system, but also being a good solution to solve the common problems related with the control systems implementation [2].

Continuous and discrete control are not incompatible, but complementary, when dealing with complex systems. Combination of both techniques offers new possibilities, such as the implementation of hierarchical control systems.

In the implementation of the proposed control system, it should be noted that it is very difficult to establish a direct relationship between inputs and outputs in order to perform a diagnosis of the land quality from measurements taken in the process. Therefore this diagnosis should be carried out by means of non-lineal control technique. On the other hand, the adoption of expert systems for the discrete variables control, fault diagnosis and global system supervision supposes a good solution due to its efficiency and programming easiness. Consequently, owing to the specific characteristics of this system, its implementation has been carried out according to a hierarchical model, as shown in Figure 1.

A new hierarchical architecture based on the combination of neural networks [3] and rule nets [4], is proposed for the integral greenhouse control, providing interesting characteristics such as low-cost implementation, possibility that even non-expert users could design the control system, self-learning system capacity and fault tolerance.

Nowadays this architecture is being applied to the intelligent control of an automated greenhouse. A greenhouse is a space with the appropriate microclimate and land quality for the optimum development of a specific plantation, therefore, giving the design and the automatic control system should be obtained in the greenhouse the temperature, relative humidity, adapted ventilation, shady, carbonic anhydride artificial atmosphere, PH land, accurate fertilization and land composition to reach high productivity, at low cost, in less time, with smaller environmental impact, protecting to the rains, the hail, the freezes or the wind excesses that could harm a cultivation.

2 Neural Networks

Neural networks (NN) are systems modeled after the neurons (nerve cells) in a biological nervous system. A neural network is designed as an interconnected system of processing elements, each with a limited number of inputs and outputs. Neurons receiving information from outside are called input neurons, whereas those which send information outside are given the name of output neurons. Logically, the rest of neurons are called intermediate neurons. Within the neurons, several levels can be distinguished, in such a way that the neurons get information from those of the lower level and provide an already processed information to neurons on the upper level. The number of levels, the amount of neurons on each level, and the number of feedback processes (connections between an upper level and an equal or lower one) define the topology of a neural network. On the other hand, the main feature of these type of networks consists of the fact that, rather than being programmed, these systems learn to recognize patterns. Therefore, they must be suitably trained.

These systems are highly appropriate to reflect knowledge that can be neither programmed nor justified.

3 Rule Nets

Rule nets are a formalism that seeks to express an automatism in a similar way to as would make it a human being: "IF antecedents THEN consequents". But at the same time Rule nets are a tool for the design, analysis and implementation of rule based systems (RBS), and consist on a mathematic-logical structure which analytically reflects the set of rules that the human expert has designed.

Basically, Rule Nets (RN) are a symbiosis between Expert Systems (based on rules) and Petri Nets (PN) [5], in such a way that facts resemble places and rules are close to transitions. Similarly, a fact may be true or false; on the other hand, since a place may be marked or not, a rule (like a transition) may be sensitized or not. In the first case, it can be fired, thus changing the state of the system.

Like Petri Nets, RN admit a graphic representation as well as a matricial one; additionally, it also accepts a grammatical notation as a production rule, which drastically simplifies the design, thus avoiding the typical problems associated with PN.

Every control system works with input and output variables, among other internal variables. In the case of RN, each variable has associated facts (corresponding to their possible status). For each associated fact, the set of complementary facts is defined as the one formed by the rest of facts associated with that variable.

Mathematically, a RN is a pair $RN = <F, R>$, where:

F is the set constituted by all the facts associated to all system variables (the ordinal of F is assumed to be f).

R is the set of rules. Each rule has one or several antecedents as well as one or several consequent facts (the ordinal of R is assumed to be r).

In order to represent a RN in matricial form, the following structures are defined:

Matrix A of antecedents. It is an $r \cdot f$ matrix where element A_{ij} is 1 if fact f_j is antecedent in rule r_i, or 0 if it is not.

Matrix C of consequents. It is a $r \cdot f$ matrix where element C_{ij} is 1 if fact f_j is consequent in rule r_i, or 0 if it is not.

Matrix D of dismarking. It is a $r \cdot f$ matrix where element D_{ij} is 0 if element C_{ik} is 1 and fact f_j is complementary of fact f_k; if it is not so, D_{ij} is 1.

Vector S of states. It is a vector with f components, so that each component corresponds to a fact of the system. In this vector, component S_i is 1 if fact f_i is true and 0 if it is false.

Rule r_j is sensitized in a determined states S if the two following conditions are met:

$$\bar{S} \wedge A^j = 0 \tag{1}$$

$$\bar{S} \vee C^j \neq 0 \tag{2}$$

(where from now on X^j will denote the jth row of matrix X).

The firing of rule r_j sensitized in the S_m state yields a change in the state vector into a new one, S_n, which may be calculated from the following equation (3):

$$\bar{S}_n = (\bar{S}_m \wedge C^j) \vee D^j \tag{3}$$

It can be easily demonstrated that all operations necessaries to run the RN are implemented by means of easy logical operations. It should also be remarked that there are other figures dealt with in the RN that are useful in the automation of processes, such as timers, etc., as well as the possibility of obtaining the dynamic properties (vivacity, non-cyclicity, determinism, etc.) owned by a system.

4 System description

The implemented control system is composed by two levels. Upper layer has been implemented by means of Rule Nets, which allows for planning -in a simple way- the analyses of the system, opening and closing the necessary electrovalves, and making the sensors read the values when appropriate. Also, RN performs discrete control loops [6] as fan, vent, illumination and irrigation control, executes diagnoses and evaluates the obtained values in order to detect and determine the nature of possible malfunctions in the measurement subsystem, and provides the user with a powerful, though simple interface. Nevertheless, rule-based expert systems permit the information to be handled in a conceptual way, that is to say, by assigning logical concepts to the values of the variables.

For the implementation of the land analysis subsystem proposed, an analogical treatment of the results obtained from the measurement system (as PH land, composition, humidity, sulfates, etc.) is required. Therefore it is necessary to complement the functions of the RN-based expert system with the utilization of non-linear reasoning systems, such as neural networks.

In the hierarchical proposed system, all the results obtained by the measurement subsystem are not directly handled by the expert system, but they are evaluated by the neural network, which provides a quality index of the analyzed land, upgrading the value of RN variables, that are taken into account together with others variables directly controlled by the RN, for the global diagnosis and consequently to carry out the opportune actions in order to maintain the optimum microclimate and land composition.

The implementation of the above mentioned hierarchical system is summarized in Figure 1, which also shows the already commented interaction among the different levels. By means of combination of both layers, a very powerful control system has been achieved.

The fault tolerant characteristics of the system must be highlighted. When being a hierarchical system, the upper level takes charge from the topic of fault tolerance. When a failure in the system is detected, the upper level takes the control of the system, evaluating the failure and deducing from the characteristics of it the correction actions that should be carried out for the reconfiguration of the system (operation in degraded way) or those guided to drive the system into a safe failure state.

A design, programming and training environment has been developed, providing user and designer access. It offers the user a simple and intuitive interface for

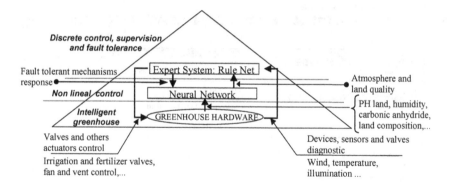

Fig. 1. Hierarchical control system architecture and interrelations.

the specification of the desired behavior of the expert system with no specific programming abilities required, by means of a rule-based language (see Figure 2). First the system variables should be specified. There are three types, *input variables* (sensors and other data acquisition devices and variables associated with neural networks results), *internal* and *output variables* (valves and other actuators). Later on, the possible values for each variable and facts should be specified. Finally to introduce the rules it is enough with selecting graphically the antecedents and consequents for each one.

For the lower layer specification (neural networks that take charge of the non-lineal control) the development tool also provides a powerful environment that allows the design and train of neural networks (see Figure 3), following the stages that are exposed in the following section. Once the user carries out the design it is possible to evaluate the behavior of the system, as well as if the coherence of the information is guaranteed.

All these features make the proposed system easily profitable for users in general, who can start without important computer knowledge to program their own control systems. This is very interesting in applications such as greenhouses automation.

5 Results and discussion

The system described has been applied to the control of a greenhouse as previously it has been indicated. For the correct identification of land quality, the neural network has been trained [7] by means of a significant number of samples covering a wide range of lands (with different composition).

The first step for setting up the neural network consists of defining the representative variables for the determination of land quality (see Section 1).

One time fixed the desired inputs and outputs of the neural network, the next stage is the selection of the most suitable network topology for solving the problem. This topology makes reference to the number of layers by which it is formed, the

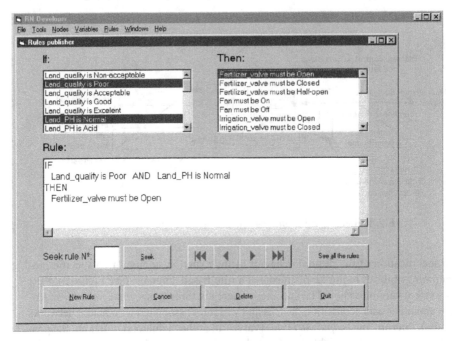

Fig. 2. Rule Nets specification tool.

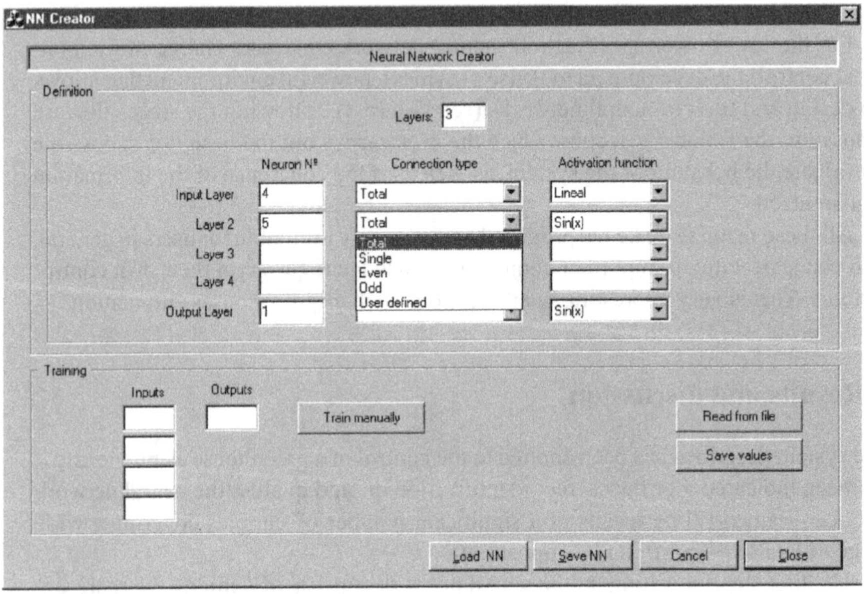

Fig. 3. Neural Networks design and train tool.

number of neurons on each layer, the type of transfer function of each neuron, and the interconnection among the neurons of each layer.

To begin with, there were two alternatives: (a) a feed-forwarded neural network; input neurons only receive external inputs, neurons of each layer are connected only with neurons of the following layer, and output neurons are not connected to any other neuron of the neural network. They are able to solve any problem that does not require the knowledge of previous inputs; (b) a recurrent neural network: there are no cycles, i.e. a neuron of a layer does not only receive inputs from lower layers but also from upper layers. These nets are capable of deciding their input as a function of previous net status.

As for the number of layers, it defines the power of the neural network, thus limiting the complexity of the problems to be solved. At least we will need two layers (an input and an output layer), with any number of layers between them. Finally, the number of neurons per layer affects the complexity of the zones that can be differentiated, as well as the complexity of the functions that can simulate. It can be considered as the degrees of freedom a certain function can take. But we had to reach a compromise in this sense. On one hand, the more neurons on each layer, the higher the risk to lose the always appreciated generalization characteristics of neural networks. In this case, the system would behave itself in a complicated, undesirable way. On the other, if our neural network does not have sufficient neurons on any of the layers, it will not be able to adequate itself to relatively complex functions, and at most we will achieve a coarse generalization of the function we wanted. To get a suitable number of neurons is somewhat troublesome and for this purpose the procedure we have adopted has consisted of performing tests until obtaining acceptable results.

Taking into account all these considerations, in the present work there was no need to resort to feedback in the system. With respect to the number of levels and the number of neurons at each level, no predetermined procedure can be used to optimize them; instead both the experience and the experimentation will be responsible for leading us to define the best values for these two parameters in each application. The starting point was four input neurons (one for each analytical parameter) and one output neuron; thereafter the number of intermediate levels was being altered and so was the number of neurons of these levels, until the most suitable configuration was obtained. The best results were provided by a single five-neuron layer, it giving rise to a 4/5/1 topology.

Once the neural network topology has been selected, the next step is its learning (training). It consists of deducting -by means of a series of learning functions- the set of weights that produce the correct outputs. For this purpose, we have exposed the neural network to an input pattern until an output has been obtained. The comparison between this output and the expected one provides a difference which is fed again in the opposite way. Each neuron modifies the weight of its inputs to get closer to this value; once the weights of the entire network have been changed, the process is repeated with another input pattern. Some iterations of the pattern set are then carried out in such a way that, with the passing of the values, weights are slightly modified, thus approximating to those desired.

However, learning is not an easy task, since an excessive training causes the specialization of the neural network, which would only be able to recognize sample elements as valid, and therefore could not generalize [9]. On the other hand, the learning process may encounter local minimum values of the function of weight calculus, which would not allow the NN to get closer to the desired solution of the problem.

In our case, the training process was carried out using a series of input value sets along with their correct evaluation provided by an expert. From these values a sufficiently big sample was taken at random to allow for a complete training of the neural network covering all possible working points (training values); another series of values (checking values) was reserved to be utilized as an excellence index in the training process of the neural network. For this purpose, a comparison has been made by studying the error committed in the responses against training values and that produced when checking values were employed.

Both errors are calculated through a least square regression, i.e. by evaluating the error as the sum of the squared difference between each one of the results obtained with the already trained neural network and the corresponding theoretical value. From this training procedure a conclusion is reached in the sense that the most adequate topology for solving this problem is the so-called 4/5/1, that is to say, four input neurons, 5 intermediate neurons, and one output neuron. From the latter, a numerical value (ranging from 0 to 5) for land quality -state- is obtained (which tends to imitate the judgement of an expert).

After the design process to determine the suitable rules for the RN, the best NN topology and the appropriate training, the system was executed in an industrial PC equipped with the required I/O module, reaching all the proposed objectives and overcoming the prepared tests.

6 Conclusions

A hierarchical architecture for the intelligent greenhouse control has been presented. Two levels can be distinguished: lower level performs non-lineal control with the information obtained from sensors and other acquisition systems by means of neural networks. In the other hand, the upper level based on discrete control techniques, not only controls the discrete variables of the process but also takes decisions over the system, supervises the system function and diagnoses failures.

The system performance was completely successful and both control techniques (Rule Nets and Neural Networks) work together in a perfect manner, guaranteeing the meeting of the proposed objectives.

This intelligent system is highly flexible due to its modularity. The choice of a different neural network does not alter the performance of the system as a whole.

Very interesting results being obtained and future developments will fulfill the automation in this field.

References

1. Yen J, Langari R, Zadeh A (1995) Industrial Applications of Fuzzy Logic and Intelligent Systems. IEEE Trans., Piscataway, NJ
2. Peris M, Bonastre A, Ors R (1998) Distributed Expert System for the monitoring and control of chemical processes. Laboratory robotics and automation, Ed. Wiley & Sons
3. Kung SY (1993) Digital Neural Networks. Prentice Hall
4. Peris M, Chirivella V, Ors R, Serrano J, Bonastre A, Martínez S (1994) Rule Nets: Application to the advanced automation of a flow injection analysis system. Chemometrics and intelligent Laboratory Systems, number 26. Ed. Elsevier Science Publishers
5. Silva M (1985) The Petri Nets in Automation and Computer science. Editorial AC, Madrid
6. Moody JO, Antsaklis, Panos J (1998) Supervisory control of discrete event systems using Petri Nets. Kluwer Academic, Boston
7. Anthony M, Bartlett P (1999) Neural Network Learning: Theoretical Foundations. Cambridge University Press
8. Pal S, Mitra SJ (1999) Neuro-Fuzzy Pattern Recognition: Methods in Soft Computing. Wiley & Sons, New York

Artificial Neural Networks Design using Evolutionary Algorithms *

P.A. Castillo[1], M.G. Arenas[1], J.J. Castillo-Valdivieso[2], J.J. Merelo[1], A. Prieto[1], and G. Romero[1]

[1] Department of Architecture and Computer Technology
University of Granada. Campus de Fuentenueva. E. 18071 Granada (Spain)
[2] Department of Technology
IES Europa. C Miguel Angel Blanco s/n. Aguilas, Murcia (Spain)
e-mail: pedro@atc.ugr.es   URL: http://geneura.ugr.es

Summary. Although a great amount of algorithms have been devised to train the weights of a neural network for a fixed topology, most of them are hillclimbing procedures, which usually fall in a local optimum; that is why results obtained depend to a great extent on the learning parameters and the initial weights as well as on the network topology.

Evolutionary algorithms have proved to be very effective and robust search methods to locate zones of the search space where finding good solutions, even if this space is large and contains multiple local optimum.

This paper intends to be an updated review of the field of design of hybrid EA/ANN methods building on previous reviews such as [1, 2, 3], and also paying special attention to aspects such as variation operators, software and applications.

Keywords: Hybrid Methods, Evolutionary Algorithms, Artificial Neural Networks, Optimization

1 Introduction

Artificial neural networks (ANN) have successfully been used in a large amount of applications [4, 5, 6], nevertheless the network design creates several problems [7] since it is necessary to establish several parameters.

These problems usually are approached using a method that optimizes these parameters that determine the architecture. In general these methods of optimization are fitted in two groups: constructive/pruning methods [8, 9, 10] and methods that make ANNs evolve (hybrid methods, since they combine the methodologies of ANN and evolutionary algorithms - EA -), [1, 2, 3].

The main problem of constructive and pruning algorithms is that they are, anyways, gradient descent based methods, so they might reach a sub-optimal solution

---

* An extended version is available at: http://geneura.ugr.es/~pedro/papers/

(local optimum). In addition, criteria to add and to remove nodes depend on the network architecture and the problem to solve.

EAs search more or less randomly the space of solutions, paying special attention to those zones in which the value of the evaluation function is maximum (higher ANN capability), whereas ANN classic training algorithms are algorithms of gradient descent, that in an iterative form reduce the error until a minimum is reached.

EAs can carry out a global search, locating the ANN near an optimal point (global). Then, by means of the ANN training algorithm (local search), the optimal point can be reached [11].

This paper reviews the different approaches in which evolutionary algorithms and artificial neural networks have been combined to optimize the different design parameters of the latter, paying special attention to the specific genetic operators used in these methods, and the main libraries to evolve artificial neural networks.

The remainder of this paper is structured as follows: Section (2) presents a comprehensive review of the approaches found in the bibliography and the design decisions made to evolve ANNs. Section 3 presents the main *genetic operators* found in the bibliography. And finally, section 4 examines available evolutionary computation libraries that combine EA and ANN.

2 ANN Design Using Evolutionary Algorithms

In the evolution of ANN, three main approaches can be found: *evolution of connection weights*, *network architecture evolution* and *learning parameters and rule evolution*. Several approaches to evolve almost any kind of ANN can be found in the bibliography, however they can be divided in two broad fields: 1) evolving generic ANN, searching for the best ANN, despite its structure (avoiding to restrict the search to an specific area of the search space); 2) select a prefixed architecture easily evolutionable (it is simpler to evolve the network having some previous knowledge on the problem and well-known training algorithms can be used).

2.1 Choosing the coding and representation

In the evolution of the connection weights as well as in the evolution of network architecture, genetic operators are fundamental in the operation of the EA. However, the form in which they are applied depends on *how networks are represented* (binary or real **representation**) and on the *amount of information each individual of the population will codify* (direct or indirect **coding**).

Binary vs. real representation

Genetic algorithms usually use binary *strings* to codify candidate solutions and they use a representation scheme so that each weight is represented using several bits. An advantage of the binary representation [12, 13] is its simplicity and generality. As a disadvantage, it is necessary to make a balance between the precision and the length of the individual.

On the other hand, *floating point representation* has been proposed; each weight is represented by a real number [14, 15]. Although floating point representation is more precise, a disadvantage is that the search space is extremely large [16].

Some researchers present **hybrid methods that avoid the task of codifying the network** in a population individual. In these papers, the EA does not need to codify the network, instead of that, the data structure that represents the neural net is evolved using several specific genetic operators [17, 18, 19].

Direct vs. indirect coding

Architecture evolution takes two phases: first consists on deciding the coding of the network in the genotype of the individual. Second one is related to the genetic operators used for searching (see section 3). The key when deciding the coding of an architecture is in the amount of information that each individual will codify. It is possible either to use the maximum detail, if each connection and node of the architecture is codified (*direct coding*), or to codify only the more important parameters of the architecture, such as the number of nodes in each layer (*indirect coding*).

Using *direct coding*, each connection is specified by means of its binary representation [20, 21]. In general, to represent a network with N neurons, a matrix of $N \times N$ is used.

In order to reduce the architecture representation length, many authors have used the *indirect coding*, [22, 21], codifying only some of the characteristics of the architecture in the individual. Several indirect coding methods can be found, although most of them are based on using of *parametric representation* [22], or on *construction rules* [23, 24].

2.2 Evolving connection weights

The initial weights of a neural network are very important to obtain a fast convergence in backpropagation and other ANN training algorithms, since, depending on the point of the search space from which it has started (and that point is determined by the set of weights) better or worse solutions can be obtained when carrying out the network training [25].

Weight initialization

In the bibliography different methods of weight initialization can be found [25]. Simplest of all is based on making a random initialization [26], other methods require a statistical analysis of the training data, which make them less efficient, although probably more powerful.

Training weights

Training ANN usually is formulated as a problem of minimization of an error function between the desired outputs and the actual network output, while those weights are adjusted. BP and their variants have successfully been applied in several areas [4, 5], although it presents certain problems because it is based on gradient descent [7].

A way to solve the problems that present these methods is to directly evolve the connection weights. Thus, an EA can be used to globally search for an optimal set of weights, without using gradient information [14, 15, 27, 28]. In general, EAs are inefficient carrying out a local search, whereas they are very efficient making global searches. For that reason, evolutionary training can be improved if a local search method is used to tune the solution found by the EA.

2.3 Evolving network architecture

The network architecture includes the *network topology*, the *connectivity* and the *transference function* of each neuron. Up to now, the architecture design had been made manually, by an expert with the sufficient experience, by means of a process of trial and error. Automatic approaches more widely studied are incremental and pruning algorithms [8, 9, 10].

The design of an optimal network architecture can be formulated as a search problem in the architecture space, where each point represents a possible network architecture. This optimization problem can be solved more easily using an EA than using incremental or pruning methods.

Most of the research has been focused in the task of making evolution of the network topology [20, 29, 30, 31], leaving the transference function predefined. Several works demonstrate the importance of these functions in the ability of ANN [32, 33].

2.4 Evolving the learning rule

ANN training algorithms or learning rules depend on the type of architecture used. Due to the lack of knowledge on the network architecture, it is better to develop an automatic system to adapt the learning rule to the architecture and the problem to solve. Several models have been proposed [34, 35, 36, 37, 38], although most of them are focused on how the learning can modify or guide the evolution, and in the relation between the evolution of the architecture and the connection weights [34]. Few of them focus on the evolution of the learning rules [35, 36, 38, 31].

The current problem has been approached in different ways, first of which is based on the **search of the BP algorithm parameters** (learning rate and momentum) [22, 39]. On the other hand, *evolution of the learning rule* is oriented to provide a dynamic behavior to ANN [35].

2.5 Evolving the input vector dimensionality

Defining the ANN optimal input vector dimensionality can be formulated as a search problem, where we have a potential input vector and want to find a sub-set that contains the minimum number of elements (inputs) and simultaneously the network does not produce worse results than those produced using the complete vector. The variables selection has been faced using other methods such as Kohonen's *SOM* [6, 41] and *multidimensional scaling* [42]. Several authors have faced this problem by means of an evolutionary approach [43, 44, 45].

3 Specific Genetic Operators

One of the important issues when evolving ANNs is the selection of the genetic operators used in the EA, since the search accuracy will depend on these.

3.1 Mutation

Mutation operator have two main roles: either 1) tuning solutions or 2) in case another tuning operator is being used, changing the area in the search space.

Some authors [46] conclude that, in certain problems, to avoid falling and being trapped in local optimums, mutation operators should make big variations to move the population away from these areas. These big mutations move the population from local optimum, which degrade the value of the evaluation function in the first generations of the EA. Later on, population will be around the global optimum, which compensates the initial negative effects of these mutations. Then, using a tuning method (mutation operator that makes small changes or a local search operator), the global optimum can be reached.

3.2 Crossover

Purpose of crossover operator is to recombine those useful parts of the population individuals to form new solutions with the characteristics of both parents. The ability of the operator to do this depends on the problem at hand and on the way solutions are represented. A binary representation of the individuals makes suitable the use of a generic crossover operator. According to some authors, using crossover may cause negative effects in the ANN construction, since this operator operate correctly when it is possible to interchange construction blocks, and in ANN it is not clear what can be defined as a construction block due to the distributed nature of the information representation [47]. In RBF networks, in which the information is not distributed between all the network weights, the crossover operator is a very useful operator [48].

3.3 Incremental and decremental operators

Incremental operators start with small networks and increase them adding new randomly initialized units to the hidden layers. That increase of size can cause the MLP to have an excessive size: big networks are faster in the learning phase [49, 50], although the way to obtain a good generalization is using the simplest network [3, 51]. At the same time, the increase of size can lead in the problem of overfitting: small sized networks generalize well, although they are slow in the learning phase, while big networks (high number of weights) are faster in the learning phase but obtain a poor generalization [50, 49]. This is the reason why this operator is applied together with the following one. *Decrement operators* remove hidden units to obtain a smaller networks [52, 53, 50]. This method avoids the networks to grow too much.

In brief, these operators are efficient tools for searching the ANN architecture space, but a balance must be kept between incremental and decremental operators to aboid bloat or collapse of diversity through excessive decremental operator application.

3.4 Local search operators

In general, using local search operators is faster and lead to obtain more precise results than using bit-manipulation genetic operators. BP as local search operator acts (as proposed in [14, 54]) refining solutions so that those solutions cannot be improved locally. Compared to EA, this operator is much more efficient searching for the local optimum in an area. The EA could, eventually, reach the local optimum, although time to reach it would be higher, since tuning in EA is based on small mutations. A problem of these search operators is that some tuned individuals remain the best during the simulation (evolution stops), and tend to dominate the population due to its high fitness. Thus, the genetic search is altered, suffering a loss of diversity. When a local search operator is used, population converge quickly, since individuals tend to have many characteristics in common (reduction of the diversity).

4 Neurogenetic software

In the vast majority of cases, when a new hybrid algorithm wants to be tested by a researcher, he/she has to create his/her own neural net *and* evolutionary algorithm library, which means that coding basic classes and methods takes usually much more time than implementing the novel algorithm itself. There exist commercial software which implements genetic neural nets for a particular solution, such as the *Neuro-Genetic Optimizer* or GENETICA; however they do not offer the possibility of implementing new neurogenetic paradigms. There are free tools such as **EO+GPROP** [54, 55], **INANNA** [56], **SNNS/ENZO** [57] (C++ libraries, available under a license that allows free use by researchers).

5 Conclusions

This work presents an exhaustive review of the different approaches to design ANN using EA, paying special attention to the specific genetic operators used in these methods.

In the evolution of ANN, three main approaches can be found: *connection weights evolution*, *network architecture evolution*, and *learning rule evolution*.

Main advantage of the design of ANN using EA is in the ability of this to optimize the network parameters (initial architecture, connectivity, weights, learning rule), and its inherent parallelism (different networks can be trained simultaneously in different processors).

Nowadays, most of the methods commented in this work implement two or three of the approaches. They are only partial attempts of optimization of ANN, so it is not guaranteed to obtain the global optimum. In the short term, it would be interesting using several approaches at the same time, mainly in those cases in which there is little a priori knowledge about the problem, since in these circumstances, the use of trial and error or heuristic methods are not effective. Thus, a method that optimizes different, or most, ANN parameters (network size, initial weights, learning parameters, network input vector and learning/validation/test sets) for a prefixed architecture would be very useful.

It would be very useful to combine the training process and visualization of the evolutionary process [58]. If this idea is extended, using visualization the way the ANN operation will be better understood, and the evolutionary search and method speed will be improved, making possible fast evaluation of ANN.

References

1. X. Yao. A review of evolutionary artificial neural networks. *International Journal of Intelligent Systems, vol. 8, no. 4, pp. 539-567*, 1993.
2. X. Yao. Evolutionary artificial neural networks. *in Encyclopedia of Computer Science and Technology (A. Kent and J.G. Williams, eds.), vol. 33, pp. 137-170, New York, NY 10016: Marcel Dekker Inc.*, 1995.
3. X. Yao. Evolving artificial neural networks. *Proceedings of the IEEE, 87(9):1423-1447*, 1999.
4. S.S. Fels and G.E. Hinton. Glove-talk: a neural network interface between a data-glove and a speech synthesizer. *IEEE Trans. Neural Networks,vol.4,pp.2-8*, 1993.
5. S. Knerr, L. Personnaz, and G. Dreyfus. Handwritten digit recognition by neural networks with single-layer training. *IEEE Trans. on Neural Networks, vol. 3, pp. 962-968*, 1992.
6. L. Prechelt. PROBEN1 — A set of benchmarks and benchmarking rules for neural network training algorithms. Technical Report 21/94, Fakultät für Informatik, Universität Karlsruhe, D-76128 Karlsruhe, Germany, September 1994.
7. R.s. Sutton. Two problems with backpropagation and other steepest-descent learning procedures for networks. *In Proceedings of the 8th Annual Conference of the Cognitive Science Society, pp.823-831. Erlbaum, Hillsdale, NJ*, 1986.
8. S.E. Fahlman and C. Lebière. The Cascade-Correlation Learning Architecture. *Neural Information Systems 2. Touretzky, D.S. (ed) Morgan-Kauffman, 524-532*, 1990.
9. Y. Le Cun, J.S. Denker, and S.A. Solla. Optimal brain damage. *Neural Information Systems 2. Touretzky, D.S. (ed) Morgan-Kauffman, pp. 598-605*, 1990.
10. B. Hassibi, D.G. Stork, G. Wolff, and T. Watanabe. Optimal Brain Surgeon: extensions and performance comparisons. *In NIPS6, pp. 263-270*, 1994.
11. J.M Renders and S.P. Flasse. Hybrid methods using genetic algorithms for global optimization. *IEEE Transactions on Systems, Man, and Cybernetics. Part B: Cybernetics, Vol.26, No.2, pp.243-258*, 1996.
12. M. Srivinas and L.M. Patnaik. Learning neural network weights using genetic algorithms - improving preformance by search-space reduction. *in Proc. of 1991 IEEE International Joint Conference on Neural Networks (IJCNN'91 Singapore), vol. 3, pp. 2331-2336, IEEE Press, New York*, 1991.

13. D.J. Janson and J.F. Frenzel. Application of genetic algorithms to the training of higher order neural networks. *Journal of Systems Engineering, vol. 2, pp. 272-276*, 1992.
14. D.J. Montana and L. Davis. Training feedforward neural networks using genetic algorithms. *Proc. 11th Int. Joint Conf. on Artificial Intelligence, 762-767*, 1989.
15. V.W. Porto, D.B. Fogel, and L.J. Fogel. Alternative neural network training methods. *IEEE Expert, vol. 10, no. 3, pp. 16-22*, 1995.
16. Z. Michalewicz. *Genetic Algorithms + Data Structures = Evolution Programs , Third, Extended Edition.* Springer-Verlag, 1996.
17. P.A. Castillo, J. González, J.J. Merelo, V. Rivas, G. Romero, and A. Prieto. SA-Prop: Optimization of Multilayer Perceptron Parameters using Simulated Annealing. *Lecture Notes in Computer Science, ISBN:3-540-66069-0, Vol. 1606, pp. 661-670, Springer-Verlag*, 1999.
18. V.M. Rivas, P.A. Castillo, and J.J. Merelo. Evolving RBF Neural Networks. *Lecture Notes in Computer Science. J. Mira and A. Prieto (Eds.). ISSN:0302-9743, Volume I, LNCS 2084, pp.506-513.*, 2001.
19. P.A. Castillo, J.J. Merelo, V. Rivas, G. Romero, and A. Prieto. G-Prop: Global Optimization of Multilayer Perceptrons using GAs. *Neurocomputing, Vol.35/1-4, pp.149-163*, 2000.
20. E. Alba, J.F. Aldana, and J.M. Troya. Fully Automatic ANN Design: A Genetic Approach. *Lecture Notes in Computer Science, Vol. 686, pp. 399-404, Springer-Verlag*, 1993.
21. S. Roberts and M. Turega. Evolving neural networks: an evaluation of enconding techniques. *in Artificial Neural Nets and Genetic Algorithms, Pearson,Steele and Albrecht Eds., pp.96-99, Springer-Verlag, ISBN 3-211-82692-0*, 1995.
22. S.A. Harp, T. Samad, and A. Guha. Towards the genetic synthesis of neural networks. *in Proc. of the 3th Int. Conf. on Genetic Algorithms and Applications (Schaffer, ed.), pp. 360-369, Morgan Kaufmann, San Mateo, CA*, 1989.
23. H. Kitano. Empirical studies on the speed of convergence of neural network training using genetic algorithms. *in Proc. of the Eighth Nat'l Conf. on AI (AAAI-90), pp. 789-795, MIT Press, Cambridge, MA*, 1990.
24. F. Gruau and D. Whitley. Adding learning to the cellular development of neural networks: Evolution and the Baldwin efect. *Evolutionary Computation, Volume I, No. 3, pp. 213-233*, 1993.
25. G. Thimm and E. Fiesler. Neural network initialization. *Lecture Notes in Computer Science, Vol. 930, pp. 535-542, Springer-Verlag*, 1995.
26. J.F. Kolen and J.B. Pollack. Back Propagation is Sensitive to Initial Conditions. *Technical Report TR 90-JK-BPSIC. Laboratory for Artificial Intelligence Research, Computer and Information Science Department*, 1990.
27. P. Osmera. Optimization of neural networks by genetic algorithms. *Neural Network World, vol. 5, no. 6, pp. 965-976*, 1995.
28. M. Koeppen, M. Teunis, and B. Nickolay. Neural network that uses evolutionary learning. *in Proceedings of the 1997 Internaational Conference on Evolutionary Computation, ICEC'97, (Piscataway, NJ, USA), pp. 1023-1028, IEEE Press*, 1997.
29. D. White and P. Ligomenides. GANNet: A Genetic Algorithm for Optimizing Topology and Weights in Neural Network Design. *Lecture Notes in Computer Science, Vol. 686, pp. 322-327, Springer-Verlag*, 1993.
30. Y. Liu and X. Yao. A population-based learning algorithm which learns both architectures and weights of neural networks. *Chinese Journal of Advanced Software Research (Allerton Press, Inc., N.Y. 10011), vol. 3, no. 1, pp. 54-65*, 1996.

31. A. Ribert, E. Stocker, Y. Lecourtier, and A. Ennaji. Optimizing a Neural Network Architecture with an Adaptive Parameter Genetic Algorithm. *Lecture Notes in Computer Science, Vol. 1240, pp. 527-535, Springer-Verlag,* 1994.

32. G. Mani. Learning by gradient descent in function space. *in Proc. of IEEE int'l Conf. on System, Man, and Cybernetics, (Los Angeles, CA), pp. 242-247,* 1990.

33. B. DasGupta and G. Schnitger. Efficient approximation with neural networks: a comparison of gate functions. *tech. rep., Delp. of Computer Sci., Pennsylvania State Univ., University Park, PA 16802,* 1992.

34. J. Paredis. The evolution of behavior: some experiments. *in Proc. of the First Int'l Conf. on Simulation of Adaptive Behavior: From Animals to Animats (J. Meyer and S.W. Wilson, eds.), MIT Press, Cambridge, MA,* 1991.

35. D.J. Chalmers. The evolution of learning: an experiment in genetic connectionism. *in Proc. of the 1990 Connectionist Models Summer School (Touretzky, Elman, and Hinton, eds.), pp.81-90, Morgan Kaufmann, San Mateo, CA,* 1990.

36. S. Bengio, Y. Bengio, J. Cloutier, and J. Gecsei. On the optimization of a synaptic learning rule. *in Preprints of the Conference on Optimality in Artificial and Biological Neural Networks, (Univ. of Texas, Dallas),* 1992.

37. D.H. Ackley and M.S. Littman. Interactions between learning and evolution. *in Artificial Life II, SFI Studies in the Sciences of Complexity, vol. X (C.G. Langton, C. Taylor, J.D. Farmer, and S. Rasmussen, eds.), (Reading, MA), pp. 487-509, Addison-Wesley,* 1991.

38. J. Baxter. The evolution of learning algorithms for artificial neural networks. *in Complex Systems (D. Green and T. Bossomaier, eds.), pp. 313-326, IOS Press, Amsterdam,* 1992.

39. H.B. Kim, S.H. Jung, T.G. Kim, and K.H. Park. Fast learning method for backpropagation neural network by evolutionary adaptation of learning rates. *Neurocomputing, vol. 11, no. 1, pp. 101-106,* 1996.

40. T. Kohonen. The Self-Organizing Map. *Procs. IEEE, vol. 78, no.9, pp. 1464-1480,* 1990.

41. T. Kohonen. Self-organizing maps. *Segunda Edición, Springer,* 1997.

42. T.F. Cox and M.A.A. Cox. Multidimensional scaling. *London: Chapman and Hall,* 1994.

43. Z. Guo and R.E. Uhrig. Using genetic algorithms to select inputs for neural networks. *in Proc. of the Int'l Workshop on Combinations of Genetic Algorithms and Neural Networks (COGANN-92)(D. Whitley and J.D. Schaffer, eds.), pp. 223-234, IEEE Computer Society Press, Los Alamitos, CA,* 1992.

44. F.Z. Brill, D.E. Brown, and W.N. Martin. Fast genetic selection of features for neural network classifiers. *IEEE Transctions on Neural Networks, vol. 3, pp. 324-328,* 1992.

45. P.R. Weller, R. Summers, and A.C. Thompson. Using a genetic algorithm to evolve an optimum input set for a predictive neural network. *in Proceedings of the 1st IEE/IEEE International Conference on Genetic Algorithms in Engineering Systems: Innovations and Applications (GALESIA'95), (Stevenage, England), pp. 256-258, IEE Conference Publication 414,* 1995.

46. M. Land. Evolutionary algorithms with local search for combinatorial optimization. *PhD thesis, Computer Science and Engr. Dept. - Univ. California. San Diego,* 1998.

47. D.E. Rumelhart and J.L. McClelland. Parallel Distributed Processing: Explorations in the Microstructures of Cognition. *Cambridge, MA: MIT Press,* 1986.

48. P.J. Angeline. Evolving basis functions with dynamic receptive fields. *in Proceedings of the 1997 IEEE International Conference on Systems, Man, and Cybernetics. Part 5 (of 5), (Piscataway, NJ, USA), pp. 4109-4114, IEEE Press,* 1997.

49. I. Bellido and G. Fernandez. Backpropagation Growing Networks: Towards Local Minima Elimination. *Lecture Notes in Computer Science, Vol. 540, pp. 130-135, Springer-Verlag,* 1991.

50. G. Bebis, M. Georgiopoulos, and T. Kasparis. Coupling weight elimination with ge-
    netic algorithms to reduce network size and preserve generalization. *Neurocomputing 17
    (1997) 167-194*, 1997.
51. R.D. Reed. Pruning algorithms – a survey. *IEEE Transactions on Neural Networks, 4(5):
    740-744*, 1993.
52. T. Jasic and H. Poh. Analysis of Pruning in Backpropagation Networks for Artificial
    and Real World Mapping Problems. *Lecture Notes in Computer Science, Vol. 930, pp.
    239-245, Springer-Verlag*, 1995.
53. M. Pelillo and A. Fanelli. A Method of Pruning Layered Feed-Forward Neural Networks.
    Lecture Notes in Computer Science, Vol. 686, pp. 278-283, Springer-Verlag, 1993.
54. P.A. Castillo, J. Carpio, J.J. Merelo, V. Rivas, G. Romero, and A. Prieto. Evolving Multi-
    layer Perceptrons. *Neural Processing Letters, vol. 12, no. 2, pp.115-127. October*, 2000.
55. M. Schoenauer, M. Keijzer, J.J. Merelo, and G. Romero. Eo: Evolving objects. *Available
    from http://eodev.sourceforge.net*, 2000.
56. M. Grönroos. INANNA. *Available from http://inanna.sourceforge.net*, 2000.
57. H. Braun and T. Ragg. Enzo evolutionary network optimizing system. *Available from
    http://i11www.ira.uka.de/fagg*, 2000.
58. G. Romero, P.A. Castillo, J.J. Merelo, and A. Prieto. Using SOM for Neural Network Vi-
    sualization. *Lecture Notes in Computer Science. J. Mira and A. Prieto (Eds.). ISSN:0302-
    9743, Volume I, LNCS 2084, pp.629-636.*, 2001.

TACDSS: Adaptation Using a Hybrid Neuro-Fuzzy System

Cong Tran[1], Ajith Abraham[2], and Lakhmi Jain[1]

[1] School of Electrical and Information Engineering
University of South Australia
Adelaide, Australia
Email: Tramcm001@students.unisa.edu.au.
lakhmi.jain@unisa.edu.au
[2] Department of Computer Science, Oklahoma State University
700 N Greenwood Avenue, Tulsa, USA
Email: ajith.abraham@ieee.org

Summary. Normally an intelligent decision support system is build to solve complex problems involving multi-criteria decisions. The knowledgebase is the vital part of the decision support system containing the knowledge or data that is used for decision-making. Several works have been done where engineers and scientists have applied intelligent techniques and heuristics to obtain optimal decisions from imprecise information. In this paper, we present a hybrid neuro-fuzzy technique for the adaptive learning of Takagi-Sugeno type fuzzy *if-then* rules for the Tactical Air Combat Decision Support System (TACDSS). Experiment results clearly demonstrate the efficiency of the proposed technique. Some simulation results demonstrating the difficulties to decide the optimal number and shape of the membership functions are also provided.

1 Introduction and Related Research

Several decision support systems have been applied mostly in the fields of medical diagnosis [7], business management, control system, command and control of defence and air traffic control [3]. In most cases, decision support systems are designed for a particular application [4]. Several decision support techniques have been developed using artificial intelligence techniques. These techniques try to mimic the human way of reasoning and interpretation using expert systems, fuzzy inference systems, rough sets and neural network learning methods. We require a system that is able to learn from the data/information and automatically provide the *if-then* decision rules as desired by the user. The decision rules obtained will be further evaluated by computer simulation or by a human expert (knowledge). If the obtained decision rules are not satisfactory, the learning process is to be fine tuned to get the desired results. Figure 2 shows the database learning process in the decision support system. Our adaptive database-learning framework using evolutionary algorithms [4] could deliver fuzzy *if-then* rules. The two disadvantages of the evolutionary approach are the requirement of expert knowledge to model the objective function (*decision score* to set up fuzzy rules) and the computational complexity leading to huge amount of simulation time. There are several techniques for extracting fuzzy decision rules but most of them use "training data" to create the fuzzy rules. Cattral developed a technique of RAGA (rule acquisition using genetic algorithm) based data mining system suitable for supervised and unsupervised knowledge extraction from large and possible noisy database [2]. Jagielska used the neural networks to extract the fuzzy rules [6]. Hung combined unsupervised learning technique (self organizing maps) and the supervised technique (learning vector quantisation) to create the two stage-training network for generating the fuzzy decision rules [5]. In this paper, we propose a neural network learning technique for the automatic adaptation of fuzzy inference system. In Section 2, we introduce the some theoretical concepts on neuro-fuzzy systems followed by the complexity of the problem in tactical air combat environment (decision-making process) in Section 3. Experimentation results are provided in Section 4 and some conclusions are also provided towards the end.

2 Hybrid Neuro-Fuzzy Model

The advantage of neural network (NN) is the adaptive learning capability. The fuzzy sets introduced by Zadeh [11] has provided an inference methodology that enables approximate human reasoning capabilities, which could be applied to knowledge-based systems. The main disadvantage of Fuzzy Logic (FL) is the requirement of expert knowledge to set up the knowledge base (*if-then* rules). FL cannot learn the data and then adaptive the fuzzy rules. Neural networks can learn from the data and automatically adjust their connection weight between the layers. An analysis reveals that the drawbacks pertaining to these approaches seem complementary and therefore it is natural to consider building an integrated system combining the concepts. While the learning capability is an advantage from the viewpoint of fuzzy inference system the formation of linguistic rule base will be advantage from the viewpoint of neural networks. During the last decade, several neuro-fuzzy systems have been developed and a summary could be obtained from [1].

3 Decision Making in Tactical Air Combat

We considered a case study based on a tactical environment problem. The air operation division of Defence Science Technology Organization (DSTO) and our research team has a collaborative project to develop a tactical environment decision support system for a pilot or mission commander in tactical air combat. In Figure 1 a typical scenario of air combat tactical environment is presented. The Airborne Early Warning and Control (AEW&C) is performing surveillance in a particular area of operation. It has two hornets (F/A-18s) under its control at the ground base as shown "+" in the left corner of Figure 1. An air-to-air fuel tanker (KB707) "?" is on station and the location and status are known to the AEW&C. Two of the hornets are on patrol in the area of combat air patrol (CAP). Sometime later, the AEW&C on-board sensors detects 4 hostile aircrafts (Mig-29) that is shown as "O". When the hostile aircrafts enter the surveillance region (shown as dashed circle) the mission system software is able to identify the enemy aircraft and its distance from the Hornets in the ground base or in the CAP.

The mission operator has few options to make a decision on the allocation of hornets to intercept the enemy aircraft.

- Send the Hornet directly to the spotted area and intercept.
- Call the Hornet in the area back to ground base and send another Hornet from the ground base.
- Call the Hornet in the area for refuel before intercepting the enemy aircraft.
- The mission operator will base his decisions on a number of decision factors, such as:
  - Fuel used and weapon status of hornet in the area.
  - Interrupt time of Hornet in the ground base in the Hornet at the CAP to stop the hostile.
  - The speed of the enemy fighter aircraft and the type of weapons it possesses.
  - The information of enemy aircraft (number and type of aircrafts, weapon, etc.)

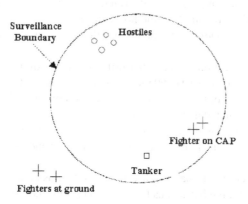

Fig. 1. A simple scenario of the air combat.

From the above simple scenario, it is evident that there are several important decision factors to be taken into account to make the overall decision. For easy demonstration of our proposed approach, we will simplify the problem by handling only a few important decision factors such as "*fuel status*", "*weapon possession status*" and "*interrupt time*" (Hornet in the ground base and the Hornet in the area of CAP) and the "*danger situation*" (friend or hostile), which is commonly known as situation awareness in a battlefield.

3.1 The Knowledge of Tactical Air Combat Data

How human knowledge could be extracted to a database? Very often people express knowledge as natural language (spoken language) or using letters or symbolic terms. The human knowledge can be analysed and converted into an information table. There are several methods to extract human knowledge. DSTO researchers use the Cognitive Work Analysis (CWA) [8] and the Cognitive Task Analysis (CTA) [9]. The CWA is a technique to analyse, design and evaluate the human computer interactive systems. The CTA is a method to identify cognitive skill, mental demands and needs to perform task proficiency. The CTA focuses on describing the representation of the cognitive elements that defines goal generation and decision making. The CTA is a reliable method to extract the human knowledge because it is based on the observations or an interview. Militallo has clearly explained the interview method to analyse the different tasks of a problem and to extract the human knowledge [9].

We made use of the CTA technique to make an interview with the expertise in the AEW&C of DSTO in Australia to set up the expert knowledge base to formulate a key knowledge for building up the complete decision support system. For the simple TACS discussed before, we have four decision factors that could affect the final decision options of "hornet in the CAP" or "hornet at the ground base". These are "fuel status" that is the quantity of fuel available to perform the intercept, the "weapon possession status" presenting the state of available weapons inside the hornet, the "interrupt time" which is required for the hornet to fly and interrupt the hostile and the "danger situation" providing information whether the aircraft is a friend or hostile.

Each of the above-mentioned factors has difference range of unit such as the fuel (0 to 1000 litres), interrupt time (0 to 60 minutes), weapon status (0 to 100 %) and the danger situation (0 to 10 points). The following are two important decision selection rules, which were formulated using expert knowledge:

The decision selection will have small value if the fuel used being too low, the interrupt time is too long, the hornet has low weapon status and the danger situation of FOE is high value.

The decision selection will have high value if the fuel used being full, the interrupt time is fast enough, the hornet has high weapon status and the danger situation of FOE is low value.

In the TACS environment, decision-making is always based on all states of all the decision factors. But sometime, a mission operator/commander can make a decision based on an important factor, such as the fuel used of the hornet is too low, enemy has more power weapon, quality and quantity of enemy aircraft etc.

3.2 Modeling TACDSS Using Adaptive Network based Fuzzy Inference System

This section will explain the modelling aspects of the TACDSS using the hybrid neuro-fuzzy learning technique ANFIS. The ANFIS is the hybrid neural-fuzzy system, which was developed by Jang [10]. This ANFIS is based on the architecture of the Takagi-Sugeno fuzzy inference system. The six-layered architecture of ANFIS is shown in Figure 3.2.

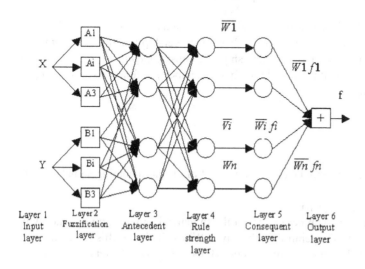

Fig. 2. Architecture of ANFIS

Suppose there are two Input Linguistic Variables (ILV) X and Y and each ILV have three membership functions (MF) A_1, A_2 and A_3 and B_1, B_2 and B_3 respectively.

Takagi-Sugeno type fuzzy *if-then* rule is set up as following:

Fuzzy Rule$_i$: *If* x isA_i and$y$is$B_i$then $f_i = p_ix + q_iy + r_i$

Where i is an index $i = 1,2,3$ and p, q and r are linear parameters of function f

Some layers of the ANFIS have the same number of nodes and the nodes in the same layer have similar functions. If the output of nodes in layer 1 is denoted as $O_{1,,i}$ with l as the layer number and i is neuron number of next layer. The function of each layer is described as follows.

Layer 1

The output of this node is the input values of the ANFIS

$O_{1,x} = x$

$O_{1,y} = y$

For TACS the four inputs are *fuel status, weapons inventory levels, time intercept* and the *danger situation*.

Layer 2

The output of nodes in this layer are presented as $O_{1,ip,i}$ where ip is the ILV and m is the degree of membership function of particular MF.

$O_{2,x,i} = \mu_{Ai(x)}$ or $O_{2,y,i} = \mu_{Bi(y)}$ for $i = 1,2$ and 3

With three MFs for each input variable, "fuel status" has 3-membership functions: *full, half* and *low,* "time intercept" has 3 membership functions: *fast, normal* and *slow,* "weapon status" has 3 membership functions: *sufficient, enough* and *insufficient* and the "danger situation of FOE" has 3 membership functions: *very danger, danger* and *endanger.*

Layer 3

The output of nodes in this layer is the product of all the incoming signals which denotes

$O_{3,n} = W_n = \mu_{Ai}(x) \times \mu_{Bi}(y),$

where $i = 1,2$ and 3, n is number of fuzzy rule. In general, any T-norm operator will perform the fuzzy AND operation in this layer. With 4 ILV and MFs for each input variable the TACS decision support system will have 81 fuzzy *if-then* rules.

Layer 4

The node in this layer is an adaptive node with the node function calculates the ratio of the ith fuzzy rules firing strength (RFS) to the sum of RFS.

$O_{4,n} = \overline{wn} = \dfrac{wn}{\sum\limits_{n=1}^{81} wn}$ where $n = 1,2,..,81$

The number of neuron in this layer is the same as the number of neuron in layer 3 that is 81 neurons. The output of this layer is also called *normalized firing strength.*

Layer 5

The node in this layer is an adaptive node being defined as

$O_{5,n} = \overline{wn} fn = \overline{wn}(p_n x + q_n y + r_n)$

$\{p_n, q_n, r_n\}$ is the parameter set of the particular node and is referred to as *consequent parameters.* This layer also has the same number of nodes in layer 4 (81 nos).

Layer 6

The single node in this layer is responsible for the defuzzification process using the center of gravity technique to computes the overall output as the summation of the incoming signal.

$$O_{6,1} = \sum_{n=1}^{81} \overline{wn} \, fn = \frac{\sum_{n=1}^{81} wn\,fn}{\sum_{n=1}^{81} wn}$$

ANFIS uses a hybrid-learning rule with a combination of gradient descent (to learn the membership function parameters) and least squares estimate algorithm to learn the rule consequent parameters [10].

4 Experimentations result

Our experiments were simulated using Matlab. In addition to the development of the decision support system, we also investigated the behaviour of TACDSS decision support system for different membership functions (shape and quantity per ILV), and learning techniques.

4.1 Comparison of TACDSS MF tuning before and after training

Figure 4.1(a) and (b) shows the three MFs for the ILV "fuel used" before and after training. The consequent parameters of fuzzy rule before training was set to zero and the parameters were learned using the hybrid learning approach.

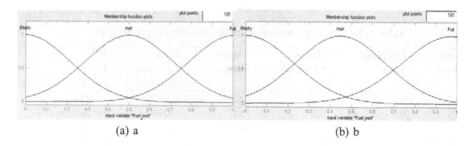

(a) a   (b) b

Fig. 3. The membership functions of the ILF "fuel used" (a) before and (b) after learning

4.2 Testing of the developed TACDSS decision support system

We tested the capabilities of the developed fuzzy inference system. The TACDSS has 4 inputs and each input has three MFs. The input and its MF were named as *"fuel used*": full, half and low, "*intercept time*": fast, normal and slow, "*weapon efficiency*": insufficient, enough and sufficient, "*danger situation*": endanger, danger and very dangerous. We have tested the decision making ability of the developed model by changing only one input variable and all other input variable were set to 0.5.

- When the "*fuel used*" was set at 0.2, the solution obtained was 0.0922 and when the fuel tank was set at 0.9 then the solution was 0.965.
- When the "*interrupt time*" was set to 0.2, the solution was 0.421 and when the setting was increased to 0.9, the solution obtained was 0.399.
- For "*weapon efficiency*" set at 0.1, the decision score obtained was 0.434 and the decision core increased to 0.524 when the setting was increased to 0.9.
- When the "*danger situation*" is set at 0.2, the solution obtained was 0.471 and for danger situation set at 0.9, the score was reduced to 0.154.

The simulation results clearly demonstrate that the developed TACDSS fuzzy inference system could provide the decision scores as same as a tactical air combat expert.

4.3 Comparison between the learning methods of FIS

We also investigated the different learning methods for learning the fuzzy inference system. Keeping the consequent parameters constant, we fine-tuned the membership functions alone using the gradient descent technique (backpropagation). Further, we used the hybrid learning method wherein the consequent parameters were also adjusted according to the least squares algorithm. Even though backpropagation is faster than the hybrid technique, learning error and the decision scores were better for the latter technique. We used 3 Gaussian MFs for each ILV. Table 1 illustrates the performance of the two learning methods for different input variable settings.

Table 1. Performance comparison for different learning techniques.

Fuel Used	Intercept Time	Weapon Efficiency	Danger Situation	Hybrid Learning Solution	BP Learning Solution
0.2	0.5	0.5	0.5	0.0920	0.0045
0.5	0.5	0.5	0.5	0.0100	0.4990
0.9	0.5	0.5	0.5	0.9650	0.0040
0.5	0.2	0.5	0.5	0.4210	0.0052
0.5	0.9	0.5	0.5	0.3990	0.0029
0.5	0.5	0.2	0.5	0.4580	0.0042
0.5	0.5	0.9	0.5	0.5240	0.0093
0.5	0.5	0.5	0.2	0.4710	0.0059
0.5	0.5	0.5	0.9	0.1540	0.0023

4.4 Comparison of the shape of membership functions of FIS

In this section, we will demonstrate the importance of the shape of membership functions. We used the hybrid-learning technique and each ILV has three MFs. Table 2 shows the training error value during the 15 epochs learning using different membership functions. We considered Generalised Bell (Gbell), Gaussian (Gaus), Gaussian 2 (Gaus2), Trapezoidal (Trap), Isosceles Triangular (Isoc) and Different Sigmoidal (Diff) membership functions. Figure 4.6 illustrates the training convergence curve for different MF's.

Table 2. The performance with different shape of membership functions

No. of epoch	Root Mean Square Error (e^{-005})					
	Gbell	Gausian	Gaus2	Trap	Isoc	Diff
1	1.393	1.183	2.198	2.183	0.783	1.942
2	1.348	1.157	2.137	2.073	1.558	1.929
3	1.315	1.137	2.039	2.275	1.558	1.903
4	1.296	1.127	1.970	2.275	1.558	1.894
5	1.288	1.108	1.912	2.275	1.558	1.902
6	1.291	1.097	1.813	2.275	1.558	1.902
7	1.291	1.085	1.673	2.275	1.558	1.902
8	1.291	1.074	1.495	2.275	1.558	1.902
9	1.291	1.062	1.412	2.275	1.558	1.902
10	1.291	1.052	1.479	2.275	1.558	1.902
11	1.291	1.043	1.479	2.275	1.558	1.902
12	1.291	1.038	1.479	2.275	1.558	1.902
13	1.291	1.039	1.479	2.275	1.558	1.902
14	1.291	1.039	1.479	2.275	1.558	1.902
15	1.291	1.039	1.479	2.275	1.558	1.902

As evident from Figure 4.6, the lowest training error was for Gaussian MF. We also tested the TACDSS models for different MFs. The input of TACDSS were set to *"fuel used"* 0.9, *"interrupt time"* 0.1, *"weapon efficiency"* 0.9 and *"danger situation"* 0.1. TACDSS gave a decision score of 0.864 (Gaussian), 0.88 (Trapezoidal), 0.861 (Gaussian Bell), 0.874 (Gaussian 2), 0.849 (Isosceles triangle) and 0.870 (Different Sigmoidal). The best decision score was obtained using the Gaussian MF and the worst using the trapezoidal MF.

4.5 Effect of Increasing Number of Membership Functions of Inputs

We used the hybrid learning and the Gaussian membership function for each input variable. The number of MF were increased from 3 to 4 for each ILV. The observation from the simulation results for 3 MFs and 4 MFs are as follows:

With a setting of *fuel used* 0.9, *interrupt time* 0.1, *weapon efficiency* 0.9 and *danger situation* 0.1, the decision score for 3 MFs was 0.865 being greater than for 4 MFs 0.857. More MFs per ILV could improve the accuracy of the decision scores.

4.6 Testing the TACDSS

We passed a randomly extracted test data set through the developed TACDSS. For example: when the ILVs are *fuel used* 0.9, *interrupt time* 0.0833, *weapon efficiency* 0.96, *danger situation* 0.2 the required decision score should be 0.9. TACDSS was simulated using different MFs such as Gaussian, Gaussian bell, and Gaussian 2 and so on. We obtained a decision score 0.90 for all the different simulations (same as the required solution). We also extracted 20 percent of master data set to form the test data and remain of master data set being the training data of the TACDSS. The shape MFs of the ILV and OLV is the Gaussian. The comparison between the actual and the desired output is shown in figure 5, the RMSE for test data is 1.498 e-5.

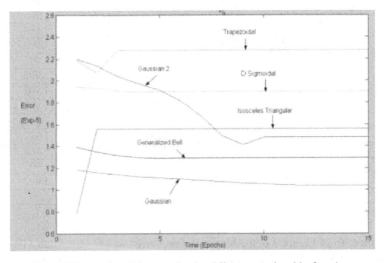

Fig. 4. Effect on learning error for the different membership functions

5 Conclusion and future research

In this paper, we have proposed the automatic construction of TACDSS using a Takagi-Sugeno neuro-fuzzy system. We used ANFIS algorithm for the automatic construction of *if-then* fuzzy rules using neural network learning techniques. Empirical results clearly reveal the developed TACDSS model could perform

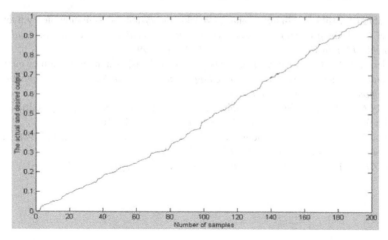

Fig. 5. Comparison between the actual and desired output of TACDSS

as efficiently as a tactical air combat expert. Our experiments also demonstrate the importance of the shape and number of membership functions for each input variable for obtaining the best performance. Compared to our previous work using evolutionary algorithms, neuro-fuzzy approach is more efficient in terms of less computational complexity and modelling simplicity. Our future research will be oriented to develop other adaptive fuzzy inference systems for the TACDSS, using decision tree analysis or unsupervised learning techniques and compare the results with current and previous works.

References

1. Abraham A. *Neuro-Fuzzy Systems: State-of-the-Art Modeling Techniques, "Connectionist Models of Neurons, Learning Processes, and Artificial Intelligence.* Springer-Verlag Germany, Granada, Spain, 2001.
2. Oppacher F. Cattral R. and Deogo D. Rule acquisition with a genetic algorithm. In *Proceedings of the congress on Evolution computation, CEC99*, volume 1, pages 125–129, 1999.
3. A. R. Chappell and J. W. McManus. Trial maneuver generation and selection in the paladin tactical decision generation system. *AIAA Guidance, Navigation, and Control Conference Paper # 92-4541.*, 1992.
4. L. Cong Tran, Jain and A. Abraham. Adaptive database learning in decision support system using evolutionary fuzzy systems: A generic framework. In *First international workshop on Hybrid Intelligent System, HIS01.*, 2001.
5. C. C. Hung. Building a neuro-fuzzy learning control system. In *AI Expert*, pages 40–49, November, 1993.
6. Jagielska I. Linguistic rule extraction from neural networks for descriptive datamining. In *The proceedings of second conference on knowledge-based intelligent electronic systems, KES'98*, volume 2, pages 89–92, 1998.

7. M. Fahimi J. Adibi, A. Ghoreishi and Z. Maleki. Fuzzy logic information theory hybrid model for medical diagnostic expert system. In *Proceedings of the Twelfth Southern Biomedical Engineering Conference*, pages 211–213, 1993.

8. Sanderson P. M. Cognitive work analysis and the analysis, design, evaluation of human computer interactive systems. In *Proceeding of the Australian/New Zealand conference on Computer-Human Interaction (OzCHI981)*, 1998.

9. R. J. B. Militallo, L. G. Hutton. Applied cognitive task analysis (acta): A practitioner's toolkit for understanding cognitive. *Ergonomics*, 41(11):1618–1642, 1998.

10. Jang J. S. R. Anfis-adaptive network based fuzzy inference system. *IEEE Trans. on Systems, Man, and Cybernetics*, 23:665–685, May 1993.

11. L. A. Zadeh. Fuzzy sets. In *Information Control*, volume 1, pages 338–353, 1965.

Visual Inspection of Fuzzy Clustering Results

Frank Klawonn, Vera Chekhtman, and Edgar Janz

Department of Computer Science
University of Applied Sciences Braunschweig/Wolfenbuettel
Salzdahlumer Str. 46/48
D-38302 Wolfenbuettel, Germany
f.klawonn@fh-wolfenbuettel.de

1 Introduction

Clustering is an explorative data analysis method applied to data in order to discover structures or certain groupings in a data set. Therefore, clustering can be seen as an unsupervised classification technique. Fuzzy clustering accepts the fact that the clusters or classes in the data are usually not completely well separated and thus assigns a membership degree between 0 and 1 for each cluster to every datum.

The most common fuzzy clustering techniques aim at minimizing an objective function whose (main) parameters are the membership degrees and the parameters determining the localisation as well as the shape of the clusters. The algorithm will always compute a result that might represent an undesired local minimum of the objective function. Even if the global minimum is found, it might correspond to a bad result, when the cluster shapes or the number of the clusters are not chosen properly. Since the data are usually multi-dimensional, the visual inspection of the data is very limited. Methods like multi-dimensional scaling are available, but lead very often to unsatisfactory results. Nevertheless, it is important to evaluate the clustering result. Although cluster validity measures try to solve this problem, they tend to reduce the information of a large data set and a number of cluster parameters to a single value.

We propose in this paper to use the underlying principles of validity measures, but to refrain from the simplification to a single value and instead provide a graphical representation containing more information. This enables the user to identify inconsistencies in the clustering result or even in single clusters.

Section 2 briefly reviews the necessary background in objective function-based fuzzy clustering. The concept of validity measures is discussed in section 3. The techniques for visualisation are introduced in section 4 and in the final conclusions we outline that the visualisation techniques tailored for fuzzy clustering are even useful in the case of crisp clustering.

2 Objective Function-Based Fuzzy Clustering

Fuzzy clustering is suited for finding structures in data. A data set is divided into a set of clusters and – in contrast to hard or deterministic clustering – a datum is not assigned to a unique cluster. In order to handle noisy and ambiguous data, membership degrees of the data to the clusters are computed. Most fuzzy clustering techniques are designed to optimise an object function with constraints. The most common approach is the so called probabilistic clustering with the objective function

$$f = \sum_{i=1}^{c} \sum_{j=1}^{n} u_{ij}^m d_{ij} \qquad (1)$$

under the constraints

$$\sum_{i=1}^{c} u_{ij} = 1 \qquad \text{for all } j = 1, \dots, n. \qquad (2)$$

It is assumed that the number of clusters c is fixed. The set of data to be clustered is $\{x_1, \dots, x_n\} \subset \mathbb{R}^p$. u_{ij} is the membership degree of datum x_j to the ith cluster. d_{ij} is some distance measure specifying the distance between datum x_j and cluster i, for instance the (squared) Euclidean distance of x_j to the ith cluster centre. The parameter $m > 1$, called fuzzifier, controls how much clusters may overlap. The constraints (2) lead to the name probabilistic clustering, since in this case the membership degree u_{ij} can also be interpreted as the probability that x_j belongs to cluster i. The parameters to be optimised are the membership degrees u_{ij} and the cluster parameters that are not given explicitly here. They are hidden in the distances d_{ij}. Since this is a non-linear optimisation problem, the most common approach to minimize the objective function (1) is to alternatingly optimise either the membership degrees or the cluster parameters while considering the other parameter set as fixed. In this paper we are not interested in the great variety of specific cluster shapes (spheres, ellipsoids, lines, quadrics,…) that can be found by choosing suitable cluster parameters and an adequate distance function. (For an overview we refer to [2, 5].) Our considerations can be applied to almost all cluster shapes. However, for shell clustering there are better suited methods. Since most shell clustering algorithms are designed for image recognition, the data are usually two-dimensional so that special visualisation techniques are not required.

The visualisation methods we propose are also suited for noise clustering [3] where the principle of probabilistic clustering is maintained, but an additional noise cluster is introduced. All data have a fixed (large) distance to the noise cluster. In this way, data that are near the border between two clusters, still have a high membership degree to both clusters as in probabilistic clustering. But data that are far away from all clusters will be assigned to the noise cluster and have no longer a high membership degree to other clusters.

We do not cover possibilistic clustering [6] where the probabilistic constraint is completely dropped and an additional term in the objective function is introduced to avoid the trivial solution $u_{ij} = 0$. However, the aim of possibilistic clustering is actually not to find the global optimum of the corresponding objective function, since this is obtained, when all clusters are identical [7].

3 Validity Measures

Cluster validity refers to the problem whether a given (fuzzy) partition fits to the data at all. We emphasize again that the clustering algorithm will always try to find the best fit for a fixed number of clusters and the parameterised cluster shapes. However, this does not mean that even the best fit is meaningful at all. Either the number of clusters might be wrong or the cluster shapes might not correspond to the groups in the data, if the data can be grouped in a meaningful way at all. The cluster validity problem is a similar one as in linear regression. One can always find the best fitting line for a given data set, even if the data come from an exponential function. This does not mean that there is a linear dependence in the data.

Cluster validity measures are used to validate a clustering result in general or also in order to determine the number of clusters. In order to fulfill the latter task, the clustering might be carried out with different numbers of clusters and the one yielding the best value of the validity measure is assumed to have the correct number of clusters.

Let us briefly review some cluster validity measures. It is beyond the scope of this paper to provide a complete overview on validity measures and we refer for a more detailed discussion to [2, 5]. We also restrict our considerations to global validity measures that evaluate a whole fuzzy partition and not single clusters.

The partition coefficient [1] is defined by

$$\frac{\sum_{i=1}^{c} \sum_{j=1}^{n} u_{ij}^2}{n}.$$

The higher the value of the partition coefficient the better the clustering result. The highest value 1 is obtained, when the fuzzy partition is actually crisp, i.e. $u_{ij} \in \{0,1\}$. The lowest value $1/c$ is reached, when all data are assigned to all clusters with the same membership degree $1/c$. This means that a fuzzy clustering result is considered better, when it is more crisp.

The partition entropy [1]

$$\frac{\sum_{i=1}^{c} \sum_{j=1}^{n} u_{ij} \ln(u_{ij})}{n}$$

is inspired by the Shannon entropy. The smaller the value of the partition entropy, the better the clustering result. This means that similar to the partition coefficient crisper fuzzy partitions are considered better.

In [4] validity measures are proposed that take the volume of the clusters into account. Let

$$A_i = \frac{\sum_{j=1}^{n} u_{ij}^m (x_j - v_i)(x_j - v_i)^\top}{\sum_{j=1}^{n} u_{ij}^m}$$

denote the (fuzzy) covariance matrix of the ith cluster where

$$v_i = \frac{\sum_{j=1}^{n} u_{ij}^m x_j}{\sum_{j=1}^{n} u_{ij}^m}$$

denotes the centre of the ith cluster and the x_j are the data vectors.

Then the validity measure called the fuzzy hypervolume is defined by

$$FHV = \sum_{i=1}^{c} \sqrt{\det(A_i)}.$$

A smaller value of FHV indicates compact and therefore better clusters.

The average partition density is given by

$$\frac{1}{c} \sum_{i=1}^{c} \frac{S_i}{\sqrt{\det(A_i)}}$$

where $S_i = \sum_{j \in Y_i} u_{ij}$ and

$$Y_j = \{ j \in \{1, \ldots, n\} \mid (x_j - v_i)^\top A^{-1} (x_j - v_i) < 1 \}$$

is the set of data near to the cluster centre v_i. S_i corresponds to the number of data assigned to cluster i that are near to the cluster centre v_i. Therefore $\frac{S_i}{\sqrt{\det(A_i)}}$ is proportional to the average density of the data in cluster i. A high density value indicates good clusters.

Finally, the partition density is defined by

$$\frac{\sum_{i=1}^{c} S_i}{FHV}$$

where a larger value refers again to better clusters.

It should be noted that validity measures like the partition coefficient or the partition entropy rely solely on the membership degrees whereas fuzzy hypervolume and (average) partition density also take the distance of the data to the clusters into account. We will use these ideas to develop graphical validity measures in the next section.

4 Visualisation

Let us first use the membership degrees only for a graphical inspection, whether a fuzzy clustering result is acceptable. It should be noted that the underlying idea is always: The more crisp the fuzzy partition, the better the clustering result. Of course, decreasing the fuzzifier m always leads to crisper, but not necessarily better partitions. This must always be taken into account.

In order to illustrate our graphical validity criterions, we use the artificial data set shown in figure 1. This data set obviously contains three well separated clusters. We have clustered this data set with the well known fuzzy c-means algorithm with three clusters as well as four clusters. For three clusters the cluster centres are not shown in the figure, but are more or less exactly in the middle of the corresponding circles. When we use four clusters, the marked spots in figure 1 show the cluster centres. A data point has the highest membership degree to the nearest cluster centre.

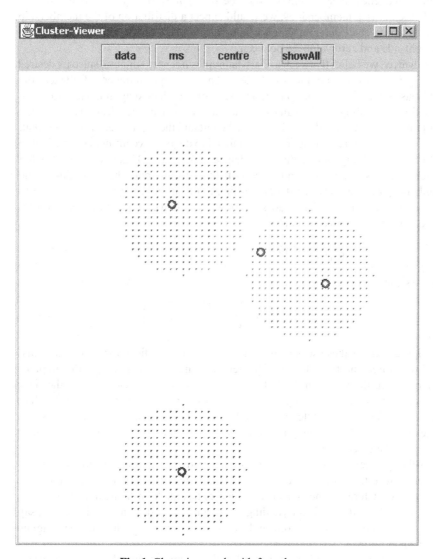

Fig. 1. Clustering result with four clusters

As a first very simple visualisation of the membership degrees we can simply look at the distribution of the membership degrees. For the ideal case of a crisp result, i.e. each datum is assigned to exactly one cluster with member ship degree 1 and to all others with membership degree 0, we would expect the following: The relative frequency of the membership degree 1 should be $n/(cn) = 1/c$ and accordingly the relative frequency of the value 0 should be $(c-1)n/(cn) = (c-1)/c$. So, for the ideal case of crisp memberships, we would expect a distribution of the membership degrees whose chart diagram shows a value of $(c-1)/c$ on the left side and $1/c$ on the right side and zero values in between.

However, we belief that such a chart diagram is not very suitable, since its desired shape depends on the number of clusters. And for larger numbers of clusters the emphasis is mainly put only on the left side of the chart diagram. Therefore, we carried out a scaling of the values in such a way that in the ideal case the chart diagram would show a value of 1 on both the left and the right side. This means that, when counting the frequencies of the membership degrees, we introduce a weighting factor. The weighting for a membership degree of 0 is $c/(c-1)$ and for a membership degree of 1 it is c. For the computation of the chart diagram, the weighting of the membership degrees between 0 and 1 is simply linear, increasing from $c/(c-1)$ to c. A single chart in the diagram does not show the relative frequency of membership degrees in a range between a and b

$$\frac{1}{n}\,\mathrm{card}\{(i,j) \in \{1,\ldots,c\} \times \{1,\ldots,n\} \mid a \leq u_{ij} < b\},$$

but the scaled frequency

$$\frac{1}{n} \sum_{(i,j):a \leq u_{ij} < b} \left(\frac{c(c-2)}{c-1} u_{ij} + \frac{c}{c-1} \right).$$

Figure 2 shows these scaled chart diagrams for our artificial data set, when clustered with three and four clusters. The height of the chart diagrams in the graphics is normalised, the value on the left side is 0.92 for the left and 0.90 for the right chart diagram. What is more significant is that we can see immediately that the chart diagrams differ in the middle and the right side. The right chart diagram shows less values near 1 and more ambiguous membership degrees. This is an indicator for a non-optimal clustering result.

Although these chart diagrams provide already interesting information on the clustering result, we propose to have a look at another diagram as well. For each datum x_j we determine the cluster for which x_j has the highest membership degree, say clusters i_1, and the cluster yielding the second highest membership degree, say i_2. Then for each x_j we plot a point at the coordinates $(u_{i_1 j}, u_{i_2 j})$ leading to a diagram as shown in figure 3.

It is clear that all points must lie within the triangle defined by the points $(0,0)$, $(0.5,0.5)$ and $(1,0)$, since the first coordinate must always be larger than the second one and according to the probabilistic constraint (2) we have $u_{i_1 j} + u_{i_2 j} \leq 1$.

Having again the ideal case of (almost) crisp membership degrees in mind, all points would be plotted near the point $(1,0)$. Points near $(0.5,0.5)$ indicate ambiguous

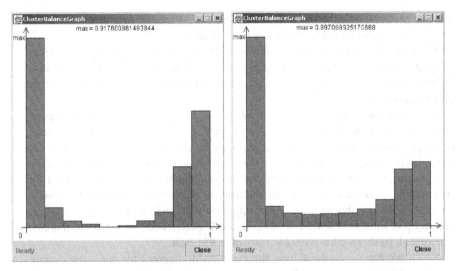

Fig. 2. Scaled membership distributions for three (left) and four (right) clusters

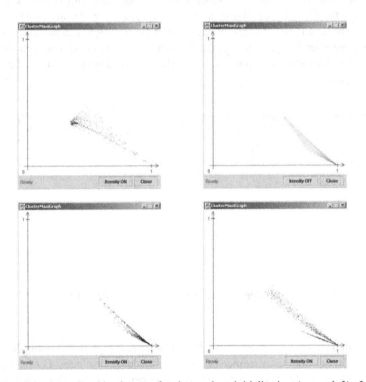

Fig. 3. Maximum membership degrees for the random initialisation (upper left), for three (lower left) and four (lower right) clusters after convergence and an intensity plot for a large data set (upper right)

data that are shared by two clusters. Points near (0,0) usually originate from noise data that have a low membership degree to all clusters.

The upper left graph in figure 3 shows this diagram at the beginning of the clustering algorithm, when the cluster centres are initialised randomly. Of course, with random cluster centres very few data are near to a cluster centre and we find only few points near (1,0).

The lower part of figure 3 shows this diagram of maximum membership degrees after we have carried out the clustering completely with three and four clusters. Since for the partition with three clusters more points are concentrated near (1,0) than for four clusters, we see from the diagram that three clusters should be preferred.

In the case of very large data sets, we recommend not to plot a single point for each datum. A colour plot where the intensity of the colour represents the density of points should be chosen instead. The upper right graph in figure 3 shows such an intensity plot for a similar data set as in figure 1, except that the density of the data is increased, so that we have 192231 instead of 951 data.

So far we have considered only the membership degrees for our visualisation. We have already seen in our short review of validity measures that this is a suitable approach, but we can use more information from the clustering output. Validity measures like the fuzzy hypervolume or the (average) partition density also take the distances of the data to the clusters into account. A lot of insight can be gained from a plot of the membership degrees over the distances for each cluster. For each cluster i we plot for every datum x_j a point at (d_{ij}, u_{ij}). This leads to diagrams as they can be seen in figure 4.

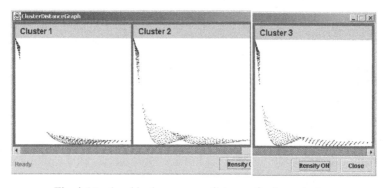

Fig. 4. Membership degrees over distances for three clusters

How should an ideal graph look like? We would expect high membership degrees for small distances and low membership degrees for large distances. Let us briefly discuss what kind of effects can occur, when cluster centres are not chosen appropriately, although there are valid clusters in the data. Typical problems that occur in this case are the following.

- One cluster has to cover two or more data clusters. This occurs especially, when the number of clusters is chosen too small. In this case, we would see almost no data with small distances in the upper left part of the diagram for this cluster.
- Two or more clusters compete and share the same data cluster. This usually occurs, when the number of clusters is chosen to high. In this case, there will occur small membership degrees even for small distances. This means, we find points in the lower left part of the diagrams of the corresponding clusters.

Figure 4 shows the corresponding diagrams for our data set, when we use three clusters. For all three clusters the diagrams look quite well. It can even be seen that cluster 1 (covering the lower data cluster in figure 1) is best separated from the other clusters.

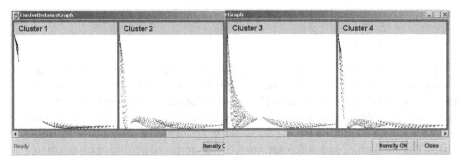

Fig. 5. Membership degrees over distances for four clusters

In figure 5 we have the diagrams for (the inappropriate choice of) four clusters. Cluster 1 corresponds to the lower data cluster in figure 1 and the diagram coincides more or less with the first cluster in figure 4. The diagram for cluster 4 (the upper left cluster centre in figure 1) is still more or less acceptable. Indeed, we can see from figure 1 that it covers the upper left data cluster almost correctly. However, compared to the diagrams in figure 4 we see a more continuous slide from high to low membership degrees, indicating that the cluster is not very well separated from the others. Cluster 3, represented by the middle cluster centre in figure 1, is the worst one. We have low membership degrees even for small distances, since it shares data with cluster 2 (the right one). There is also a gap in medium values for the membership degrees arising from the fact that cluster 3 actually covers data from two different data clusters. Cluster 2 has also low membership degrees for small distances, because of the competing cluster 3.

Analogously to the diagrams showing the maximum membership degrees, we recommend to replace the point plots by intensity plots for larger data sets as shown in figure 6, where we have used again the previously described data set with 192231 data points.

As another example we have used the artificial data set with some noise added (see figure 7).

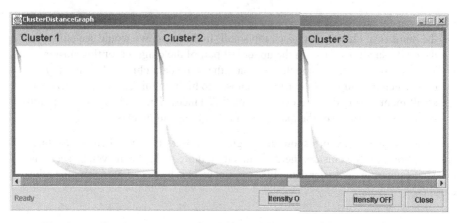

Fig. 6. Intensity plot for membership degrees over distances for a large data set

Figure 8 shows the chart diagram for the membership degree distribution and the maximum membership degree diagram for the noisy data when clustered with three clusters. These diagrams should be compared to the diagrams in figure 2 and 3, respectively, for the data set without noise. The diagrams for the membership degrees over the distances for the noisy data are shown in figure 9 to be compared to figures 4 and 5. The similarities and differences are obvious and need no further explanation.

5 Conclusions

We have a proposed a number of diagrams that support the visual validation of a fuzzy clustering result. Classical validity measures can be used to carry out clustering completely unsupervised. But this means that we rely on a very strict information compression performed by these validity measures. With our visualisation techniques much more information is available and even single good or bad clusters can be identified by inspecting the diagrams.

Our methods can also be applied in the context of crisp clustering. Of course, all our methods are based on membership degrees. But even if we have carried out a crisp clustering, we can afterwards compute membership degrees by the formulae known from fuzzy clustering and then apply our visualisation methods.

References

1. Bezdek JC (1981) Pattern recognition with fuzzy objective function algorithms. Plenum Press
2. Bezdek JC, Keller J, Krishnapuram R, Pal NR (1999) Fuzzy models and algorithms for pattern recognition and image processing. Kluwer, Boston

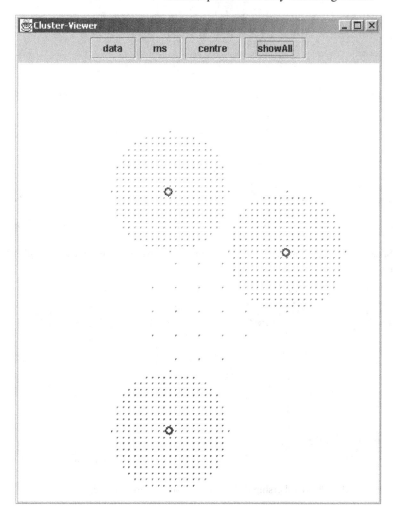

Fig. 7. Clustering result with three clusters for noisy data

3. Davé, RN (1991) Characterization and detection of noise in clustering. Pattern Recognition Letters 12: 657–664
4. Gath I, Geva AB (1989) Unsupervised optimal fuzzy clustering. IEEE Trans. Pattern Analysis and Machine Intelligence 11: 773–781
5. Höppner F, Klawonn F, Kruse R, Runkler T (1999) Fuzzy cluster analysis. Wiley, Chichester
6. Krishnapuram R, Keller J (1993) A possibilistic approach to clustering. IEEE Trans. on Fuzzy Systems 1: 98-110
7. Timm H, Borgelt C, Kruse R (2002) A modification to improve possibilistic cluster analysis. IEEE Intern. Conf. on Fuzzy Systems, Honululu (2002)

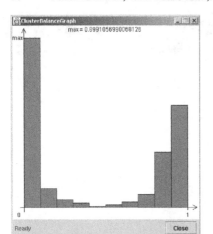 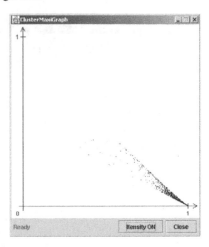

Fig. 8. Scaled chart diagram (left) and maximum membership degrees diagram (right) for noisy data

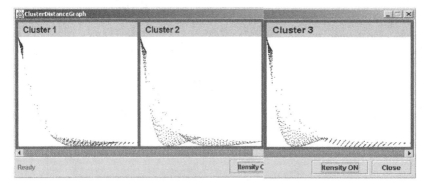

Fig. 9. Membership degrees over distances for noisy data

Benchmark Between Three Controllers in IFOC: PI, IP and New Fuzzy-PI Regulator

Jean-Camille de Barros

LAGEM - Service A51 - U3 Saint Jérôme, Avenue Escadrille Normandie Niemen - 13397
Marseille Cedex 20
Laboratory email : debarros@ms251u08.u-3mrs.fr / Home :
jc.debarros@free.fr

Summary. When the system to be ordered is complex and/or non-linear, fuzzy regulators, which are on the way to be standardised in industry, bring an original solution compared to the conventional *PID* regulators [5]. In this paper, we present benchmark test between three types of speed controller in *Indirect Field Oriented Control* (*IFOC*) of asynchronous machine: classical *PI,IP* and new *Fuzzy-PI* structure [13]. *IFOC* has permitted fast transient response by decoupled torque and flux control separately. However, *IFOC* detuning is caused by parameters variations. In particular it is highly sensitive to uncertainties in the rotor resistance of the AC motor. This sensitivity can be reduced by using fuzzy controller. New optimal *Fuzzy-PI* proposed structure is based on optimal *Ziegler-Nichols* and *Broida* classical approaches [4]. For optimising this fuzzy controller we used *Sequential Quadratic Programming* (*SQP*) optimisation method [8]. We show in computer simulation the principal results from these different regulators. We point out the interest of proposed fuzzy controllers: the fuzzy command permits good improvement of the performance. Moreover we show also that the fuzzy controller is more robust against large parameters variations (±20%) of the process.

Key words: Fuzzy controller, Fuzzy-PI, asynchronous machine, AC motor, IFOC.

1 Introduction

Vector controlled induction motor drive systems are increasingly employed in industrial applications. In these applications performances requirements become severe. Decoupled torque and flux control in *IFOC* of induction machines permits high dynamic response [2]. Because of its simplicity, *PI* or *IP* controller is widely used in asynchronous machine speed control in *IFOC* strategy. But parametric variations (essentially rotor/stator resistor and inductance variations), and load disturbances cause *IFOC* detuning and degrade severely *PI* and *IP* performances [11]. To overcome the above problems, some approaches have been developed like adaptive *PI* controller, *Kalman* observer, on-line parameters identification etc. One of the modern techniques is fuzzy logic controller [9], [10], [11], [12]. This paper presents new

Fuzzy-PI [13] regulator structure. In first approach parameters auto tuning is performed like *Ziegler-Nichols* and *Broida* principles [13]. To achieve high dynamic performances, we used *SQP* method [8] to tune in-loop parameters of our *Fuzzy-PI*. Finally, computer simulation in *Matlab/Simulink* [6] environment gives benchmark between these three speed controllers. Results are analysed and commented.

2 Definition of IFOC Induction Motor Drive

2.1 Asynchronous Machine Park's model

In *Park* (d,q) reference frame, electrical equations of AC machine are [2]:

1) Park's Flux

$$\begin{cases} \phi_{s_d} = L_s \cdot i_{s_d} + L_m \cdot i_{r_d} \\ \phi_{s_q} = L_s \cdot i_{s_q} + L_m \cdot i_{r_q} \end{cases} \& \begin{cases} \phi_{r_d} = L_r \cdot i_{r_d} + L_m \cdot i_{s_d} \\ \phi_{r_q} = L_r \cdot i_{r_q} + L_m \cdot i_{s_q} \end{cases}$$

2) Park's Voltage

$$\begin{cases} v_{s_d} = R_s \cdot i_{s_d} + \frac{d\phi_{s_d}}{dt} - \omega_s \cdot \phi_{s_q} \\ v_{s_q} = R_s \cdot i_{s_q} + \frac{d\phi_{s_q}}{dt} + \omega_s \cdot \phi_{s_d} \end{cases} \& \begin{cases} v_{r_d} = R_r \cdot i_{r_d} + \frac{d\phi_{r_d}}{dt} - \omega_{sl} \cdot \phi_{r_q} = 0 \\ v_{r_q} = R_r \cdot i_{r_q} + \frac{d\phi_{r_q}}{dt} + \omega_{sl} \cdot \phi_{r_d} = 0 \end{cases}$$

Orientation of the direct d axis according to rotor flux's leads to use this following expression of the torque [2]:

$$C_e = n_p \cdot \frac{L_m}{L_r} \cdot \left(\phi_{r_d} \cdot i_{s_q} - \phi_{r_q} \cdot i_{s_d} \right)$$

and slip angular speed is: $\omega_{sl} = \omega_s - \omega_r$

2.2 IFOC of Asynchronous machine

To have an instantaneous scalar expression of the electromagnetic torque C_e this last relation shows that the rotor flux in q axis must be set to zero [2]:

$$\phi_{r_q} = 0$$

and the preceding relation of the torque becomes:

$$C_e = n_p \cdot \frac{L_m}{L_r} \cdot \phi_{r_d} \cdot i_{s_q}$$

By maintaining the rotor flux ϕ_{r_d} constant, we will have a scalar relation for the torque C_e depending to the stator direct axis current ß$_{s_q}$ [2]:

$$C_e = K_c \cdot i_{s_q} \text{ with} K_c = n_p \cdot \frac{L_m}{L_r} \cdot \phi_{r_d}$$

Finally, *IFOC* is given by these equation systems [2]:

1) IFOC's Flux

$$\begin{cases} \phi_{s_d} = L_s \cdot i_{s_d} + L_m \cdot i_{r_d} \\ \phi_{s_q} = L_s \cdot i_{s_q} + L_m \cdot i_{r_q} \end{cases} \& \begin{cases} \phi_{r_d} = L_r \cdot i_{r_d} + L_m \cdot i_{s_d} \\ \phi_{r_q} = 0 \end{cases}$$

2) IFOC's Rotor Voltage

$$\begin{cases} v_{r_d} = R_r \cdot i_{r_d} + \frac{d\phi_{r_d}}{dt} = 0 \\ v_{r_q} = R_r \cdot i_{r_q} + \omega_{sl} \cdot \phi_{r_d} = 0 \end{cases}$$

3) Magnetizing Current and slip angular speed

$$i_\phi = \frac{\phi_{r_d}}{L_m} = \frac{1}{1 + \tau_r \cdot p} \cdot i_{s_d} \quad with \quad \tau_r = \frac{L_r}{R_r}$$

4) Slip angular speed and stator phase

$$\omega_{sl} = \frac{L_m}{\tau_r} \cdot \frac{1}{\phi_{r_d}} \cdot i_{s_q} = \frac{1}{\tau_r} \cdot \frac{1}{i_\phi} \cdot i_{s_q}$$

$$\omega_s = \omega_{sl} + \omega_r \Rightarrow \theta_s = \int_0^t \omega_s \cdot d\tau = \theta_r + \frac{L_m}{\tau_r} \cdot \int_0^t \frac{i_{s_q}}{\phi_{r_d}} \cdot d\tau$$

5) IFOC's Stator Voltage

$$\begin{cases} v_{s_d} = R_s \cdot i_{s_d} + p \cdot \left[\sigma \cdot L_s + \frac{L_s \cdot (1-\sigma)}{1 + \tau_r \cdot p} \right] \cdot i_{s_q} - e_d \\ v_{s_q} = R_s \cdot i_{s_q} + p \cdot \sigma \cdot L_s \cdot i_{s_q} - e_q \end{cases}$$

$$with \begin{cases} e_d = \sigma \cdot L_s \cdot \omega_s \cdot i_{s_q} \\ e_q = -L_s \cdot \omega_s \cdot (1 + \sigma \cdot \tau_r \cdot p) \cdot i_\phi \end{cases}$$

2.3 IFOC's usual regulators

To avoid performance degradation caused by noise (*PWM* switching), we use only *PI* regulator for different regulators used after. It must be noticed that flux and torque regulators are identical whatever the speed regulator (usual or fuzzy) may be.

1) Flux Regulator - $R_\phi(p)$

Transfer function $H_{i_\phi}(p)$ between magnetising current ß$_\phi$ and *d* axis voltage v_{s_d} is:

$$H_{i_\phi}(p) = \frac{K_\phi}{(1 + \tau_{\phi_1} \cdot p) \cdot (1 + \tau_{\phi_2} \cdot p)} \quad with \quad K_\phi = \frac{1}{R_s}$$

$$And \quad \tau_{\phi_{1,2}} = \frac{2 \cdot \sigma \cdot \tau_s \cdot \tau_r}{\tau_s + \tau_r \mp \sqrt{(\tau_s + \tau_r)^2 - 4 \cdot \sigma \cdot \tau_s \cdot \tau_r}}$$

The *PI* type flux regulator $R_\phi(p)$ is defined by:

$$R_\phi(p) = K_{P_\phi} + \frac{K_{I_\phi}}{p} = \frac{K_{I_\phi}}{p} \cdot (1 + \tau_\phi \cdot p) \quad with \quad \tau_\phi = \frac{K_{P_\phi}}{K_{I_\phi}}$$

To control flux response time, *PI*'s gains are:

$$K_{I_\phi} = \frac{\tau_{\phi 2}}{K_\phi \cdot \tau'^2_\phi} \quad \& \quad K_{P_\phi} = \tau_\phi \cdot K_{I_\phi} = \frac{\tau_{\phi 1} \cdot \tau_{\phi 1}}{K_\phi \cdot \tau'^2_\phi} \Rightarrow \xi'_\phi = \frac{\tau'_\phi}{2 \cdot \tau_{\phi 2}}$$

where τ'_ϕ is the desired response time and ξ'_ϕ is obtained damping factor. To control flux over-shoot, *PI*'s gains are:

$$K_{I_\phi} = \frac{1}{4 \cdot K_\phi \cdot \xi'^2_\phi \cdot \tau_{\phi 2}} \quad ; \quad K_{P_\phi} = \frac{\tau_{\phi 1}}{4 \cdot K_\phi \cdot \xi'^2_\phi \cdot \tau_{\phi 2}} \Rightarrow \tau'_\phi = 2 \cdot \tau_{\phi 2} \cdot \xi'_\phi$$

ξ'_ϕ is desired damping factor and τ'_ϕ is obtained response time.

2) Torque Regulator - $R_{C_e}(p)$

Transfer function $H_{i_{sq}}(p)$ between stator q axis current i_{sq} and voltage v_{sd} is [2]:

$$H_{i_{sq}}(p) = \frac{i_{sq}}{v_{sq} + e_q} = \frac{K_{i_{sq}}}{1 + \sigma \cdot \tau_s \cdot p} \quad with \quad \tau_s = \frac{L_s}{R_s} \, and \, K_{i_{sq}} = \frac{1}{R_s}$$

and $\sigma \cdot \tau_s$ is the torque time constant. The *PI* type torque regulator $R_{C_e}(p)$ is defined by:

$$R_{C_e}(p) = K_{P_{C_e}} + \frac{K_{I_{C_e}}}{p} = \frac{K_{I_{C_e}}}{p} \cdot (1 + \tau_{C_e} \cdot p) \quad with \quad \tau_{C_e} = \frac{K_{P_{C_e}}}{K_{I_{C_e}}}$$

To control torque response time, *PI*'s gains are:

$$K_{I_{C_e}} = \frac{K_{r_{C_e}}}{K_{i_{sq}}} \quad ; \quad K_{P_\phi} = \frac{K_{r_{C_e}}}{K_\phi \cdot \tau'_{C_e}} \quad and \quad K_{r_{C_e}} = \frac{\tau_{C_e}}{\tau'_{C_e}} \geq 10$$

where $K_{r_{C_e}}$ is the swiftness of torque loop.

3) Speed Regulator - $R_{\omega_r}(p)$

Physical relation between rotor speed ω_r and torque C_e is:

$$\omega_r = \frac{n_p}{f + J \cdot p} \cdot (C_e - C_r)$$

where C_r is the loaded torque [2]. *PI* or *IP* is used to control speed.

The *PI* type rotor speed regulator $R_{\omega_r}(p)$ is defined by:

$$R_{\omega_r}(p) = K_{P_{\omega_r}} + \frac{K_{I_{\omega_r}}}{p} = \frac{K_{P_{\omega_r}} \cdot p + K_{I_{\omega_r}}}{p}$$

Closed-loop of rotor speed is given by:

$$H_{\omega_r}^{C-L}(p) = \frac{\omega_r}{\omega_r^*} = \frac{f \cdot K_{\omega_r} \cdot (K_{P_{\omega_r}} \cdot p + K_{I_{\omega_r}})}{f \cdot K_{\omega_r} \cdot K_{I_{\omega_r}} + f \cdot (1 + K_{\omega_r} \cdot K_{P_{\omega_r}}) \cdot p + J \cdot p^2}$$

$$\Rightarrow H_{\omega_r}^{C-L}(p) = \frac{1+\tau_{\omega_r} \cdot p}{1+\frac{2 \cdot \xi'_{\omega_r}}{\omega'_{o\omega_r}} \cdot p + \frac{1}{\omega'^2_{o\omega_r}} \cdot p^2} \quad and \quad \omega'_{o\omega_r} = \frac{1}{\tau'_{\omega_r}}$$

where ξ'_{ω_r} and τ'_{ω_r} are respectively desired damping factor and closed-loop response time. *PI* regulator gains are given by:

$$K_{I\omega_r} = \frac{J}{f \cdot K_{\omega_r} \cdot \tau'^2_{\omega_r}} \quad and \quad K_{P\omega_r} = \frac{2 \cdot J \cdot \xi'_{\omega_r} - f \cdot \tau'_{\omega_r}}{f \cdot K_{\omega_r} \cdot \tau'_{\omega_r}}$$

and

$$\tau_{\omega_r} = \frac{K_{P\omega_r}}{K_{I\omega_r}} = \frac{\tau'_{\omega_r} \cdot (2 \cdot J \cdot \xi'_{\omega_r} - f \cdot \tau'_{\omega_r})}{J}$$

First closed-loop transfer function $H_{(1)\omega_r}^{C-L}(p)$ is written by:

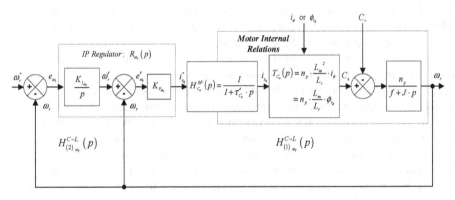

Fig. 1. IFOC rotor speed controlled by*IP* regulator

$$H_{(1)\omega_r}^{C-L}(p) = \frac{K_{P\omega_r} \cdot K_{\omega_r}}{(1+K_{P\omega_r} \cdot K_{\omega_r})} \cdot \frac{K_{I\omega_r}}{1+\frac{J}{f \cdot (1+K_{P\omega_r} \cdot K_{\omega_r})} \cdot p} = \frac{K_{1\omega_r}}{1+\tau_{1\omega_r} \cdot p}$$

where $\tau_{1\omega_r}$ is the desired response time of this loop and $K_{1\omega_r}$ is resulting static gain. Then *IP* proportional gain $K_{P\omega_r}$ is:

$$K_{P\omega_r} = \frac{J - f \cdot \tau_{1\omega_r}}{f \cdot K_{\omega_r} \cdot \tau_{1\omega_r}} \Rightarrow K_{1\omega_r} = \frac{K_{P\omega_r} \cdot K_{\omega_r}}{(1+K_{P\omega_r} \cdot K_{\omega_r})}$$

Second closed-loop transfer function $H_{(2)\omega_r}^{C-L}(p)$ is written by:

$$H_{(2)\omega_r}^{BF}(p) = \frac{1}{1+\frac{1}{K_{1\omega_r} \cdot K_{I\omega_r}} \cdot p + \frac{\tau_{1\omega_r}}{K_{1\omega_r} \cdot K_{I\omega_r}} \cdot p^2} = \frac{1}{1+\frac{2 \cdot \xi'_{\omega_r}}{\omega'_{o\omega_r}} \cdot p + \frac{1}{\omega'_{o\omega_r}} \cdot p^2}$$

and $\omega'_{o_{\omega_r}} = 1/\tau'_{\omega_r}$. Finally, *IP* integral gain $K_{I_{\omega_r}}$ is:

$$K_{I_{\omega_r}} = \frac{\tau_{1_{\omega_r}}}{K_{1_{\omega_r}} \cdot \tau'^2_{\omega_r}} \Rightarrow \xi'_{\omega_r} = \frac{\tau'_{\omega_r}}{2 \cdot \tau_{1_{\omega_r}}}$$

Figure 2 gives the complete structure of *IFOC* strategy with usual and *Fuzzy-PI* regulators.

3 PROPOSED OPTIMAL FUZZY-PI

3.1 Definition of Proposed FUZZY-PI

The *Fuzzy-PI* regulator proposed exhibits an overall trend very closed to the traditional *PI* while keeping the non-linear characteristics of the fuzzy regulator. Its structure is given in *Figure 3*. *Fuzzy-PI* proposed is based on the structure of traditional parallel *PI* , [3] combined of *Fuzzy-P*: the idea was to make fuzzy the proportional. It has 7 triangular membership functions for the input and 7 singleton membership functions type for the output. We used the Max-Min *Mamdani*'s inference engine and [5]. The defuzzification used is the method of the center of gravity applied to singleton membership functions of the output (*Sugeno*'s fuzzy controller). For the *"fuzzy proportional gain"* we use the table of inference proposed by *H. Buhler* [1]: Besides its internal structure, fuzzy regulator has many parameters of adjustment (around one hundred parameters for a regulator with seven membership functions in input and output), each one having a more or less significant impact on its answer [4]. In particular, the factors of normalisation and denormalisation, the position of the modal values of the membership functions and their distribution play a major role [7]. Nevertheless, the table of inference has only one local role on the transfer function of the regulator [3].

In *Figure 5* we give input and output membership functions suggested.

3.2 Analysis of proposed FUZZY-PI

Proposed Reference *Fuzzy-PI* has equidistant and regular distribution input and output membership functions. Normalisation and denormalisation factors are: $D_u = K_I = 1$. For the proposed *Fuzzy-PI* input membership functions of e are tightened around 0. Output membership functions of u^* are tightened around 1 and mobile (modal value depends on e - it is one of the originalities of the proposed *Fuzzy-PI*). The first essential issue is the possibility to carry out a *Fuzzy-PI* identical to the traditional *PI* [7]: in fact, that permits to directly replace *PI* by *Fuzzy-PI* (*Figure 6*). Then, the behavior of *Fuzzy-PI* can be completely modified while exploiting his parameters. Indeed, as shown the ramp and sine responses (*Figure 6*), this regulator has the property to superimpose the response of the fuzzy regulator to that of a traditional *PI* [7]. So the structure of *Fuzzy-PI* proposed combines at the same time the characteristics of a traditional *PI* with those of a fuzzy regulator.

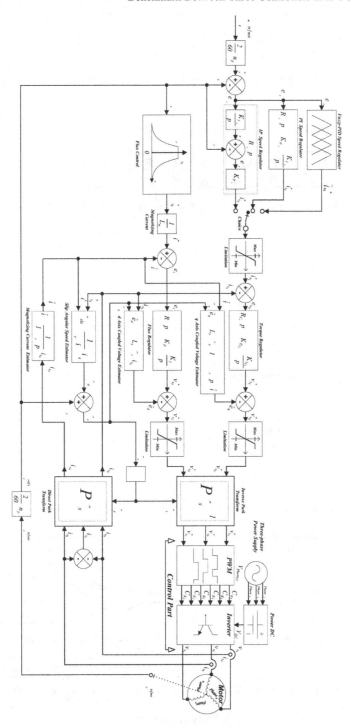

Fig. 2. IFOC rotor speed controlled by *IP* regulator

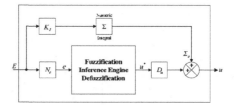

Fig. 3. IFOC rotor speed controlled by *IP* regulator

e	NB	NM	NS	Z	PS	PM	PB
u^*	NB	NM	NS	Z	PS	PM	PB

Fig. 4. Suggested membership functions for proposed Fuzzy-PI

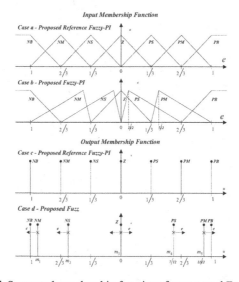

Fig. 5. Suggested membership functions for proposed Fuzzy-PI

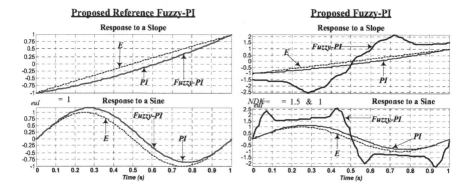

Fig. 6. Study of Fuzzy-PI

A value near unit saturates the output of the *Fuzzy-PI* (these results are not showed at the current study). The factor of denormalisation D_u acts like a total gain on output u [5]. The gain K_I has a direct integral action [7]. Tightening of the u^* mobile output's membership functions around 1, allows to relocate the response of this controller according to the error e . Lastly, the response to a sine gives the dynamic behavior of proposed*Fuzzy-PI* controller [13].

Proposed *Fuzzy-PI* is very interesting because it preserves the general behavior of a *PI* while having non-linearity in its entire characteristic. A priori, that is interesting because physical systems often present a *'globally'* linear behavior, while having low no linearity in their response curve: one can think to compensate them with this *Fuzzy-PI* .

4 APPLICATION IN IFOC AND RESULTS

In *IFOC*, *Fuzzy-PI* controller gains are:

$$N_e = \frac{\gamma_{N_e}}{c} \; ; \; K_I = K_{I_{\omega_r}}^{PI} = \frac{\gamma_{K_I} \cdot J}{f \cdot K_{\omega_r} \cdot \tau_{\omega_r}'^2} \; ; \; D_u = \gamma_{D_u} \cdot c$$

where γ_{N_e}, γ_{K_I} and γ_{D_u} are adjustable parameters, and c is the speed reference in rd/s. By using *SQP* method [8], optimal value of these adjustable parameters are:

$$\gamma_{N_e} = 6.1734 \; ; \; \gamma_{K_I} = 0.7710 \; ; \; \gamma_{D_u} = 0.1648$$

with $\tau_{\omega_r}' = 0.15\,s$

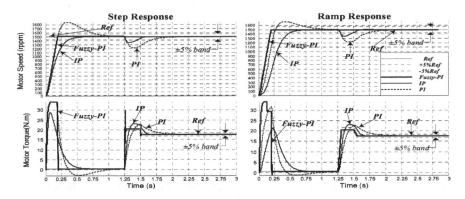

Fig. 7. Regulators response: 1500 rpm speed and 20.2 Nm torque

In *Figure 7* and in *Figure 10* we give over-shoot: $\delta^\%$ and Response time: $\tau_r^{5\%}$. As given in the following tables in more cases, *Fuzzy-PI* is better than traditional controller. In particular, *Fuzzy-PI* is insensitive with the torque load in nominal point!

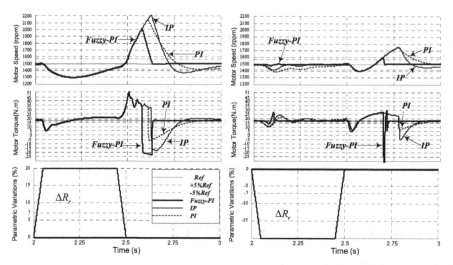

Fig. 8. Regulators response: 1500 rpm, 20.2 Nm torque step and $\pm20\%\Delta R_r$

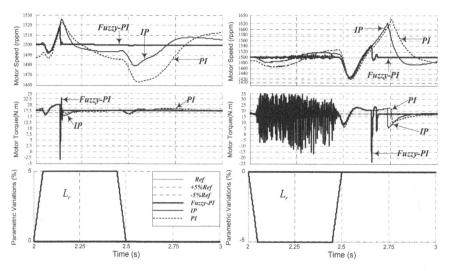

Fig. 9. Regulators response: 1500 rpm, 20.2 Nm torque step and $\pm5\%\Delta L_r$

PI controller is overall most sensitive than *IP* and *Fuzzy-PI*. *PI* is sensitive to R_s variations, *IP* is less and *Fuzzy-PI* not or very little. These three controllers are sensitive to R_r variations, but it is *Fuzzy-PI* which is less and *PI* more (*Figure 8*). To L_s variations, *IP* is more sensitive than *PI* and *Fuzzy-PI* which also less. And to L_r variations (*Figure 9*) the most sensitive controller is *PI* and the less is *Fuzzy-PI*. In conclusion, *Fuzzy-PI* is the less sensitive of the three controllers and *PI* the more [13].

Comparative study of the speed response

	%		t_r % s	
	Step	Ramp	Step	Ramp
PI	24.4947	12.3401	0.7949	0.7318
IP	0.2646	0.2604	0.3768	0.4962
Fuzzy-PI	0.4167	0.0179	0.1820	0.1057
Fuzzy-PI / PI	-98.3%	-99.9%	-77.1%	-85.6%
Fuzzy-PI / IP	+57.5%	-93.1%	-51.7%	-78.7%

Comparative study of the torque response

	%		t_r % s	
	Step	Ramp	Step	Ramp
PI	12.4014	11.0811	0.5632	0.5889
IP	13.3251	13.1118	0.3362	0.3631
Fuzzy-PI	44.8093	0	0.0173	0.0227
Fuzzy-PI / PI	+261.3%	-100%	-96.9%	-96.1%
Fuzzy-PI / IP	+236.3%	-100%	-94.9%	-93.8%

Fig. 10. Comparative Study of Speed and Torque Response

5 Conclusion

It has been shown the impact of the parameters of the fuzzy regulators proposed on their response to various references (the ramp allows to know the static behavior and the sine, the dynamic behavior). Among the number of possible parameters, we privileged normalisation and denormalisation coefficients, and input and output membership functions [4]. Our study for the regulators is made on Matlab/Simulink ® [6]. Thus, we showed that proposed *Fuzzy-PI* under some conditions, is perfectly equal to a traditional *PI*, and it can have really interesting non linear characteristics.

On the basis of work of *Hissel* [4], in my first works [13] and SQP optimisation techniques [8], we gave, in 4th part, the optimal parameters of the fuzzy regulators proposed within the meaning of criterion *IAE* [7]. Then a comparative study in *IFOC* shows the excellent static and dynamic behavior of the fuzzy regulators proposed compared to traditional *PI* and *IP* regulator. In this last case, the proposed *Fuzzy-PI* improves the over shoot and the response time more than 90%.

All the results prove the great robustness [7] of the fuzzy regulators. Even if *Fuzzy-PI* presents a simpler structure than *Fuzzy-PID* [13], it brings performances quite higher than the traditional *PI* and *IP* structure.

6 Nomenclature

We use Leroy-Somer LSVMV00L motor.
Nominal stator frequency : $\omega_{s_n}^{(rd/s)} = 100\pi$; Rotor frequency : $\omega_r^{(rd/s)}$; Sleep frequency : $\omega_{sl}^{(rd/s)}$; Nominal rotor speed : $\Omega_{r_n}^{(rpm)} = 1420$; Nominal / starting torque : $C_{e_n}^{(Nm)} / C_{eStart}^{(Nm)} = 20.2/7.3$; Number of poles pairs : $n_p = 2$; Stator / rotor resistance : $R_s^{(\Omega)} / R_r^{(\Omega)} = 2.5/4.5$; Stator / rotor inductance : $L_s^{(H)} = L_r^{(H)} = 0.270$; Mutual Inductance : $L_m^{(H)} = 0.255$; Nominal stator flux : $\phi_{s_n}^{(W)} = V_{s_n}/\omega_{s_n} = 1.2096$; *Blondel*'s coefficient : $\sigma = 0.108$; Rotor time constant : $\tau_r^{(s)} = L_r/R_r = 0.06$; Stator time constant : $\tau_s^{(s)} = L_s/R_s = 0.108$; Rotor mechanics friction : $f^{(N \cdot m \cdot s)} = 2.8 \cdot 10^{-3}$; Rotor

inertia : $J^{(kg \cdot m^2)} = 37 \cdot 10^{-3}$; Load inertia : $C_r^{(kg \cdot m^2)}$; *Park* (d, q) axes parameters : $X_{d,q}$;

References

1. H. Buhler - "Réglage par logique floue" (1994) - Collection Electricité. Presses Polytechnique et Universitaires.
2. J.-P. Caron & J.-P. Hautier - "Modélisation et Commande de la Machine Asynchrone" (1995) - Collection Méthodes et Pratiques de l'Ingénieur dirigée par P. Borne, n°7 - Edition Technip.
3. D. Driankou & H. Hellendoorn & M. Reinfrauk - "An Introduction to Fuzzy Control" (1993) - Springer-Verlug Berlin Heidelberg - New York.
4. D. Hissel - "Contribution à la Commande de dispositifs électrotechniques par logique Floue" (1998) - Thèse présentée à l'INPT (Toulouse).
5. B. Kosko - "Fuzzy Engineering" (1996) - Prentice Hall International Editions.
6. MATLAB - "Fuzzy Logic Toolbox" (1998) - The MathWorks.
7. Y. Takahashi & M. J. Rabins & D.M. Auslander - "Control and dynamic system" (1969) - Addison-Wesley Publishing Company.
8. MATLAB® - "Non Linear Control Design Blockset" - (1997) - The MathWorks Inc.
9. I. Miki & N. Nagai & S. Nishiyama & T. Yamada - "Vector Control of Induction Motor with Fuzzy PI Control" (1991) - IEEE.
10. C. M. Liaw & S. Y. Cheng- "Fuzzy Two-Degrees-of-Freedom Speed Controller for Motor Drives" (April 1995) - IEEE Transaction on Industrial Electronics, Vol. 42, No 2.
11. J. B. Wang & C. M. Liaw - "Performance Improvement of a Field-Oriented Induction Motor Drive via Fuzzy Control" (1999) - Electric Machines and Power Systems.
12. B. Heber & L. Xu & Y. Tang - "Fuzzy Logic Enhanced Speed Control of an Indirect Field Oriented Induction Machine Drive" (1995) - IEEE.
13. J.-C. de Barros & M. Tholomier - "New Optimal Fuzzy-PID Controller Approach to Indirect Vector Control Loop of a Three-Phase Induction Motor" (2002) - ICEM 2002.

Determining the Model Order of Nonlinear Input-Output Systems by Fuzzy Clustering

Balazs Feil, Janos Abonyi, and Ferenc Szeifert

University of Veszprem, Department of Process Engineering,
Veszprem, P.O. Box 158, H-8201, Hungary
abonyij@fmt.vein.hu, http://www.fmt.vein.hu/softcomp

Summary. Selecting the order of an input-output model of a dynamical system is a key step toward the goal of system identification. By determining the smallest regression vector dimension that allows accurate prediction of the output, the false nearest neighbors algorithm (FNN) is a useful tool for linear and also for nonlinear systems. The one parameter that needs to be determined before performing FNN is the threshold constant that is used to compute the percentage of false neighbors. For this purpose heuristic rules can be followed. However, for nonlinear systems choosing a suitable threshold is extremely important, the optimal choice of this parameter will depend on the system. While this advanced FNN uses nonlinear input-output data based models, the computational effort of the method increases along with the number of data and the dimension of the model. To increase the efficiency of the method this paper proposes the application of a fuzzy clustering algorithm. The advantage of the generated solutions is that it remains in the horizon of the data, hence there is no need to apply nonlinear model identification tools. The efficiency of the algorithm is supported by a data driven identification of a polymerization reactor.

1 Introduction

In recent years a wide range of model-based engineering tools have been developed [1]. However, most of these advanced techniques require models of relatively low order and restricted complexity. Since most of the current data-driven identification algorithms assume that the model structure is a priori known, the structure and the order of the model have to be chosen before identification. Several information theoretic criteria have been proposed for the selection of the order of input-output models of linear dynamical systems. A technique based on prediction-error variance, the Final Prediction-Error (FPE) criterion, was developed by Akaike [2]. Akaike also proposed another well known criterion, Aikaike's Information Criterion (AIC), that is derived from information theoretic concepts, but do not yield consistent estimates of the model order. To avoid this problem, the Minimum Description Length (MDL) criterion has been developed by both Schwartz and Rissanen, and its ability to produce consistent estimates of model order has been also proven [9]. With the usage of these tools, determining the model order of linear systems is not a problematic task. While there is extensive work in determining the proper model order for linear systems, there is relatively little work in the filed of nonlinear systems. For the determination of the order of nonlinear models deterministic suitability measures [4] and false nearest neighbors (FNN) based algorithm [10] have been worked out and applied in the chemical process industry. These ideas build upon similar methods developed for the analysis of self-driven chaotic time series [8]. The idea behind the FNN algorithm is geometric in nature. If there is enough information in the regression vector to predict the future output, then any of two regression vectors which are close in the regression space will also have future outputs which are close in some sense. Hence, the model order identification is reformulated as a determination of a distance measure and the calculation a problem-specific threshold that is used to compute the percentage of false neighbors for all combinations of the possible input variables. The one parameter that needs to be determined before performing FNN is the threshold constant. For this purpose heuristic rules can be followed [4]. However, for nonlinear systems choosing a suitable threshold is extremely important, the optimal choice of this parameter will depend on the system [10]. While this advanced FNN uses nonlinear input-output data based models, the computational effort of the method increases along with the number of data and the dimension of the model. To increase the efficiency of the method this paper proposes the application of a fuzzy clustering algorithm. The advantage of the generated solutions is that it remains in the horizon of the data, hence there is no need not to apply nonlinear model identification tools.

The paper is organized as follows. Section 2 presents the idea behind the FNN algorithm. In Section 3, the application of fuzzy clustering for improvement of this algorithm is proposed. An application example is given in Section 4. Conclusions are given in Section 5.

2 FNN Algorithm

Many non-linear static and dynamic processes can be represented by the following regression model

$$y_k = f(\mathbf{x}_k) \tag{1}$$

where $f(.)$ is a nonlinear function and $\mathbf{x}$ represents its input vector. Among this class of models, the identification of discrete-time, Non-linear Auto-Regressive models with eXogenous inputs (NARX) is considered in this paper. In the NARX model, the model inputs are past values of the process outputs $y(k)$ and the process inputs $u(k)$.

$$\mathbf{x}_k = [y(k-1), \ldots, y(k-m), u(k-1), \ldots, u(k-n)]^T \tag{2}$$

while the output of the model is the one-step ahead prediction of the process, $y_k = y(k)$. The number of past outputs used is m and the number of past inputs is n. The values m and n are often referred to as model orders. The above SISO system representation can be assumed without a loss of generality since the extension to MISO and MIMO systems is straightforward.

The method of false nearest neighbors (FNN) was developed by Kennen [8] specifically for determining the minimum embedding dimension, the number of time-delayed observations necessary to model the dynamic behavior of chaotic systems. For determining the proper regression for input/output dynamic processes, the only change to the original FNN algorithm involves the regression vector itself [10]. The main idea of the FNN algorithm utilized in this article stems from the basic property of a function. If there is enough information in the regression vector to predict the future output, then any of two regression vectors which are close in the regression space will also have future outputs which are close in some sense. For all regression vectors embedded in the proper dimensions, for two regression vectors that are close in the regression space and their corresponding outputs are related in the following way:

$$y_k - y_j = df\left(\mathbf{x}_k^{m,n}\right)\left[\mathbf{x}_k^{m,n} - \mathbf{x}_j^{m,n}\right] + o\left(\left[\mathbf{x}_k^{m,n} - \mathbf{x}_j^{m,n}\right]\right)^2 \tag{3}$$

where $df\left(\mathbf{x}_k^{m,n}\right)$ is the jacobian of the function $f(.)$ at $\mathbf{x}_k^{m,n}$.

Ignoring higher order terms, and using the Cauchy-Schwarz inequality the following inequality can be obtained:

$$\left|y_k - y_j\right| \leq \left\|df\left(\mathbf{x}_k^{m,n}\right)\right\|_2 \left\|\mathbf{x}_k^{m,n} - \mathbf{x}_j^{m,n}\right\|_2 \tag{4}$$

$$\frac{\left|y_k - y_j\right|}{\left\|\mathbf{x}_k^{m,n} - \mathbf{x}_j^{m,n}\right\|_2} \leq \left\|df\left(\mathbf{x}_k^{m,n}\right)\right\|_2 \tag{5}$$

If the above expression is true, then the neighbors are recorded as true neighbors. Otherwise, the neighbors are false neighbors.

Based on this theoretical background, the outline of the FNN algorithm is the following.

1. Identify the nearest neighbor to a given point in the regressor space. For a given regressor:

$$\mathbf{x}_k^{m,n} = [y(k), \ldots, y(k-m), u(k), \ldots, u(k-n)]^T$$

find the nearest neighbor $\mathbf{x}_j^{m,n}$ such that the distance d is minimized:

$$d = ||\mathbf{x}_k^{m,n} - \mathbf{x}_j^{m,n}||_2$$

2. Determine if the following expression is true or false

$$\frac{\left|y_k - y_j\right|}{||\mathbf{x}_k^{m,n} - \mathbf{x}_j^{m,n}||_2} \leq R$$

where R is a previously chosen threshold value. If the above expression is true, then the neighbors are recorded as true neighbors. Otherwise, the neighbors are false neighbors.

3. Continue the algorithm for all times k in the data set.

4. Calculate the percentage of points in the data that have false nearest neighbors $J(m,n)$.
5. Continue the algorithm for increasing m and n using the percentage of false nearest neighbors drops to some acceptably small number.

Because the model order is determined by finding the number of past outputs m and past inputs n, the $J(m,n)$ indices become a surface in two dimensions. It is possible to find a 'global' solution (or solutions) for the model orders by computing the desired index over all values of m and n in a certain range and determining which points satisfy the order determination conditions. The smallest m and n values such that $J(m,n)$ is zero lie in the corner of this that is nearest to the origin, $\hat{m}$ and $\hat{n}$. This corner is easily identified since $J(m,n) \neq 0$ for $m \leq \hat{m}$ and $n \leq \hat{n}$. When the noise is not zero, $J(m,n)$ will not be zero if m and n are chosen as $m \geq \hat{m}$ and $n \geq \hat{n}$, but it will tend to remain relatively small and flat. Therefore, we calculate table of $J(m,n)$ and then search for the corner where $J(m,n)$ drops quickly similarly to the MDL based method suggested in [9]. A more heuristic 'local' solution is also possible. In this case, initial guesses for m and n are used, and the optimum model order is computed competitively; ate each iteration, either m or n is increased by one, depending on which reduces the index the greatest amount [4].

In cases where the available input-output data set is small, the algorithm is sensitive to the choice of the R threshold. In [4] the threshold value was selected by trial and error method based on empirical rules of thumb, $10 \leq R \leq 50$. However, choosing a single threshold that will work well for all data sets is impossible task. In this case, it is advantageous to estimate R based on 5 using the following expression $R = max_k \left\| df\left(\mathbf{x}_k^{m,n}\right) \right\|$ as it has been suggested by Rhodes and Morari [10]. While the method uses input-output data based models for the estimation of $\left\| df\left(\mathbf{x}_k^{m,n}\right) \right\|$, the computational effort of FNN increases along with the number of data and the dimension of the model. To increase the efficiency of the method this paper proposes the application of a fuzzy clustering algorithm that will be introduced in the following section.

3 Application of Fuzzy Clustering to FNN

The available input-output can be clustered. The main idea of the paper is that when the appropriate number of regressors are used, the collection of the obtained clusters will approximate the regression surface of the model of the system. In this case the clusters can be approximately regarded as local linearizations of the system and can be used to estimate R.

Clusters of different shapes can be obtained by different clustering algorithms by using an appropriate definition of cluster prototypes (e.g., points vs. linear varieties) or by using different distance measures. The Gustafson–Kessel clustering algorithm [7] has been often applied to identify Takagi–Sugeno fuzzy systems that are based on local linear models [3]. The main drawbacks of this algorithm are that only clusters with approximately equal volumes can be properly identified which constrain makes the application of this algorithm problematic for the task of this paper. To circumvent this problem, in this paper Gath–Geva algorithm is applied [6] that will be described in the following subsection.

3.1 Gath-Geva Clustering of the Data

The objective of clustering is to partition a data set $\mathbf{Z}$ into c clusters, where the available identification data, $\mathbf{Z}^T = [\mathbf{X}\mathbf{y}]$ formed by concatenating the regression data matrix $\mathbf{X}$ and the output vector $\mathbf{y}$

$$
\mathbf{X} = \begin{bmatrix} \mathbf{x}_1^T \\ \mathbf{x}_2^T \\ \vdots \\ \mathbf{x}_N^T \end{bmatrix} \qquad \mathbf{y} = \begin{bmatrix} y_1 \\ y_2 \\ \vdots \\ y_N \end{bmatrix} \tag{6}
$$

This means, each observation consists of $m+n+1$ variables, grouped into an $m+n+1$-dimensional column vector $\mathbf{z}_k = [x_{1,k},\ldots,x_{n+m,k},y_k]^T = [\mathbf{x}_k^T\, y_k]^T$. Through clustering, the fuzzy partition matrix $\mathbf{U} = [\mu_{i,k}]_{c \times N}$ is obtained, whose element μ_{ik} represents the degree of membership of the observation $\mathbf{z}_k$ in the cluster $i = 1,\ldots,c$. In this paper, c is assumed to be known, based on prior knowledge, for instance. For methods to estimate or optimize c in the context of system identification refer to [3].

The GG algorithm is based on the minimization of the sum of the weighted squared distances between the data points, $\mathbf{z}_k$ and the cluster centers, $\mathbf{v}_i, i = 1,\ldots,c$.

$$
J(\mathbf{Z},\mathbf{U},\mathbf{V}) = \sum_{i=1}^{c} \sum_{j=1}^{N} \mu_{i,k}^m D_{i,k}^2 \tag{7}
$$

where $\mathbf{V} = [\mathbf{v}_1,\ldots,\mathbf{v}_c]$ contains the cluster centers and $m \in [1,\infty)$ is a weighting exponent that determines the fuzziness of the resulting clusters and it is often chosen as $mw = 2$. The fuzzy partition matrix has to satisfy the following conditions:

$$
U \in \mathbf{R}^{c \times N} \quad \text{with} \quad \mu_{i,k} \in [0,1], \, \forall i,k; \quad \sum_{i=1}^{c} \mu_{i,k} = 1, \, \forall k; \quad 0 < \sum_{k=1}^{N} \mu_{i,k} < N, \, \forall i \tag{8}
$$

The minimum of 1 is sought by the alternating optimization (AO) method given below:

Initialization Given a set of data $\mathbf{Z}$ specify c, choose a weighting exponent $m > 1$ and a termination tolerance $\varepsilon > 0$. Initialize the partition matrix such that 3 holds.
Repeat for $l = 1,2,\ldots$
Step 1 Calculate the cluster centers.

$$
\mathbf{v}_i^{(l)} = \frac{\sum_{k=1}^{N} \mu_{i,k}^{(l-1)} \mathbf{z}_k}{\sum_{k=1}^{N} \mu_{i,k}^{(l-1)}}, \, 1 \le i \le c \tag{9}
$$

Step 2 Compute the distance measure D_{ik}^2.

The distance to the prototype is calculated based the fuzzy covariance matrices of the cluster

$$\mathbf{F}_i^{(l)} = \frac{\sum_{k=1}^{N} \mu_{ik}^{(l-1)} \left(\mathbf{z}_k - \mathbf{v}_i^{(l)}\right) \left(\mathbf{z}_k - \mathbf{v}_i^{(l)}\right)^T}{\sum_{k=1}^{N} \mu_{ik}^{(l-1)}}, 1 \leq i \leq c \tag{10}$$

The distance function is chosen as

$$D_{i,k}^2(\mathbf{z}_k, \mathbf{v}_i) = \frac{(2\pi)^{\left(\frac{n+1}{2}\right)} \sqrt{\det(F_i)}}{\alpha_i} \exp\left(\frac{1}{2}\left(\mathbf{z}_k - \mathbf{v}_i^{(l)}\right)^T \mathbf{F}_i^{-1}\left(\mathbf{z}_k - \mathbf{v}_i^{(l)}\right)\right) \tag{11}$$

with the *a priori* probability α_i

$$\alpha_i = \frac{1}{N} \sum_{k=1}^{N} \mu_{i,k} \tag{12}$$

Step 3 Update the partition matrix

$$\mu_{i,k}^{(l)} = \frac{1}{\sum_{j=1}^{c} \left(D_{ik}(\mathbf{z}_k, \mathbf{v}_i)/D_{jk}(\mathbf{z}_k, \mathbf{v}_j)\right)^{2/(mw-1)}}, \quad 1 \leq i \leq c, 1 \leq k \leq N. \tag{13}$$

until $||\mathbf{U}^{(l)} - \mathbf{U}^{(l-1)}|| < \varepsilon$.

3.2 Estimation of the R Threshold Coefficient

The collection of c clusters approximates the regression surface as it is illustrated in Figure 1. Hence, the clusters can be approximately regarded as local linear subspaces.

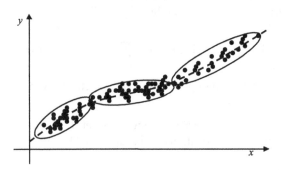

Fig. 1. Example for clusters approximating the regression surface.

This is reflected by the smallest eigenvalues $\lambda_{i,m+n+1}$ of the cluster covariance matrices $\mathbf{F}_i$ that are typically in orders of magnitude smaller than the remaining eigenvalues [3] (see Figure 2).

The eigenvector corresponding to this smallest eigenvalue, $\mathbf{t}^i_{m+n+1}$, determines the normal vector to the hyperplane spanned by the remaining eigenvectors of that cluster

$$(\mathbf{t}^i_{m+n+1})^T (\mathbf{z}_k - \mathbf{v}_i) = 0 \qquad (14)$$

Similarly to the observation vector $\mathbf{z}_k = [\mathbf{x}^T_k \, y_k]^T$, the prototype vector and is partitioned as $\mathbf{v}_i = \left[(\mathbf{v}^x_i)^T \, v^y_i \right]$ into a vector $\mathbf{v}^x$ corresponding to the regressor $\mathbf{x}$, and a scalar v^y_i corresponding to the output y. The smallest eigenvector is partitioned in the same way, $\mathbf{t}^i_{m+n+1} = \left[\left(\mathbf{t}^{i,x}_{m+n+1} \right)^T \, t^{i,y}_{m+n+1} \right]^T$. By using this partitioned vectors 14 can be written as

$$\left[\left(\mathbf{t}^{i,x}_{m+n+1} \right)^T \, t^{i,y}_{m+n+1} \right]^T \left([\mathbf{x}^T_k \, y_k]^T - \left[(\mathbf{v}^x_i)^T \, v^y_i \right] \right) = 0 \qquad (15)$$

from which the parameters of the hyperplane defined by the cluster can be obtained:

$$y_k = \underbrace{\frac{-1}{t^{i,y}_{m+n+1}} \left(\mathbf{t}^{i,x}_{m+n+1} \right)^T}_{\mathbf{a}^T_i} \mathbf{x}_k + \underbrace{\frac{1}{t^{i,y}_{m+n+1}} \left(\mathbf{t}^i_{m+n+1} \right)^T \mathbf{v}_i}_{b_i} = \mathbf{a}^T_i \mathbf{x}_k + b_i \qquad (16)$$

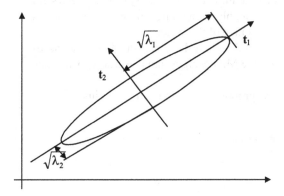

Fig. 2. Example for clusters approximating the regression surface.

Although the parameters have been derived from the geometrical interpretation of the clusters, it can be shown [3] that 16 is equivalent to the weighed total least-squares estimation of the consequent parameters, where each data point is weighed by the corresponding $\sqrt{\mu_{ik}}$.

The main contribution of this paper is that it suggests the application of an adaptive threshold function that takes into account the nonlinearity of the system. This means, based on the result of the fuzzy clustering, for all input-output data pairs different R_k values are calculated. Since, the optimal value of R_k is $R_k = \left\| df \left(\mathbf{x}^{m,n}_k \right) \right\|$

and the $df\left(\mathbf{x}_k^{m,n}\right)$ partial derivatives can be estimated based on the shape of the clusters from 16

$$df\left(\mathbf{x}_k^{m,n}\right) \approx \sum_{i=1}^{c} \mu_{ik} \frac{-1}{t_{m+n+1}^{i,y}} \left(\mathbf{t}_{m+n+1}^{i,x}\right)^T \tag{17}$$

the threshold can be calculated as

$$R_k = \left\| \sum_{i=1}^{c} \mu_{ik} \frac{-1}{t_{m+n+1}^{i,y}} \left(\mathbf{t}_{m+n+1}^{i,x}\right)^T \right\|_2 \tag{18}$$

4 Application for Continuous Polymerization Reactor

The following example illustrates identification using data from a model of a continuous polymerization reactor. The model describes the free-radical polymerization of methyl methacrylate with azobisisobutyronitrile as an initiator an toluene as a solvent. For further information on the details of this model and how it is derived, see [5]. The reaction takes place in a jacketed CSTR. The dimensionless state variable x_1 refers to the monomer concentration, and x_4/x_3 is the number-average molecular weight (an also the output y). The input u is the dimensionless volumetric flow rate of the initiator. Since a model of the system is known, large amounts of data can be collected for analysis. For this example we apply a uniformly distributed random input over the range 0.007 to 0.015 with a sampling time of 0.2. By driving the system with this input signal, an output that is roughly in the range of 26,000 to 34,000 is produced, which is the desired operating range of the system.

The model with output order $m = 1$ and input order $n = 2$ should give an accurate estimate of future outputs, because the MARS algorithm constructs an accurate model for this problem [10].

The Table 1 shows the results of the proposed algorithm with c=6 cluster.

Table 1. FNN results for polymerization data when R is obtained by fuzzy clustering

Input Delays (n)	Output Delays (m)				
% FNN	0	1	2	3	4
0	100.00	99.87	62.40	38.14	17.25
1	99.19	69.14	8.76	0.54	0.54
2	73.45	3.77	1.89	0.67	0.00
3	8.76	0.40	0.14	0.00	0.00
4	0.14	0.40	0.13	0.00	0.00

The number with m=2 and l=1 is enough small, but larger input and output orders are acceptable, too [10]. The clustering algorithm has an parameter: c, the number of the clusters. The increasing of this parameter increases the accuracy of the model as a general rule. For the purpose to avoid the overfitting and the increasing of the calculation requirement it is recommended to determine the number of the clusters automatically. For this purposes the method of Gath and Geva can be applied [6] and [3].

For comparisons the next table shows the results when constant threshold has been used. In this case the value of R has been estimated based on the parameters of a linear ARX model identified based on the data used for clustering purposes.

Table 2. FNN results for polymerization data when R is obtained based on the parameters of a linear ARX model

Input Delays (n)	Output Delays (m)				
% FNN	0	1	2	3	4
0	100.00	99.73	97.04	92.32	84.64
1	99.60	68.19	9.97	1.48	0.94
2	73.72	5.52	4.45	2.02	0.14
3	9.57	2.16	1.08	1.48	0.27
4	1.48	1.08	1.35	0.94	0.27

We can allocate that this linear model based method does not give conspicuously incorrect results, but induces larger error for high nonlinear systems, because it results in more inaccurate approximation. Hence, the application of the proposed clustering based approach is much more advantageous.

5 Conclusions

By determining the smallest regression vector dimension that allows accurate prediction of the output, the FNN algorithm is a useful tool for linear and also nonlinear systems. It reduces the overall computational effort, simplifies and makes more effective the nonlinear identification which becomes difficult and gives not certainly accurate results by false regression vector.

To increase the efficiency of FNN this paper proposed the application of clustering algorithm. The advantage of our approach is that it remains in the horizon of the data and there is need not to apply nonlinear model identification tools to determine the threshold parameter of the FNN algorithm.

Acknowledgement

The financial support of the Hungarian Ministry of Culture and Education (FKFP-0073/2001) and the Hungarian Science Foundation (T037600) is greatly acknowledged. Janos Abonyi is grateful for the financial support of the Janos Bolyai Research Fellowship of the Hungarian Academy of Science.

References

1. J. Abonyi. *Fuzzy Model Identification for Control.* Birkhauser, Cambridge, MA, 2003.
2. H. Akaike. A new look at the statistical model identification. *IEEE Trans. on Automatic Control*, 19:716–723, 1974.
3. R. Babuška. *Fuzzy Modeling for Control.* Kluwer Academic Publishers, Boston, 1998.
4. J.D. Bomberger and D.E. Seborg. Determination of model order for NARX models directly from input–output data. *Journal of Process Control*, 8:459–468, Oct–Dec 1998.
5. F.J. Doyle, B.A. Ogunnaike, and R. K. Pearson. Nonlinear model-based control using second-order volterra models. *Automatica*, 31:697, 1995.
6. I. Gath and A.B. Geva. Unsupervised optimal fuzzy clustering. *IEEE Transactions on Pattern Analysis and Machine Intelligence*, 7:773–781, 1989.
7. D.E. Gustafson and W.C. Kessel. Fuzzy clustering with fuzzy covariance matrix. In *Proceedings of the IEEE CDC, San Diego*, pages 761–766. 1979.
8. M.B. Kennel, R. Brown, and H.D.I. Abarbanel. Determining embedding dimension for phase-space reconstruction using a geometrical construction. *Physical Review*, A:3403–3411, 1992.
9. G. Liang, D.M. Wilkes, and J.A. Cadzow. Arma model order estimation based on the eigenvalues of the covariance matrix. *IEEE Trans. on Signal Processing*, 41(10):3003–3009, 1993.
10. C. Rhodes and M. Morari. Determining the model order of nonlinear input/output systems. *AIChE Journal*, 44:151–163, 1998.

Fuzzy Self-Organizing Map based on Regularized Fuzzy c-means Clustering

Janos Abonyi, Sandor Migaly and Ferenc Szeifert

University of Veszprem, Department of Process Engineering,
Veszprem, P.O. Box 158, H-8201, Hungary
abonyij@fmt.vein.hu, http://www.fmt.vein.hu/softcomp

Summary. This paper presents a new fuzzy clustering algorithm for the clustering and visualization of high-dimensional data. The cluster centers are arranged on a grid defined on a small dimensional space that can be easily visualized. The smoothness of this mapping is achieved by adding a regularization term to the fuzzy c-means (FCM) functional. The measure of the smoothness is expressed as the sum of the second order partial derivatives of the cluster centers. Coding the values of the cluster centers with colors, regions with different colors evolve on the map and the hidden relation between the variables reveal. Comparison to the existing modifications of the fuzzy c-means algorithm and several application examples are given.

1 Introduction

Clustering based computational intelligence methods are becoming increasingly popular in the pattern recognition community. They are able to learn the mapping of functions and systems [1, 3], and can perform classification from labeled training data as well as explore structures and classes in unlabeled data [5]. If there are lot of clusters, it is hard to evaluate the results of the clustering. To avoid this problem, it is useful to arrange the clusters on a low dimensional grid and visualize them. The aim of this paper is to develop an algorithm for this visualization task.

The visualization of high-dimensional data is an important pattern recognition task. Advanced visualization tools should be able to convert complex, nonlinear statistical relationships between high-dimensional data items into simple geometric relationships on a low-dimensional display and compress information while preserving the most important topological and metric relationships of the primary data items.

Among the wide range of possible tools, the self-organizing map (SOM) is one of the most effective [6]. The Self-Organizing Map as a special clustering tool provides a compact representation of the data distribution, has been widely applied in the visualization of high-dimensional data. SOM implements an ordered mapping of the high-dimensional distribution of the data onto a low-dimensional grid.

The SOM algorithm can be considered as a regularized version of the hard c-means clustering algorithm. Hard clustering methods are based on classical set theory, and it requires an object that either does or does not belong to a cluster. Fuzzy clustering methods operate with fuzzy sets allowing the objects to belong several clusters simultaneously with different degrees of membership [5]. The data set is thus partitioned into c fuzzy subsets. In many real situations, fuzzy clustering is more natural than hard clustering, as objects on the boundaries between several classes are not forced to fully belong to one of the classes. Recently, several approaches have been worked out to increase the performance of SOM by the incorporation of fuzzy logic. In the study of Vuorimaa [9], a modified SOM algorithm is introduced by the replacement of the neurons with fuzzy rules that allows an efficient modelling of continuous valued functions. In [2] fuzzy clustering combined with SOM is used to project the data to lower dimensions. Chen-Kuo Tsao et al. integrate some aspects of the fuzzy c-means model into the classical Kohonen-type hard clustering framework [8]. The most interesting approach has been presented in [7], where a fuzzy self-organizing map is developed based on the modifications of the fuzzy c-means functional. In this approach, similarly to SOM, the cluster centers are distributed on a regular low-dimensional grid, and a regularization term is added to guarantee a smooth distribution for the values of the code vectors on the grid. The idea of the ordering of the clusters in a smaller dimensional space can be also found in [10], where the fuzzy c-means functional has been modified to detect smooth lines.

The aim of this paper is to generalize these ideas to develop a clustering algorithm that can be simultaneously used for the clustering and the visualization of high-dimensional data. With the introduction of a new regularization term into the standard fuzzy clustering algorithm it is possible to arrange the cluster prototypes on a grid. This type of arrangement of the code vectors helps greatly in the visualization of the results, as the regularization orders the similar cluster centers closer to each other. The values of the cluster centers are separately plotted on maps, where each map represents a variable. The color-coding of the numerical values of the clusters results in regions on the map that shows the relations among the variables. Hence, the proposed algorithm can be used as an alternative of SOM.

The paper is organized as follows. Section 2 presents the idea behind the proposed Regularized Fuzzy c-means Clustering algorithm. This section also shows how the recently published modifications of the fuzzy c-means algorithms can be derived from this algorithm. Some illustrative application examples are given in Section 3, and the conclusions are drawn in Section 4.

2 Regularized Fuzzy c-means Clustering

2.1 Regularized Fuzzy c-means Clustering Algorithm

In this paper the clustering of quantitative data is considered. The data are typically observations of some physical phenomenon. Each observation consists of n measured variables, grouped into an n-dimensional column vector $\mathbf{z}_k = [z_{1,k}, \ldots, z_{n,k}]^T$. The objective of clustering is to partition the set of N observations into c clusters, where the available identification data, $\mathbf{Z}$ is formed as $\mathbf{Z} = [\mathbf{z}_1^T, \cdots, \mathbf{z}_N^T]^T$.

Through clustering, the fuzzy partition matrix $\mathbf{U} = [\mu_{i,k}]_{c \times N}$ is obtained, whose element $\mu_{i,k}$ represents the degree of membership of the observation $\mathbf{z}_k$ in the cluster ß $= 1, \ldots, c$. In this paper, c is assumed to be known, based on prior knowledge, for instance. For methods to estimate or optimize c in the context of system identification refer to [3].

Most of the fuzzy clustering algorithms are based on the minimization of the sum of the weighted squared distances between the data points, $\mathbf{z}_k$ and the cluster centers, $\mathbf{v}_i, i = 1, \ldots, c$.

$$J(\mathbf{Z}, \mathbf{U}, \mathbf{V}) = \sum_{i=1}^{c} \sum_{j=1}^{N} \mu_{i,k}^m D_{i,k}^2(\mathbf{z}_k, \mathbf{v}_i) \tag{1}$$

where $\mathbf{V} = [\mathbf{v}_1, \ldots, \mathbf{v}_c]$ contains the cluster centers, $D_{i,k}^2$ represents the distance between the i-th cluster and the k-th data, and $m \in [1, \infty)$ is a weighting exponent that determines the fuzziness of the resulting clusters and it is often chosen as $m = 2$. $D_{i,k}^2(\mathbf{z}_k, \mathbf{v}_i)$ can be determined by any appropriate distance measure, e.g., by an $\mathbf{A}$-norm:

$$D_{i,k}(\mathbf{z}_k, \mathbf{v}_i) = \|(\mathbf{z}_k - \mathbf{v}_i)\|_\mathbf{A} = \sqrt{(\mathbf{z}_k - \mathbf{v}_i)^T \mathbf{A}(\mathbf{z}_k - \mathbf{v}_i)} \tag{2}$$

Usually spherical clusters are applied when $\mathbf{A}$ is an identity matrix, $\mathbf{A} = I$.

The fuzzy partition matrix has to satisfy the following conditions: $U \in \mathcal{R}^{c \times N}$ with:

$$\mu_{i,k} \in [0,1], \ \forall i,k; \quad \sum_{i=1}^{c} \mu_{i,k} = 1, \ \forall k; \quad 0 < \sum_{k=1}^{N} \mu_{i,k} < N, \ \forall i \tag{3}$$

The classical fuzzy c-means clustering algorithm is able to detect groups in the data, but the obtained clusters are not ordered which makes the interpretation of the model to be difficult. The aim of this paper is to increase the transparency of the result of clustering by ordering the cluster centers (prototypes) on an easily visualizable low (in this paper two) dimensional space. According to this motivation, similarly to SOM, the cluster centers (also referred as code vectors) are distributed on a two-dimensional lattice, but other topologies can be considered.

The proposed algorithm performs a topology preserving mapping from high, n, dimensional space of the cluster centers onto a smaller,$s \ll n$, dimensional map. The cluster centers of this s dimensional map are connected to the adjacent cluster centers by a neighborhood relation, which dictates the topology of the map. For instance, Figure 1 shows a regular square grid, corresponding to $c = 9$ code vectors. This kind of arrangement of the clusters makes the result of the clustering interpretable only if the smoothness of the distribution of the cluster centers on this s dimensional map is guaranteed. Since the smoothness of the grid of the cluster centers can be measured as

$$S = \int \left\| \frac{\partial^2 \mathbf{v}}{\partial^2 \mathbf{x}} \right\| d\mathbf{x}. \tag{4}$$

This smoothness can be approximated by second order differentiation, so 4 can be expressed in matrix multiplication form:

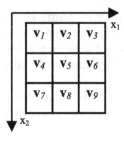

Fig. 1. Example of cluster centers arranged on a two-dimensional space

$$S \approx tr(\mathbf{VLV}^T) = tr(\mathbf{VG}^T\mathbf{GV}^T). \tag{5}$$

where $\mathbf{G} = [\mathbf{G}_1^T, \cdots \mathbf{G}_s^T]^T$, where $\mathbf{G}_i$ denotes the second order partial difference operator in the $i = 1, \ldots, s$-th direction of the map. At the boundary of the map the the second order difference of the cluster centers cannot be evaluated, hence $\mathbf{G}_i$ has to calculated taking into the borders of the map. In case of the model depicted in Figure 1, these matrices are the following:

$$\mathbf{G}_1 = \begin{bmatrix} 0 & 0\,0\,0 & 0\,0\,0 & 0\,0 \\ 1 & -2\,1\,0 & 0\,0\,0 & 0\,0 \\ 0 & 0\,0\,0 & 0\,0\,0 & 0\,0 \\ 0 & 0\,0\,0 & 0\,0\,0 & 0\,0 \\ 0 & 0\,0\,1 & -2\,1\,0 & 0\,0 \\ 0 & 0\,0\,0 & 0\,0\,0 & 0\,0 \\ 0 & 0\,0\,0 & 0\,0\,0 & 0\,0 \\ 0 & 0\,0\,0 & 0\,0\,1 & -2\,1 \\ 0 & 0\,0\,0 & 0\,0\,0 & 0\,0 \end{bmatrix} \quad \mathbf{G}_2 = \begin{bmatrix} 0\,0\,0 & 0 & 0 & 0\,0\,0\,0 \\ 0\,0\,0 & 0 & 0 & 0\,0\,0\,0 \\ 0\,0\,0 & 0 & 0 & 0\,0\,0\,0 \\ 1\,0\,0 & -2 & 0 & 0\,1\,0\,0 \\ 0\,1\,0 & 0 & -2 & 0\,0\,1\,0 \\ 0\,0\,1 & 0 & 0 & -2\,0\,0\,1 \\ 0\,0\,0 & 0 & 0 & 0\,0\,0\,0 \\ 0\,0\,0 & 0 & 0 & 0\,0\,0\,0 \\ 0\,0\,0 & 0 & 0 & 0\,0\,0\,0 \end{bmatrix} \tag{6}$$

To obtain ordered cluster centers, this smoothness measure can be added to the original cost function of the clustering algorithm, 1

$$J(\mathbf{Z}, \mathbf{U}, \mathbf{V}) = \sum_{i=1}^{c} \sum_{j=1}^{N} \mu_{i,k}^m D_{i,k}^2(\mathbf{z}_k, \mathbf{v}_i) + \varsigma tr(\mathbf{VLV}^T) \tag{7}$$

Taking into account the constraints related to the membership values 3, the minimum of 7 can be solved by using a variety of available methods [2]. The most popular method is the alternating optimization (AO) [4]. The AO algorithm is iterative. The name of AO comes from the fact that this kind of minimization algorithms alternately selects a variable, hold the other constants and moves the selected variable toward the local minimum of the objective function.

For fixed membership values, $\mathbf{U}$, the minimum of the cost function with respect to the cluster centers, $\mathbf{V}$, can be found by solving the following system of linear equations:

$$\mathbf{v}_i \sum_{k=1}^{N} \mu_{i,k}^m + \varsigma \sum_{j=1}^{c} L_{i,j} \mathbf{v}_j = \sum_{k=1}^{N} \mu_{i,k} \mathbf{z}_k \ i = 1, \dots, c \tag{8}$$

where $L_{i,j}$ denotes the i, j-th element of the $\mathbf{L} = \mathbf{G}^T \mathbf{G}$ matrix.

Based on the previous equation, the first step of the regularized FCM-AO algorithm is the following:

- **Initialization**
  Given a set of data $\mathbf{Z}$ specify c, choose a weighting exponent $m > 1$ and a termination tolerance $\varepsilon > 0$. Initialize the partition matrix such that 3 holds.
- **Repeat for** $l = 1, 2, \dots$
- **Step 1: Calculate the cluster centers.**

$$\mathbf{v}_i^{(l)} = \frac{\sum_{k=1}^{N} \left(\mu_{i,k}^{(l-1)} \right)^m \mathbf{z}_k - \varsigma \sum_{j=1, j \neq i}^{c} L_{i,j} \mathbf{v}_j}{\sum_{k=1}^{N} \left(\mu_{i,k}^{(l-1)} \right)^m + \varsigma L_{i,j}}, \ 1 \leq i \leq c \tag{9}$$

- **Step 2 Compute the distances $D_{i,k}^2$.**

$$D_{i,k}^2(\mathbf{z}_k, \mathbf{v}_i) = \left(\mathbf{z}_k - \mathbf{v}_i^{(l)} \right)^T \mathbf{A} \left(\mathbf{z}_k - \mathbf{v}_i^{(l)} \right) \ 1 \leq i \leq c, 1 \leq k \leq N \tag{10}$$

- **Step 3: Update the partition matrix**

$$\mu_{i,k}^{(l)} = \frac{1}{\sum_{j=1}^{c} \left(D_{i,k}(\mathbf{z}_k, \mathbf{v}_i) / D_{j,k}(\mathbf{z}_k, \mathbf{v}_j) \right)^{2/(m-1)}}, \tag{11}$$
$$1 \leq i \leq c, 1 \leq k \leq N.$$

until $\|\mathbf{U}^{(l)} - \mathbf{U}^{(l-1)}\| < \varepsilon$.

2.2 Relation to the Fuzzy Curve-Tracing Algorithm

It is interesting to note that the fuzzy curve-tracing algorithm [10] developed for the extraction of smooth curve from unordered noisy data can be considered as a special case of the previously proposed algorithm, where a neighboring cluster centers are linked to produce a graph according to the average membership values. After the loops in the graph have been removed, the data are then re-clustered using the fuzzy c-means algorithm, with the constraint that the curve must be smooth. In this approach, the measure of smoothness is also the sum of second order partial derivatives for every cluster centers:

$$J(\mathbf{Z}, \mathbf{U}, \mathbf{V}) = \sum_{i=1}^{c} \sum_{j=1}^{N} \mu_{i,k}^m D_{i,k}^2(\mathbf{z}_k, \mathbf{v}_i) + \varsigma \sum_{i=1}^{c} \left\| \frac{\partial^2 \mathbf{v}_i}{\partial^2 x} \right\|^2 \tag{12}$$

As this cost-function shows, the fuzzy curve-tracing applies a clustering algorithm that can be considered as a special one-dimensional case of the proposed regularized clustering algorithm 7, where $\mathbf{G}_1$ is the second order difference operator. If $c = 5$ then $\mathbf{G}_1$ and $\mathbf{L}$ are:

$$\mathbf{G}_1 = \begin{bmatrix} 0 & 0 & 0 & 0 & 0 \\ 1 & -2 & 1 & 0 & 0 \\ 0 & 1 & -2 & 1 & 0 \\ 0 & 0 & 1 & -2 & 1 \\ 0 & 0 & 0 & 0 & 0 \end{bmatrix} \quad \mathbf{L} = \begin{bmatrix} 1 & 2 & -1 & 0 & 0 \\ 2 & 5 & 4 & -1 & 0 \\ -1 & 4 & 6 & 4 & -1 \\ 0 & -1 & 4 & 5 & 2 \\ 0 & 0 & -1 & 2 & 1 \end{bmatrix} \quad (13)$$

It is interesting to note that the elements of L are identical to the regularization co-efficients derived "manually" in [10]. As the proposed approach gives compact and generalized approach to the fuzzy curve-tracing algorithm, we can call our algorithm as Fuzzy Surface-Tracing (FST) method.

2.3 Relation to the Smoothly Distributed FCM Algorithm

One possible interpretation of smoothness is that the cluster centers must be close to the average value of their nearest neighbors on the grid. Referring to Figure 1, this means that $\mathbf{G}_i$ becomes a first order gradient operator [7]:

$$\mathbf{G}_1 = \begin{bmatrix} 0 & 0 & 0 & 0 & 0 & 0 & 0 & 0 & 0 \\ 1 & 1 & 0 & 0 & 0 & 0 & 0 & 0 & 0 \\ 0 & -1 & 1 & 0 & 0 & 0 & 0 & 0 & 0 \\ 0 & 0 & 0 & 0 & 0 & 0 & 0 & 0 & 0 \\ 0 & 0 & 0 & -1 & 1 & 0 & 0 & 0 & 0 \\ 0 & 0 & 0 & 0 & -1 & 1 & 0 & 0 & 0 \\ 0 & 0 & 0 & 0 & 0 & 0 & 0 & 0 & 0 \\ 0 & 0 & 0 & 0 & 0 & 0 & -1 & 1 & 0 \\ 0 & 0 & 0 & 0 & 0 & 0 & 0 & -1 & 1 \end{bmatrix} \quad \mathbf{G}_2 = \begin{bmatrix} 0 & 0 & 0 & 0 & 0 & 0 & 0 & 0 & 0 \\ 0 & 0 & 0 & 0 & 0 & 0 & 0 & 0 & 0 \\ 0 & 0 & 0 & 0 & 0 & 0 & 0 & 0 & 0 \\ -1 & 0 & 0 & 1 & 0 & 0 & 0 & 0 & 0 \\ 0 & -1 & 0 & 0 & 1 & 0 & 0 & 0 & 0 \\ 0 & 0 & -1 & 0 & 0 & 1 & 0 & 0 & 0 \\ 0 & 0 & 0 & -1 & 0 & 0 & 1 & 0 & 0 \\ 0 & 0 & 0 & 0 & -1 & 0 & 0 & 1 & 0 \\ 0 & 0 & 0 & 0 & 0 & -1 & 0 & 0 & 1 \end{bmatrix} \quad (14)$$

This type of penalizing is not always advantageous, since it attracts the cluster centers to the center of the data, where the magnitude of this contraction is determined by the regulation parameter ς.

3 Illustrative Examples

In this section the proposed algorithm is used to trace a spiral in three-dimensional space and to show how the elastic lattice of the cluster centers can be fold a half sphere.

3.1 Tracing a Spiral in 3D

The aim of the first example is to trace a part of a spiral in 3D. For this purpose 300 points are available with noise with 0 mean and variance 0.2. The aim of the clustering is to detect seven ordered clusters that can be lined up to detect the 3D curvature. Both FCM and its regularized version are used, where the parameters of the clustering algorithm were $m = 2$ and $\varsigma = 2$.

As can be seen from the comparison of Figure 2 and Figure 3, the with the help of the proposed regularized clustering algorithm the detected cluster centers are arranged in a proper order.

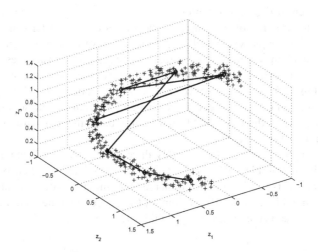

Fig. 2. Detected clusters and the obtained ordering when the standard FCM algorithm is used.

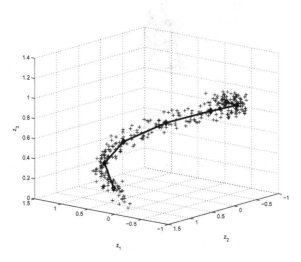

Fig. 3. Detected clusters and the obtained ordering when the regularized FCM algorithm is used.

3.2 Mapping of Spheres

In the second example we folded a 6x6 grid on a half sphere. 900 points were taken and noise with zero mean and 0.1 variance was added. The cluster centers were initialized equally distributed on a grid of the subspace defined by the first two principal components of the data. The parameters of the fuzzy clustering were $m = 2$ and $\varsigma = 0.5$.

As can be seen in Figure 4, the proposed regularized FCM algorithm successfully obtained the ordering of the clusters. The values of the cluster centers were separately plotted on maps, where each map represents a variable (see Figure 5). As on these maps black means low, white colors high values of the variables, regions with different colors evolve and the hidden relation between the variables reveal. On the maps (Figure 5 and Figure 6) we can see the effect of our regularization. In Figure 5, the values of codebook vectors are changing smoothly so these figures are easy to interpret. When the basic FCM algorithm is used, the maps are unordered (Figure 6), so it is not interpretable.

It is interesting to note that the smoothly distributed FCM [7] does not give a perfect result in this problem due to the different L regularization matrix. Based on several experiments, we have realized that as only the first order derivatives are utilized in the smoothly distributed algorithm, it tries to contract the code vectors to the center of data, that is the dissatdvatage of this algorithm.

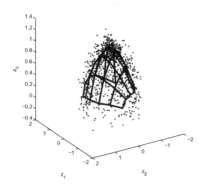

Fig. 4. Example of cluster centers arranged on a two-dimensional space

4 Conclusions

Clustering is useful tool to detect the structure of the data. If there are lot of clusters, it is hard to evaluate the results of the clustering. To avoid this problem, it is useful to arrange the clusters on a low dimensional grid and visualize them. The aim of this

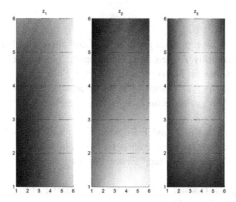

Fig. 5. The map of the input variables when the regularized FCM algorithm is used.

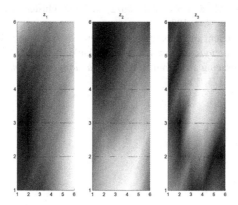

Fig. 6. The map of the input variables when the standard FCM algorithm is used.

paper was to develop an algorithm for this visualization task. With the introduction of a new regularization term into the standard fuzzy clustering algorithm it is possible to arrange the cluster prototypes on a grid. This type of arrangement of the code vectors helps greatly in the visualization of the results, as the regularization orders the similar cluster centers closer to each other. The color-coding of the numerical values of the clusters results in regions on the map that shows the relations among the variables. Hence, the proposed algorithm can be used as an alternative of SOM.

Acknowledgement

The financial support of the Hungarian Ministry of Culture and Education (FKFP-0073/2001) and the Hungarian Science Foundation (T037600) is greatly acknowledged. Janos Abonyi is grateful for the financial support of the Janos Bolyai Research Fellowship of the Hungarian Academy of Science.

References

1. J. Abonyi. *Fuzzy Model Identification for Control*. Birkhauser, Cambridge, MA, 2003.
2. A. Ahalt, A.K. Khrisnamurthy, P.Chen, and D.E. Melton. Competitive learning algorithms for vector quantization. *Neural Networks*, (3):277–290, 1990.
3. R. Babuška. *Fuzzy Modeling for Control*. Kluwer Academic Publishers, Boston, 1998.
4. J.C. Bezdek. *Pattern Recognition with Fuzzy Objective Function Algorithms*. Plenum, New York, 1981.
5. F. Hoppner, F. Klawonn, R. Kruse, and T. Runkler. *Fuzzy Cluster Analysis – Methods for Classification, Data Analysis and Image Recognition*. John Wiley and Sons, 1999.
6. T. Kohonen. The self-organizing map. *Proceedings of the IEEE*, (78(9)):1464–1480, 1990.
7. R.D Pascal-Marqui, A.D. Pascual Montano, K. Kochi, and J.M.Carazo. Smoothly distributed fuzzy c-means: a new self organizing map. *Pattern Recognition*, (34):2395–2402, 2001.
8. E.C.K. Tsao, J.C. Bezdek, and N.R. Pal. Fuzzy kohonen clustering networks. *Pattern Recognition*, (27(5)):757–764, 1994.
9. P. Vuorimaa, T. Jukarainen, and E. Karpanoja. A neuro-fuzzy system for chemical agent detection. *IEEE Transactions On Fuzzy Systems*, (4):403–414, 1995.
10. H. Yan. Fuzzy curve-tracing algorithm. *IEEE Transactions on Systems, Man, and Cybernetics, Part B*, (5):768–773, 2001.

Outline Representation of Fonts Using Genetic Approach

Muhammad Sarfraz

Department of Information & Computer Science
King Fahd University of Petroleum & Minerals
KFUPM # 1510, Dhahran 31261, Saudi Arabia
Email: sarfraz@kfupm.edu.sa

Summary. An Algorithm has been proposed that models the outlines of handprinted and electronic fonts. By capturing the input through a scanning device, the bitmapped image is converted to a more flexible format by means of a mathematical description. This, in turn, reflects the input through a series of non-uniform cubic spline segments, pieced together via an acceptable level of continuity. An optimal solution is realised by the method being applied using a genetic approach.

Keywords: Font, Significant Points, Contour, Outline, Spline

1 Introduction

For the Roman like languages, although a great amount of work [11] has been done as far as font designing is concerned. But, still the work is in progress to improve the existing techniques as well as to find out new and better techniques. However, non-Roman languages need considerable amount of work [10] in the designing of their fonts. It is important to mention here that more than half of the world population reads, speaks and writes non-Roman Languages. Therefore it is desirable to work in this dimension.

There are two fundamental techniques for storing fonts in computers [3, 9, 15, 17]: the bitmap technique and the outline technique. The outline representation of fonts has many advantages over bitmap representation [11]. In an outline representation, multiple sizes may be derived from a single stored representation by suitable scaling. In addition, different typefaces can also be derived using geometric transformations. The fonts, using outline methodology, can be easily translated, rotated, scaled and clipped without any distortion and jaggy appearance. Therefore, most of the contemporary desktop publishing systems are based on outline based methodology.

The method proposed in this study, mainly differs to the traditional approaches [10] in various ways. It is based on various algorithms as its essential components. These algorithms include boundary detection, spline fitting, corner detection and genetic algorithms.

Some times corners are not detected precisely and some times only corner points are not sufficient to fit the outline which represents the original character. Therefore, in addition to corner points, some more points are needed to achieve a best fit. This paper, in addition to the discovery of corner points, also proposes another family of points called significant points. These significant points are dynamically discovered using a genetic algorithm while the corner points remain static throughout the genetic process. The significant points play important role in the overall shape of the final outline of a character. The union of set of corner points with the set of characteristic points will be called set of characteristic points. Optimal curve fitting is done by fitting a global spline model at the characteristic points in a genetic way. The spline model used, in this work, is a non-uniform cubic spline (NUCS), which has attracting geometric features.

The organization of the paper is made as follows. The discussion summary of scanning the image and filtering the noise is made in Section 2. This Section is also meant for the boundary detection algorithm to the scanned images. The issue of detecting the characteristic points is summarized in Section 3. The Spline model is explained in Section 4. Genetic Algorithm is discussed in Section 5. The Section 6 discusses the issue of best possible boundary fit. The Section 7 concludes the paper.

2 Image Outline Extraction

The extraction of the contour points, from the gray-level image, is the first step of the whole process of Font recognition. During the digitization process (converting the gray-level image to a bilevel image), some noise may arise. This adds irregularities to the outer boundary of the image and, hence, may have some undesired effects. The algorithm of Avrahami and Pratt [2] is a reasonable solution to convert the gray-level image to a bilevel image. Although, this algorithm minimizes the error during the conversion process but it requires some modifications, which have been incorporated in this theory. The algorithm returns number of boundary points and their values:

$$\mathbf{F}(i) = (x(i), y(i)), i = 1, ..., N \tag{1}$$

3 Detection of Characteristic Points

After finding out boundary points, the next step in preprocessing is the detection of *characteristic points*. We can categorize them into two classifications: *corner points* and the *significant points*.

3.1 Corner Points

The corner points are those points, which partition the outline into various segments. There have been a good amount of work done for the detection of corner points given the boundary of an image. A number of approaches has been proposed by researchers[4, 5]. These include curvature analysis with numerical techniques and some signal processing methods. In [4] some of the possible ways to detect corners in an image are presented. The curvature can be analyzed using some numerical approaches. The algorithm, in [10], has used the numerical approach. But it presents a problem of scaling. The detection of corner actually depends on the actual resolution of the image and processing width to calculate the curvature.

For the detection of corners, in this paper, the author has adopted the simple technique presented in [4]. The algorithm depicts the corners intelligently. The algorithm has been adopted with certain specific degrees of freedom in its description so that best possible results are achieved in order to obtain the objectives. The details of the algorithm are explained in the following paragraphs.

The proposed two-pass algorithm defines a corner in a simple and intuitively appealing way, as a location where a triangle of specified size and opening angle can be inscribed in a curve (see Figure 1). A curve is represented by a sequence of points p_i in the image plane. The ordered points are densely sampled along the curve, but contrary to the other algorithms, no regular spacing between them is assumed. A chain-coded curve can also be handled if converted to a sequence of grid points. In the first pass the algorithm scans the sequence and selects candidate corner points. The second pass is post-processing to remove superfluous candidates.

The First pass is as follows. In each curve point p the detector tries to inscribe in the curve a variable triangle

$$(\mathbf{p}^-, \mathbf{p}, \mathbf{p}^+) \tag{2}$$

Constrained by a set of simple rules:

$$d_{min}^2 \leq |\mathbf{p} - \mathbf{p}^+| \leq d_{max}^2 \tag{3}$$

where $|\mathbf{p}\text{-}\mathbf{p}^+| = |a| = a$ is the distance between $\mathbf{p}$ and $\mathbf{p}^+$ and α is the opening angle of the triangle. The later is computed as follows:

$$\alpha = arccos \frac{a^2 + b^2 - c^2}{2ab} \tag{4}$$

Variations of the triangle that satisfy the above mentioned conditions are called admissible. Search for the admissible variations starts from p outwards and stops if any of the conditions is violated. Among the admissible variations, the least opening

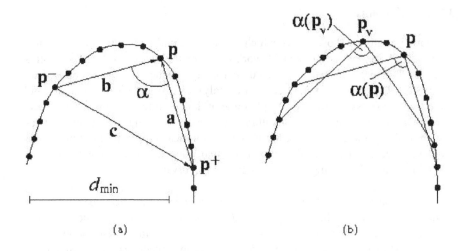

Fig. 1. Detecting High Curvature Points. (a) Determining if **p** is a candidate point. (b) Testing **p** for sharpness non-maxima suppression

angle $\alpha(\mathbf{p})$ is selected. $\pi - |\alpha(\mathbf{p})|$ is assigned to p as the sharpness of the candidate. If no admissible triangle can be inscribed, **p** is rejected and no sharpness is assigned.

The Second pass of the algorithm is as follows. A corner detector can respond to the same corner in a few consecutive points. Similarly to edge detection, a post-processing step is needed to select the strongest response by discarding the non-maxima points. A candidate point **p** is discarded if it has a sharper valid neighbor $\mathbf{p}_v : \alpha(\mathbf{p}) > \alpha(\mathbf{p}_v)$. A candidate point $\mathbf{p}_v$ is a valid neighbor of **p** if $|\mathbf{p} - \mathbf{p}_v| \le d_{max}^2$. The d_{min}, d_{max} and α_{max} are the parameters of the algorithm. The upper limit d_{max} is necessary to avoid false sharp triangles formed by distant points in highly varying curves. α_{max} is the angle limit that determines the minimum sharpness accepted as high curvature. In practice, we often set $d_{max} = d_{min} + 2$ and tune the remaining two parameters, d_{min} and α_{max}. The default values are $d_{min} = 7$ and $\alpha_{max} = 150^0$.

3.2 Significant Points

Since corner points are not always enough to produce the outline of the character [17]. Therefore, in addition to corner points, we identify some more points, called significant points. A genetic approach is proposed, for this purpose, which searches for the best significant points so that the curve fit is optimal to the contour of the character. The points on the contour, which owe highest error for the corresponding spline points, are taken as significant points.

After the set of significant points is achieved, it is then merged in an orderly manner with the set of corner points. This super set would be named as set of characteristic points. This set of characteristic points is updated after each iteration of the genetic algorithm, which is explained in Section 5 so much so that an optimal curve is obtained. This phenomena can be explained mathematically as follows:

Let C be the set of Corner Points and $S^{[i]}$ denote the set of points in the ith iteration. Now, if $R^{[i]}$ denote the set of characteristic points in the ith iteration, then we define

$$R^{[0]} = C, S^{[0]} = \Phi \text{ and}$$

$$R^{[i+1]} = R^{[0]} S^{[i]}, i = 0, 1, 2, ...$$ (5)

Remark: The set of characteristic points is a subset of the points on the image contour. It is obligatory to maintain the order of the points in the union process of [5] according to the order of the path of the contour. One will not be able to achieve the required output without this restriction.

4 Spline Model

Since the advent of computers, curve designing is mainly done by fitting an appropriate spline model [6, 13] to the given data. The spline model used, in this work, is a non-uniform cubic spline (NUCS), which has attracting geometric features. The theory of the proposed study is based upon the NUCS having cubic degree. The points on a parametric NUCS curve are represented as ordered set of values.

The NUCS consist of curve segments, which are cubic polynomial in each segment. Let the *knot* vector be a non-decreasing sequence of real numbers, i.e., $t_i \leq t_{i+1}$, $i=0,1,2,...,n-1$.

Let $F_i \in R^m, i \in Z$, be data values given at the distinct knots $t_i \in R_i \in Z$, with interval spacing $h_i = t_{i+1} - t_i > 0$. Let $D_i \in R^m, i \in Z$, denote the first derivative values defined at the knots. Then the NUCS, in the form a parametric piecewise cubic Hermite function $P:R \to R^m$, is defined by:

$$P|_{(t_i, t_{i+1})}(t) = (1 - \theta_i)^3 F_i + 3\theta_i(1 - \theta_i)^2 V_i + 3\theta_i^2(1 - \theta_i)W_i + \theta_i^3 F_{i+1}$$ (6)

where

$$\theta_i \equiv \theta_i(t) = \theta|_{(t_i, t_{i+1})}(t) = \frac{(t - t_i)}{h_i}$$ (7)

$$V_i = F_i + \frac{1}{3}h_i D_i \text{ and } W_i = F_{i+1} - \frac{1}{3}h_i D_{i+1}$$ (8)

It can be simply observed that the NUCS is C^1 as the following Hermite Interpolation Conditions

$$P(t_i) = F_i \ and \ P^{(1)}(t_i) = D_i, i \in Z, \tag{9}$$

are very obvious. In most of the applications, the tangent information is not provided. We define a distance based choice for tangent vectors D_i at F_i as follows.

5 A Brief Overview of Genetic Algorithm

Genetic Algorithms (GAs), are search techniques based on the concept of evolution [7, 8, 12, 14]. Given a well-defined search space in which each solution is represented by a bit string, called a chromosome, a GA is applied with its three genetic search operators (selection, crossover and mutation) to transform a population of chromosomes with the objective of improving the quality of the chromosomes. The individual bits of a chromosome are called genes. Before the search starts, a set of chromosomes is randomly chosen from the search space to form the initial population. The three genetic search operations are then applied one after the other to obtain a new generation of chromosomes in which the expected quality over all the chromosomes is better than that of the previous generation. The process is repeated until stopping criterion is met. Finally the best chromosome of the last generation is reported as a final solution

6 Outline Capture

The scheme used, in this study, converts the original continuous problem into a discrete optimization problem. Each data point (boundary point or contour point) corresponds to a single gene in the bit string of a chromosome. In this formulation if a gene is equal to 1, we select the corresponding data point as a knot point of the spline. If the gene is equal to 0, the corresponding point is rejected (see Figure 3). All the selected knot points will be considered as the characteristic points.

The initial population consists of K individuals of genelength L. The genes are randomly set to 0 and 1. However, the corner points are determined before the creation of initial population and the genes corresponding to those points are intentionally set to 1 in the initial population and in the population of the subsequent generations. The idea behind this scheme is not to lose those points as they are important in determining the outlines of the shapes.

For the best optimality as well as for fast computation, some additional parameters are proposed. Their description is made in the following sections.

6.1 The Knot Ratio

In addition to the conventional genetic control parameters (crossover and mutation), another control parameter knot ratio R is also proposed. Akaike's Information Criterion (AIC) [1] is used as a fitness measure. By using AIC we can choose the best model among the candidate models automatically. AIC is given by

$$AIC = Nlog_e^Q + 2(2n+4),$$ (10)

where N is the number of data, n is the number of interior knots, the number 4 is the order of the spline to be fitted on the given data, and Q is given by

$$Q = \sum_{j=1}^{N}\{\{Sx_j(t) - x_j(t)\}^2 + \{Sy_j(t) - y_j(t)\}^2\}.$$ (11)

Here $Sx(t)$ and $Sy(t)$ are the x and y components respectively of the approximated spline over the data F. It should be noted that the smaller value of AIC gives better fitness.

6.2 Decimation

The parameter, named as decimation, enables the data to be selected interval wise, without loosing the contour of the font as well as the significant points determined by the algorithm. This has been used in order to decrease the genelength of the chromosomes.

In the context of genetic algorithm, a Roulette wheel selection and a double point crossover has been used. The probability of crossover P_C is taken to be 0.7 and the probability of mutation P_M is taken to be 0.001, while the value of R has been ranged as $0 \leq R < 0.5$.

6.3 Algorithm

The summary of the algorithm is given as follows:

1. Input the control parameters.
2. Find the boundary using boundary detection algorithm.
3. Find corner points using corner detection algorithm
4. Create an initial population by using random numbers.
5. For each individual in the population make the bits corresponding to the significant points as 1.
6. For each individual compute data fitting and obtain the fitness value.
7. If total number of generations exhausted, stop the computation, otherwise go to step 9.
8. Do selection by using the fitness values.
9. Do crossover and make the individuals of the next generation.
10. Do mutation and go back to step 6.

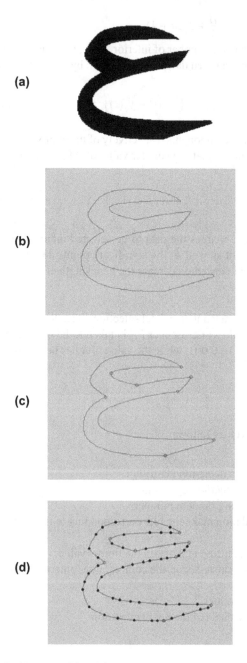

Fig. 2. (a) An Arabic Alphabet 'Ain', (b) Detected Boundary, (c) Corner Points (Circles) detected on the outline boundary, (d) The Genetic Algorithm converged to a desired outline with detected corner points (Circle) and significant points (bullets)

7 Demonstration

The results, in Figure 2, are obtained by applying the algorithm on an Arabic alphabet 'Ain', see Figure 2(a). This letter has been approximated with the NUCS. For the letter 'Ai', 311 points were determined by the boundary detection algorithm (see Fig 2(b) to see the outline obtained). The corner detection algorithm detected 8 corner points shown as circles in Figure 2(c). The genetic algorithm was run for 120 generations with a population size of 30. The selection was based on Roulette wheel. The knot Ratio was 0.3 and the crossover and mutation rates of 0.7 and 0.001, respectively, were used. The algorithm converged at 75th generation (see Fig 2(d)). The bullets are the significant points detected by the algorithm.

8 Concluding Remarks

A method for font designing has been proposed which is suitable for both Roman like as well as non-Roman languages like Arabic, Gothic, etc. In addition to the detection of corner points, a strategy to detect a set of significant points is also explained to optimize the outline. A NUCS model, through a genetic approach, has been utilized to identify the significant points and hence capture the outline of the fonts. The proposed approach minimizes the human interaction in obtaining the outline of original character. This research, in addition to the automatic capture of Fonts, is equally good to capture hand-drawn images. The author feels that the proposed approach has the potential to be enhanced and make more automated and robust treatment using parallelism towards the computing aspect. Therefore, such a work is still in progress and the authors are expecting some more elegant results. The author also feels that the methodology using genetic algorithm and NUCS, proposed in this research, will prove to be quite reasonable as compared to the existing methods.

References

1. Karow P. Font technology (methods and tools). *Springer - Verlag*, 1994. Berling.
2. Itoh and Koichi. A curve fitting algorithm for character fonts. *Electronic Publishing*, pages 195–205, 1993.
3. Chang H. and Yan H. Vectorization of hand-written image using piece wise cubic bezier curve fitting. *Pattern Recognition*, 31(11):1747–1755, 1998. Elsevier Science.
4. Hersch R. D. Towards a universal auto-hinting system for typographic shapes. *Electronic Publishing*, 7(4):251–260, 1994.
5. Sarfraz M and Khan M A. Automatic outline capture of arabic fonts. *Journal of Information Sciences*, 140(3-4):269–281, 1998. Elsevier Science Inc.
6. Sarfraz M. and Khan M. Towards automation of capturing outlines of arabic fonts. *Proceedings of the third KFUPM Workshop on Information and Computer Science:Software Developement for the new Millenium (WICS' 2000)*, pages 83–98, 2000. Saudi Arabia.
7. Avrahami G. and Pratt V. Sub - pixel edege detection in character digitization. *Raster Imaging and Digital Typrography II,*, pages 54–64, 1991. Eds. Morris, R. and Andre, J., Cambridge University Press.

8. Chetverikov D. and Szabo Z. A simple and efficient algorithm for detection of high curvature points in planar curves. *Proc. 23rd Workshop of the Australian Pattern Recognition Group*, pages 175–184, 1999.

9. Davis L. Shape matching using relaxation technique. *IEEE Transactions PAMI*, 1:60–72, 1979.

10. Farin G.E. Curves and surfaces for computer aided geometric design. *Academic Press*, 1974. New York.

11. Sarfraz M. Cubic spline curves with shape control. *Computers and Graphics*, 18(5):707–713, 1994. Elsevier Science.

12. GoldBerg D. E. *Genetic Algorithms in search, Optimization and Machine Learning*. Addison Wesley, 1989.

13. Harada T, Yoshimoto F, and Aoyama Y. Data fitting using a genetic algorithm with real number genes. *Proc. of the IASTED International Conference on Computer Graphics and Imaging*, 2000.

14. Moriyama M, Yoshimoto F, and Harada T. A method of plane data fitting with a genetic algorithm. *Proceedings of the IASTED international conference on computer graphics and Imaging*, pages 21–31, 1998. Elsevier Science.

15. Sarfraz M and Raza A. Visualization of data using genetic algorithm. *Proceedings of the Fourth KFUPM Workshop on Information and Computer Science:Internet Computing (WICS 2002)*, pages 253–265, 2002. Saudi Arabia ISBN:9960-07-187-1.

16. Sarfraz M. and Raza S.A. Capturing outline of fonts using genetic algorithm and splines. *Proceedings of IEEE International Conference on Information Visualization - IV*, pages 738–743, 2001. IEEE Computer Society Press.

17. Akaike H. A new look at the statistical model identification. *IEEE Transactions Automatic Control*, 19(6):716–723, 1974.

Process-Oriented Plant Layout Design using a Fuzzy Set Decomposition Algorithm

Erel Avineri[1] and Prakhar Vaish[2]

[1] Technion - Israel Institute of Technology, Faculty of Civil and Environmental Engineering, Haifa 32000, Israel, E-mail: avineri@internet-zahav.net

[2] Oracle Software India Ltd., 5th Floor, Cyber Gateway, Hi-Tec City, Madha Pur; Hyderabad - 500081, India, E-mail: vaishprakhar@yahoo.com

Summary. In this paper a novel application of the concept of fuzzy sets theory to process-oriented plant design and layout is being discussed. The introduced algorithm uses the reasoning of fuzzy sets and judicially applies mass transfer from one machine to another as the criterion for deciding the appropriate position of the machine. As a thumb rule it considers that a machine from which maximum mass exchange occurs with other machines should be placed in the most easily accessible position, which is easily approachable from all associated machines. The algorithm also considers indirect mass transfer between machines.

Key words: Facility Layout Planning

1 Introduction

Efficient plant layout is critical to achieve and maintain manufacturing competitiveness. In U.S., between 20% to 50% of total operating cost is spent on *Material Handling (MH)* and an appropriate facilities design can reduce this cost by at least 10% to 30% [6]. An improved layout design often brings reduction in the cost of MH, in terms of cost of transportation, congestion and work-in-process [9]. Reducing the movement of material on the factory floor is highly important in *Just In Time (JIT)* layouts. These kind of movements are considered as waste movements, since they do not add value to the factory products. Consequently, we want flexible layouts that reduce the movement of both people and material. When a layout reduces distance, the firm also saves space and eliminates potential areas for unwanted inventory [3]. Here we discuss a process-oriented layout, method which was applied to physical arrangement of given number of machines (or facilities) within a given configuration. When designing a process layout, the main aim is to arrange machines so as to minimize MH costs. In other words, machines with large flow of material among them should be placed closer to one another. MH costs, in this approach, depends on (i) the amount of load to be carried between machines during some period of time and (ii) the distance-related costs of moving loads between machines. *Facility Layout Problem (FLP)* considers the amount of load or number of trips multiplied with distance-related costs, and looks for a solution which would minimize the value of function F [5]:

$$F = \sum_{i=1}^{M} \sum_{j=1}^{M} (\text{traffic})_{ij} \cdot (\text{distance})_{f(i)f(j)} \tag{1}$$

Where $(\text{traffic})_{ij}$ is the traffic between machines i and j; $(\text{distance})_{f(i)f(j)}$ is the distance between machine locations and M is the number of machines.

Note here that in solving the FLP we need to specify an appropriate *distance model*, i.e., a general expression for calculating the distance between two locations. If (x_i, y_i) and (x_j, y_j) represent the coordinates of two locations i and j, then the distance model can be:

- *Rectilinear*. Distance between i and j is:

$$d_{ij} = |x_i - x_j| + |y_i - y_j| \tag{2}$$

This method may be useful for building layouts where all the rooms are rectangular and corridors are situated along the walls of the rooms. In this work the rectilinear distance model is used.
- *Euclidean*. Distance between i and j is:

$$d_{ij} = \sqrt{(x_i - x_j)^2 + (y_i - y_j)^2} \tag{3}$$

- *Squared Euclidean*. Distance between i and j is:

$$d_{ij} = (x_i - x_j)^2 + (y_i - y_j)^2 \tag{4}$$

In all of the above methods, the distance is usually measured from the location's center of gravity.

The number of different layout patterns, in which M machines are placed in N locations or cells $(N \geq M)$ is given by:

$$\frac{N!}{(N-M)!} \tag{5}$$

Because there are $M!$ ways of placing M machines into M locations, the above function can take $M!$ different values at most. Accordingly, in order to get the best layout (i.e. the least F value), all the $M!$ patterns should be estimated. Nevertheless, since $M!$ grows extremely large if M grows large, it is impossible to search for all patterns in polynomial time. That is, the FLP consisting of many facilities is a NP-complete problem [8].

Owing to its NP-completeness of the FLP it is impractical to search for optimal solutions. Therefore, in order to find reasonably good solutions, many sub-optimal methods such as CRAFT [1] and MAT [2] have been suggested. Fuzzy sets theory [10] was found to be a powerful method of approximate reasoning. The fuzzy set decomposition algorithm used in this article was first introduced as an application to integrated circuits design [7].

2 Mathematical Formulation

Consider a connection matrix $[M_{ij}]$ which has elements $m_{ij} \geq 0$, where a material of mass m_{ij} is moved from machine i to machine j. If $m_{ij} > 0$, $m_{jk} > 0$, and $m_{ik} = 0$ then mass m_{ij} moves from machine i to machine j and mass m_{jk} moves from machine j to machine k, but no mass moves from machine i to machine k. Moreover, indirectly mass moves from machine i to machine k via machine j (but no processing is done on this mass in machine j).

This also means that if we have mass transfer from machine i to machine j, from machine j to machine k and from machine k to machine l, but none from machine i to machine l, then to find appropriate placement we need to take the combination of mass transfer from machine i to machine j to machine k to machine l. We may also consider any higher-level combination of mass transfer between machines.

The distance between machines i and j, d_{ij}, is inversely proportional to m_{ij} because the more is the mass, the closer the machines should be. Similarly d_{jk} is inversely proportional to m_{jk}. Thus:

$$d_{ik} \leq d_{ij} + d_{jk} \tag{6}$$

(this formulation is known as the *triangle inequality*).

So, we can write:

$$1/m_{ik} \leq 1/m_{ij} + 1/m_{jk} \tag{7}$$

or:

$$m_{ik} \geq \frac{m_{ij} \cdot m_{jk}}{m_{ij} + m_{jk}} \tag{8}$$

Hence, even with no direct mass transfer occurs between machine i and machine k, m_{ik} is more (or equal) than $\frac{m_{ij} \cdot m_{jk}}{m_{ij} + m_{jk}}$, which has a positive value. This forms the main principle of the fuzzy set algorithm: Considering non-direct transfer of mass between machines.

Let $\sum_j m_{ij}$ be the sum of all mass transported from machine i to other machines, and let $\sum_i m_{ij}$ be the sum of mass transported to machine j from all other machines. Let us define:

$$P_{\max} = \max\{\sum_j m_{ij}, \sum_i m_{ij}\} \tag{9}$$

Thus:

$$\frac{m_{ik}}{P_{max}} \geq \left(\frac{m_{ij}}{P_{max}} \cdot \frac{m_{jk}}{P_{max}}\right) / \left(\frac{m_{ij}}{P_{max}} + \frac{m_{jk}}{P_{max}}\right) \quad \forall i,k \mid m_{ij} + m_{jk} > 0 \tag{10}$$

we can extend Eq. 10 to:

$$\frac{m_{ik}}{P_{max}} \geq \left(\frac{m_{ij}}{P_{max}} \cdot \frac{m_{jl}}{P_{max}} \cdot \frac{m_{lk}}{P_{max}}\right) / \left(\frac{m_{ij}}{P_{max}} + \frac{m_{jl}}{P_{max}} + \frac{m_{lk}}{P_{max}}\right) \quad \forall i,k \mid m_{ij} + m_{jl} + m_{lk} > 0$$

$$\tag{11}$$

We may also extend Eq. 10 to any higher-level combination of mass transfer between machines.

A fuzzy relation matrix M' can be defined. Each of its elements represents the membership value of a pair of machines to the fuzzy set "*machines that should be placed close to each other*", as calculated by Eq. 10 and its extensions. In other words, The higher the membership value is, the nearer the machines should be placed. It is assumed that mass may be transferred between machines in both flow directions. The element $\max\{m'_{ij}, m'_{ji}\}$ represent the mass transferred in the dominant flow between machines, and the element $\min\{m'_{ij}, m'_{ji}\}$ represents the mass transferred against the dominant flow between machines. The total mass transfer between machines i and j is $m'_{ij} + m'_{ji}$. By adding the element $\min\{m'_{ij}, m'_{ji}\}$ to the element $\max\{m'_{ij}, m'_{ji}\}$ we will get the matrix M''_{ij}. The larger the value of $m'_{ij} + m'_{ji}$, the closer the machines i and j should be placed.

Applying the fuzzy decomposition algorithm to M''_{ij}, the decomposition of fuzzy relation may be summarized as:

$$R = \sum_\alpha \alpha \cdot R_\alpha \qquad 0 < \alpha \le 1; \qquad \alpha_1 > \alpha_2 \to R_{\alpha_1} \subset R_{\alpha_2} \qquad (12)$$

where R_α denotes a relation between all machine pairs with $m''_{ij} > \alpha$, and $\alpha \cdot R_\alpha$ means all the elements of relation R_α multiplied with α [4, 11]. The basic structure of the algorithm is as follows:

Step 1: Construct the modified fuzzy relation matrix M''_{ij}. Each of its elements represents the membership value of a pair of machines to the fuzzy set "*machines that should be placed close to each other*".

Step 2: Set a high value of the decomposition variable α; Set the interval size to decrease the decomposition variable, $\Delta\alpha$;

Step 3: Let R_1 be the set of all pairs of machines; Let $R_2 = \dot{O}$[*empty set*] be the set of pairs which already considered and placed;

Step 4: Set $R_3 = R_1 \setminus R_2$ be the set of pairs that haven't been considered yet in the plant layout;

Step 5: Set R_2 be a subset of R_3, including all the pairs of machines to be considered for placing in the current iteration $(m''_{ij} \ge \alpha)$. This way, only the pairs with a membership value between α and $\alpha + \Delta\alpha$ are included;

Step 6: Place the machine pairs that are included in set R_2; If more than one pair is included in the set R_1, place them sequentially with pairs having high membership value to pairs with lower membership value;

Step 7: Set $R_1 = R_1 \cup R_2$; Set $R_3 = R_3 \setminus R_2$;

Step 8: If R_3 is an empty set (no more machines to be placed) then stop, otherwise - decrease the value of the decomposition variable α: Set $\alpha = \alpha - \Delta\alpha$;

Step 9: Go to Step 4.

Hence, the value of the decomposition variable α is gradually decreased until all the machines are included in the plant layout.

3 Example

Consider a workshop where 6 machines namely A to F are used. We regard the factory floor to be divided into 2 rows and 3 columns, i.e. six individual cells, as described in figure 1. For Simplicity, let all the horizontal distances and all the vertical distances between neighboring cells be the same.

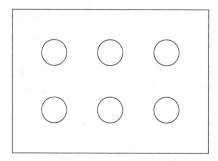

Fig. 1. The factory floor

The flow mass (in Kg./Hour) between machines is presented in the following *From-To* matrix M_{ij}:

From i To j	A	B	C	D	E	F	Row_Sum $\sum_i m_{ij}$
A	—	10	0	0	0	0	10
B	0	—	15	0	0	0	15
C	20	0	—	20	0	0	40
D	40	0	0	—	25	5	70
E	0	0	32	0	—	0	28
F	0	0	0	0	15	—	15
Column_Sum $\sum_j m_{ij}$	60	10	43	20	40	5	

Thus:

$$P_{\max} = \max\{\sum_j m_{ij}, \sum_i m_{ij}\} = \max\{60, 70\} = 70 \qquad (13)$$

Applying the fuzzy decomposition algorithm we get the matrix $M_{ij}/P_{\max}$. Since machine i cannot have mass flow from/to itself, the element $(M_{ii}/P_{\max})$ has no meaning. Alternatively, one may consider $M_{ii}/P_{\max}$ to have the highest membership value possible, since machine i should be placed as close as possible to itself.

From i To j	A	B	C	D	E	F
A	—	0.143	0	0	0	0
B	0	—	0.214	0	0	0
C	0.286	0	—	0.286	0	0
D	0.571	0	0	—	0.357	0.071
E	0	0	0.457	0	—	0
F	0	0	0	0	0.214	—

Path-Length is defined as the product of mass flowing between pairs of machines along alternate route. We populate matrix element (M'_{ij}) with the maximum value of the path length along these alternate routes. For example, the maximum value of the path length between machines A and B is calculated using Eq. 10 and its extensions:

$$m'_{AB} \geq \left(\frac{m_{AC}}{P_{max}} \cdot \frac{m_{CB}}{P_{max}} \right) / \left(\frac{m_{AC}}{P_{max}} + \frac{m_{CB}}{P_{max}} \right)$$
$$= \frac{0.286 \cdot 0.214}{0.286 + 0.214} = 0.122 \tag{14}$$

$$m'_{AB} \geq \left(\frac{m_{AD}}{P_{max}} \cdot \frac{m_{DC}}{P_{max}} \cdot \frac{m_{CB}}{P_{max}} \right) / \left(\frac{m_{AD}}{P_{max}} + \frac{m_{DC}}{P_{max}} + \frac{m_{CB}}{P_{max}} \right)$$
$$= \frac{0.571 \cdot 0.286 \cdot 0.214}{0.571 + 0.286 + 0.214} = 0.033 \tag{15}$$

(All the other values of path length between machines A and B are zeros). Thus:

$$m'_{AB} \geq \max\{0.122, 0.033\} = 0.122$$

Similarly, the maximum value of the path lengths between all pairs of machines is calculated. These values are summarized in the following matrix:

From i To j	A	B	C	D	E	F
A	—	0.143	0.086	0.014	0.003	0.001
B	0.122	—	0.214	0.122	0.026	0.008
C	0.286	0.095	—	0.286	0.159	0.057
D	0.571	0.114	0.201	—	0.357	0.071
E	0.176	0.021	0.457	0.176	—	0.011
F	0.029	0.004	0.146	0.029	0.214	—

There are pairs of machines in which there is mass flow for both directions (for example machines A and B). The total mass transfer between machines i and j is $m'_{ij} + m'_{ji}$. By adding the element $\min\{m'_{ij}, m'_{ji}\}$ to the element $\max\{m'_{ij}, m'_{ji}\}$ we will get the matrix M''_{ij} as following:

From i / To j	A	B	C	D	E	F	Row_Sum $\sum_i m_{ij}''$
A	—	0.265	0	0	0	0	0.265
B	0	—	0.310	0.237	0.047	0.011	0.604
C	0.371	0	—	0.486	0	0	0.858
D	0.585	0	0	—	0.533	0.101	1.219
E	0.179	0	0.616	0	—	0	0.795
F	0.030	0	0.203	0	0.226	—	0.459
Column_Sum $\sum_j m_{ij}''$	1.166	0.265	1.128	0.723	0.805	0.112	

When the total mass flow from/to a specific machine x, $\sum_i m_{ix}'' + \sum_j m_{xj}''$, is large, we should consider to place it in the most approachable position. The next table describes the total mass flow from/to each of the machines:

	A	B	C	D	E	F
Row_Sum $\sum_i m_{ij}''$	1.166	0.265	1.128	0.723	0.805	0.112
Column_Sum $\sum_j m_{ij}''$	0.265	0.604	0.858	1.219	0.795	0.459
$\sum_i m_{ij}'' + \sum_j m_{ij}''$	**1.431**	**0.869**	**1.986**	**1.942**	**1.600**	**0.571**

Let us assume a decomposition variable α. For $\alpha = 0.6$ the only subset obtained is: $\{(C,E), 0.616\}$. Machine C has the maximal value of *Row_Sum* and *Column_Sum* ($\sum_i m_{iC}'' + \sum_j m_{Cj}'' = 1.986$). Thus, we select machine C to be placed in the middle of a row (most approachable position) and place machine E next to it. There are two alternatives for such a placement, which are described in figure 2 (actually there are more alternatives which are symmetric to those and therefore are equivalent).

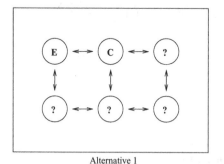
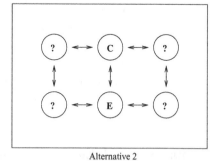

Alternative 1   Alternative 2

Fig. 2. First arrangement ($\alpha = 0.6$)

Next, the value of the decomposition variable α is decreased. For $\alpha = 0.5$ the new subsets obtained are: $\{(A,D), 0.585\}$ and $\{(E,D), 0.533\}$. Machine E was already placed, and we should place machine D close to it. Since there are two different alternatives to place machine E, there are also two different layouts to place machine D, as described in figure 3. Machine A is placed as well, next to machine D.

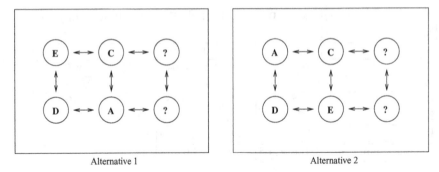

Alternative 1   Alternative 2

Fig. 3. Second arrangement ($\alpha = 0.5$)

For $\alpha = 0.4$ a new subset is obtained: $\{(D,C), 0.486\}$. Since machines D and C are already placed we move to the next value of the decomposition variable. For $\alpha = 0.3$ the new subsets obtained are: $\{(A,C), 0.371\}$ and $\{(C,B), 0.310\}$. Machines A and C are already placed. Machine B is placed as close to machine C, in both of the alternatives. Since only one cell have been remained, and only one machine (F), we will place machine F and will get the two alternatives to the final arrangement of the plant layout, described in figure 4.

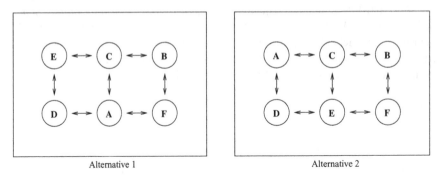

Alternative 1   Alternative 2

Fig. 4. Final arrangement at the end of the process ($\alpha = 0.3$)

4 Comparison with the Optimal Solution

The total cost (number of movements) of the final arrangement solution recommended by the fuzzy set decomposition method (described in figure 4) is 247 for the first alternative and 217 for the second one, which is therefore better. Actually, the second alternative is also the optimal solution of this problem. Thus, Fuzzy sets based algorithm gives good solution (close to the optimal solution), with a reasonable explanation of how to proceed with the layout and can be used for a layout with large number of machines.

More complicated problems, such as 3-D facility layout planning, or considering planning constraints, are easily to be solved using the same algorithm

5 Conclusion

It can be concluded that fuzzy set based algorithms can form a firm basis for the design of plant layout. They provide a basic reasoning for the machine layout in the plant which otherwise are subject to a complex process.

References

1. Buffa ES, Armor GS, Vollmann TE (1964) Allocating Facilities with CRAFT, Harvard Business Review 42:2:136–159
2. Edwards HK, Gillett BE, Hale ME (1970) Modular Allocation Technique (MAT), Management Science 17:3:161–169
3. Heizer J, Render B (1999) Operations Management, 5th Edition. Prentice Hall
4. Kaufmann A (1975) Theory of Fuzzy Subsets, Vol. 1, Fundamental Theoretical Elements. Computer Science Press
5. Kusiak A, Heragu SS (1987) The Facility Layout Problem, European Journal of Operational Research 29:229–251
6. Meller RD, Gau KY (1996) The Facility Layout Problem: Recent and Emerging Trends and Perspectives, Journal of Manufacturing Systems 15:351–366
7. Razaz M, Gan J (1990) Fuzzy Set Based Initial Placement for IC Layout. In: Proceedings of the European EDAC, European Design Automation Conference 655–659
8. Sahni S, Gonzalez T (1976) P-Complete Approximation Problem, Journal of ACM 23:555–565
9. Sule DR (1988) Manufacturing Facilities - Location, Planning and Design. PWS-KENT, Boston
10. Zadeh LA (1965) Fuzzy sets, Information Control 8:338–353
11. Zimmerman HJ (1985) Fuzzy Set Theory and its Applications. Kluwer Nijhoff Publishing

A Fuzzy Colour Image Segmentation Applied to Robot Vision

J. Chamorro-Martínez, D. Sánchez and B. Prados-Suárez

Department of Computer Science and Artificial Intelligence, University of Granada
C/ Periodista Daniel Saucedo Aranda s/n, 18071 Granada, Spain
E-mail: {jesus,daniel,belenps}@decsai.ugr.es

Summary. In this paper, a new growing region algorithm to segment colour images is proposed. A region is defined as a fuzzy subset of connected pixels and it is constructed using topographic and colour information. To guide the growing region process, a distance defined in the HSI colour space is proposed. This distance is used both to select the pixels which will be linked in each step of the algorithm, and to calculate membership degree of each point to each region. The proposed technique is applied to vision-guided robot navigation with the aim of detecting doors in indoor environments.

Keyword: Colour image segmentation, fuzzy segmentation, colour distance, robot vision.

1 Introduction

The image segmentation, view as the process of dividing the image into significant regions, is one of the most widely used step in image processing. Usually, this operation allows to arrange the image in order to make it more understandable for higher levels of the analysis (for example, in image database retrieval [4], motion estimation [10] or robot navigation [2, 8]).

Many types of segmentation techniques have been proposed in the literature. They can be grouped into three main categories corresponding to three different definitions of regions: the methods in the first group, called pixel based segmentation methods, define a region as a set of pixels satisfying a class membership function. In this category are included the histogram based techniques [5] and the segmentation by clustering algorithms [12]. The methods in the second group, corresponding to the area based segmentation techniques, consider a region like a set of connected pixels satisfying a uniformity condition. The growing region techniques [7] and the split and merge algorithms [1] are two examples of this group. The last type of segmentation methods corresponds to the edge based algorithms, where a region is defined as a set of pixels bounded by a colour contour [11].

In real images, the separation between regions is usually imprecise. This is one of the main problems of the crisp segmentation techniques, where each pixel have to belong to an unique region. To solve this problem, some approaches propose the definition of *region* as a fuzzy subset of pixels, in such a way that every pixel of the image has a membership degree to that region [3]. In this sense, many algorithms have been developed to segment grey scale images, but the fuzzy segmentation of colour images has been paid less attention.

Other important aspect to take into account is the colour information processing. The most common solutions in the literature are (i) combining the information of each band into a single value before processing (for example, the gradient), or (ii) analyzing each band separately and then combining the results (for example, histogram analysis of each band and subsequent combination). Apart from the difficulty to choose an adequate criterion to pool the data, the main problem of these approaches is the application of the same combination rule to the whole image, without considering the particularities that appear in the comparison of two colours. That problem is most significant in methods, like those based on growing region, where the decision in each step depends on the difference between pixels.

In this paper, a new growing region algorithm to segment colour images is proposed. In our approach, a region is defined as a fuzzy subset of connected pixels and it is constructed using topographic and colour information, i.e., two pixels will be assigned to the same region if they are connected through a path of similar colours. To process the colour information, a distance defined in the HSI colour space is proposed. This distance is used in the growth of a region both (i) to select the pixels which will be linked in each step of the algorithm, and (ii) to calculate the membership degree of each point to each region.

Although the proposed algorithm is a multipurpose segmentation technique, in this paper it has been applied to vision-guided robot navigation in indoor environments. Concretely, it has been used to detect doors with the aim of moving a robot to the exit of the room.

2 Colour space

Many colour spaces may be used in image processing: RGB, YIQ, HSI, HSV, etc. [9]. Although the RGB is the most used model to acquire digital images, it is well known that it is not adequate for colour image segmentation. Instead, other colour spaces based on human perception (HSI, HSV or HLS) seem to be a better choice for this purpose [9]. In these systems, hue (H) represents the colour tone (for example, red or blue), saturation (S) is the amount of colour that is present (for example, bright red or pale red) and the third component (called intensity, value or lightness) is the amount of light (it allows the distinction between a dark colour and a light colour).

In this paper, the HSI colour space will be used. This model separate the colour information in ways that correspond to the human visual system. Moreover, it offers many advantages in a segmentation process (for example, the use of hue avoids the shading effects). Geometrically, this colour space is represented as a cone, in which the axis of the cone is the grey scale progression from black to white, distance from the central axis is the saturation, and the direction is the hue (figure 1).

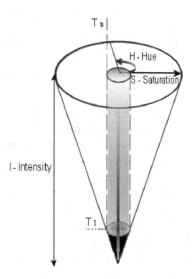

Fig. 1. HSI colour space

Since we use a digital camera to acquire images in RGB coordinates, it is necessary to define a relationship between RGB and HSI systems. In this paper, the following transform is applied [9]:

$$
\begin{aligned}
H &= \arctan\left(\frac{\sqrt{3}(G+B)}{2R-G-B}\right)\\
S &= 1 - \min\{R,G,B\}/I\\
I &= (R+G+B)/3
\end{aligned}
\tag{1}
$$

2.1 Distance between colours

Once the colour space is selected, a distance between two colour points $p_i = [H_i, S_i, I_i]$ and $p_j = [H_j, S_j, I_j]$ is defined. For this purpose, we first define the differences between components as:

$$
\begin{aligned}
d_H &= \begin{cases} |H_i - H_j| & if\ |H_i - H_j| \leq \pi \\ 2\pi - |H_i - H_j| & otherwise \end{cases}\\
d_S &= |S_i - S_j|\\
d_I &= |I_i - I_j|
\end{aligned}
\tag{2}
$$

with d_H, d_S and d_I being the distances between hue, saturation and intensity respectively. Based on the previous distances, the following equation will be used to measure the difference between colours:

$$dc(p_i,p_j) = \begin{cases} \frac{d_I}{MAXI} & \text{if } p_i \text{ or } p_j \text{ are achromatic} \\ \frac{1}{\sqrt{3}} \left[\left(\frac{d_I}{MAXI}\right)^2 + (d_S)^2 + \left(\frac{d_H}{\pi}\right)^2 \right]^{1/2} & \text{if } p_i \text{ and } p_j \text{ are chromatic} \\ \frac{1}{\sqrt{2}} \left[\left(\frac{d_I}{MAXI}\right)^2 + (d_S)^2 \right]^{1/2} & \text{otherwise} \end{cases} \quad (3)$$

with $MAXI$ being a constant equals to the maximum level of intensity (usually 255). Notice that $dc(p_i,p_j) \in [0,1]$. In the previous equation we introduce the notions of chromaticity/achromacitiy to manage two well known problems of the HSI representation: the imprecision of the hue when the intensity or the saturation are small, and the non-representativity of saturation under low levels of intensity. An often practical solution to solve this problem is to perform a partition of the colour space based on the chromaticity degree of each point. In equation (3), we propose to split the HSI space into three regions: *chromatic*, *semi-chromatic* and *achromatic* (figure 1). This partition is defined on the basis of two thresholds T_I and T_S : the first one, T_I, correspond to the level of intensity under which both hue and saturation are imprecise; the second one, T_s, defines the value of saturation under which the hue is imprecise (but not the intensity). Therefore, a point $p_i = [H_i, S_i, I_i]$ will be *achromatic* if $I_i \leq T_I$ (black zone in figure 1), *semi-chromatic* if $I_i > T_I$ and $S_i \leq T_S$ (grey zone in figure 1), and *chromatic* if $I_i > T_I$ and $S_i > T_S$ (white zone in figure 1). In this paper, the thresholds have been fixed empirically to $T_I = MAXI/5$ and $T_s = 1/5$.

3 Segmentation method

Our approach will segment the colour image in two steps:

1. Firstly, a set of "*seed points*", noted as $\Theta = \{s_1, s_2, \ldots, s_q\}$, will be calculated.
2. Secondly, a collection of fuzzy sets, noted as $\widetilde{\Theta} = \{\widetilde{S}_1, \widetilde{S}_2, \ldots, \widetilde{S}_q\}$, will be defined from Θ. For this purpose, a method to calculate the membership degree of a given pixel to a fuzzy set $\widetilde{S}_k$ will be proposed based on topographic an colour information.

In the following sections, the previous stages will be analyzed in detail.

3.1 Seed points

In the growing region methods, the selection of seeds that hopefully may correspond to structural units is a critical step. Due to this paper is focused in a particular application, i.e., the localization of doors in indoor environments, a seed point selection is proposed for this specific case.

In our approach, we have considered that a door have at least one corner point in its frame. Based on the previous assumption, the seed selection is performed in two steps. In the first one, corner points are detected in the intensity band using the Harris method [6]. Afterwards, a couple of seeds are put at a distance D from each corner in the direction of its gradient (one per each way). In this paper, D has been fixed to 7 pixels. Figure 2 shows an example where twelve seed points have been marked corresponding to six corner points (they are showed with red points and green crosses respectively).

Fig. 2. Example of seed points selection

It would be desirable for each door to contain a single seed, but this is not often the case. Indeed, it is usual to find several seeds into the same door (which implies several regions). To solve this problem, the proposed algorithm will discard seeds during the segmentation process. This point will be explained in detail in section 3.3.

Although the previous seed selection method has been designed for a particular case, a general one could be considered in order to give more generality to the technique (for example, seed regions could be located in the local minima calculated over the gradient of the image). This will be an object of future research.

3.2 Fuzzy regions

In this section, a method to fuzzify a set of "seed points" $\Theta = \{s_1, s_2, \ldots, s_q\}$ is proposed. For this purpose, a membership function for fuzzy regions is defined (section 3.2) based on a distance between pixels (section 3.2).

Distance between pixels

Let $\prod_{ij}$ be the set of possibles paths linking the pixels p_i and p_j through pixels of the image I. Given a path $\pi_{ij} \in \prod_{ij}$, its cost is defined as the greatest distance between two consecutive points on the path:

$$cost(\pi_{ij}) = max \left\{ d_C(p_r, p_{r+1}) \, / \, p_r, p_{r+1} \in \pi_{ij} \right\} \tag{4}$$

where p_r and p_{r+1} are two consecutive points of π_{ij}, and $d_C(p_r, p_s)$ is defined in equation (3) and measures the distance between colours. Let $\pi_{ij}^* \in \prod_{ij}$ be the optimum path between p_i and p_j defined as the path that link both points with minimum cost:

$$\pi^*_{ij} = \underset{\pi_{i,j} \in \Pi_{i,j}}{argmin} \ \{cost(\pi_{i,j})\} \tag{5}$$

Based on the previous definition, the distance between two pixels p_i and p_j is defined as the cost of the optimum path from p_i to p_j:

$$d_P(p_i, p_j) = cost(\pi^*_{ij}) \tag{6}$$

Let us remark that the distance defined in (6) make use of topographic information (paths linking the pixels) and distances between colours. In addition, it is sensitive to the presence of edges in the following sense: if the optimum path linking two points p_i and p_j pass through an edge (that is, a point which separate two regions), its cost, and consequently the distance between p_i and p_j, will be high. That is because of the fact that there are consecutive points, in the portion of the path that cross over the edge, with a high distance between them.

Membership function for fuzzy regions

The membership degree $\mu_{\tilde{S}_v}(p_i)$ of a pixel p_i to a fuzzy region $\tilde{S}_v$ is defined as

$$\mu_{\tilde{S}_v}(p_i) = \frac{1 - d^*_P(p_i, s_v)}{\sum_{k=1}^{q} (1 - d^*_P(p_i, s_k))} \tag{7}$$

where $s_v \in \Theta$ is the seed point of $\tilde{S}_v$, and $d^*_P(p_i, s_v)$ is defined as

$$d^*_P(p_i, s_v) = \begin{cases} 1 & if \ p_i \in \Theta, \ p_i \neq s_v \\ d_P(p_i, s_v) & otherwise \end{cases}$$

Let us remark that $\sum_{v=1}^{q} \mu_{\tilde{S}_v}(p_i) = 1$. Using the equation (7) we can calculate the membership degree of every point $p_i \in I$ to each seed $s_v \in \Theta$. That allows us to obtain the set of fuzzy regions $\tilde{\Theta} = \{\tilde{S}_1, \tilde{S}_2, \ldots, \tilde{S}_q\}$ from the set of seed points $\Theta = \{s_1, s_2, \ldots, s_q\}$.

3.3 Algorithm

Given a set of seed points $\Theta = \{s_1, s_2, \ldots, s_q\}$, a growing region algorithm is applied to each $s_v \in \Theta$ in order to obtain the set of fuzzy regions $\tilde{\Theta} = \{\tilde{S}_1, \tilde{S}_2, \ldots, \tilde{S}_q\}$. For each $s_v \in \Theta$, the membership degree $\mu_{\tilde{S}_v}(p_i)$ is obtained for every point of the image using the equation (7). For this purpose, it is necessary to calculate the distance $d_P(p_i, s_v)$ given by equation (6). The algorithm 1 calculates that distance for all the points p_i of an image I with respect to a given seed point $s_v \in \Theta$. Its computational complexity is $O(n)$, where $n = N \cdot M$ is the number of pixels of the image. Due to the algorithm have to be applied for each seed point of Θ, the overall process has a complexity of $O(qn)$, where q is the number of seeds.

As we mentioned in section 3.1, it is usual to find several seeds into the same region. To overcome this problem, given a seed point $s_v \in \Theta$, all the seeds of the subset $\{s_u\}_{u=1..K}$ verifying $d_P(s_v, s_u) < C$ will be eliminated. This new step may be introduced in the algorithm 1 without increase its complexity. In this paper, the constant C has been fixed to 0.01.

Algorithm 1 Algorithm to calculate d_P

Input:
   Image I of size $N \times M$
   Seed point s_v

Notation:
   $Contour(L)$: Pixels in the contour of a region L
   $Neighbor(p_i)$: 8-neighborhood of p_i

1.-Initialization
   $d_P(s_v, s_v) = 0$
   $L = \{s_v\}$

2.- While $Card(L) \neq N \cdot M$
   $(p_{in}, p_{out}) = \underset{(p_i, p_j)}{argmin} \{d_C(p_i, p_j) / p_i \in Contour(L), p_j \in Neighbor(p_i) \setminus L\}$
   $d_P(p_{out}, s_v) = max[d_P(p_{in}, s_s), d_C(p_{in}, p_{out})]$ (1)
   $L = L \cup \{p_{out}\}$

---

4 Results

The proposed segmentation method has been applied to different colour images. To acquire these images, a SONY EVI-401 colour camera connected to the mobile robot Nomad 200 has been used (figure 3).

Six of the images used for testing our methodology are showed in figure 4(A-F). To show the results, the segmentation has been "defuzzified" allocating each pixel to the region for which it has the highest membership degree. The examples 4(A-D) show images with a single closed door. In all the cases, the proposed algorithm generates a fuzzy region corresponding to the door (the associated crisp region is showed in blue colour in figure 4(G-J)). The number of regions obtained are 5, 10, 4 and 3 respectively, although the number of seeds detected in each case was 12, 34, 14 and 12 respectively. That reveals the goodness of the method proposed to discard seeds during the segmentation process

---

[1] Notice that $d_P(p_{in}, s_v)$ has been calculated in a previous step

Fig. 3. Mobile robot Nomad 200

The examples 4(E-F) correspond to images with two doors. Whereas in the first case both of them are closed, in the second one there are an open door in the foreground and a close one behind it. In both cases, our approach obtains one fuzzy region for each door (figure 4(K-L)). On the image 4(E), the algorithm generates four regions: one of them corresponds to the background, another one is associated to a label, and the last two regions correspond to the doors. On the example 4(F) our technique obtains eight fuzzy regions, four of them corresponding to the two doors and its frames (figure 4(L)). The number of seeds was 9 and 30 respectively.

Although in all the examples there are a degradation in the luminance and zones with different brightness (see the door in image B), the results show that our methodology detect the doors without split them in several regions. That reveals that the use of a distance which consider both colour and topographic information improves the results.

5 Conclusions

A new fuzzy methodology to segment colour images has been presented. The technique is based on the growth of regions which are treated as fuzzy subsets. To manage the colour information, a distance between pixels has been defined in the HSI colour space. Using that distance and topographic information, the membership degree of each point to each region has been calculated. Our experiments suggest the combination of the proposed colour distance and the growing region process improve the results obtained with the classical segmentation techniques.

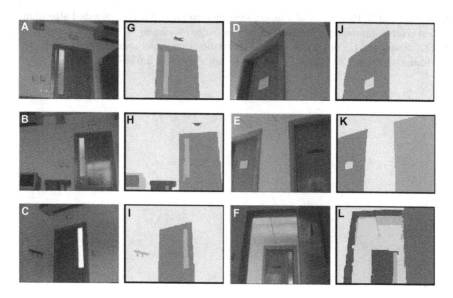

Figure 4 : Segmentation results

References

1. G.A. Borges and M.J. Aldon. A split-and-merge segmentation algorithm for line extraction in 2d range images. *Proc. 15th Inter. Conf. on Pattern Recognition*, 1:441 – 444, 2000.
2. J. Bruce, T. Balch, and M. Veloso. Fast and inexpensive color image segmentation for interactive robots. *Proc. IEEE Inter. Conf. on Intelligent Robots and Systems*, 3:2061 – 2066, 2000.
3. B.M. Cavalho, C.J. Gau, G.T. Herman, and T.Y. Kong. Algorithms for fuzzy segmentation. *Proc. Inter. Conf. on Advances in Pattern Recognition*, pages 154–63, 1999.
4. J.M. Fuertes, M. Lucena, N. Perez de la Blanca, and J. Chamorro-Martinez. A scheme of colour image retrieval from databases. *Pattern Recognition Letters*, 22:323–337, 2001.
5. A. Gillet, L. Macaire, C. Botte-Lococq, and J.G Postaire. Color image segmentation by fuzzy morphological transformation of the 3d color histogram. *Proc. 10th IEEE Inter. Conf. on Fuzzy Systems*, 2:824 –824, 2001.
6. C.G. Harris and M. Stephens. A combined corner and edge detection. *Proc. 4th ALVEY Vision Conference*, pages 147–151, 1988.
7. A. Moghaddamzadeh and N. Bourbakis. A fuzzy region growing approach for segmentation of color images. *Pattern Recognition*, 30(6):867–881, 1997.
8. E Natonek. Fast range image segmentation for servicing robots. *Proc. IEEE Inter. Conf. on Robotics and Automation*, 1:406 – 411, 1998.
9. J.C. Russ. *The Image Processing Handbook*. CRC Press and IEEE Press, third edition, 1999.
10. L. Salgado, N. Garcia, J.M. Menendez, and E. Rendon. Efficient image segmentation for region-based motion estimation and compensation. *Proc. IEEE Trans. on Circuits and Systems for Video Technology*, 10(7):1029 – 1039, 2000.

11. A. Shiji and N. Hamada. Color image segmentation method using watershed algorithm and contour information. *Proc Inter. Conf. on Image Processing*, 4:305 – 309, 1999.
12. D.X. Zhong and H. Yan. Color image segmentation using color space analysis and fuzzy clustering. *Proc. IEEE Signal Processing Society Workshop*, 2:624– 633, 2000.

Applying Rule Weight Derivation to Obtain Cooperative Rules

R. Alcalá[1], J. Casillas[2], O. Cordón[2], and F. Herrera[2]

[1] Dept. Computer Science, University of Jaén, E-23071 Jaén, Spain alcala@ujaen.es
[2] Dept. Computer Science and A.I., University of Granada, E-18071 Granada, Spain
{casillas,ocordon,herrera}@decsai.ugr.es

Summary. In this work we propose the hybridization of two techniques to improve the co-operation among the fuzzy rules: the use of rule weights and the Cooperative Rules learning methodology. To do that, the said methodology is extended to include the learning of rule weights within the rule cooperation paradigm. Considering these kinds of techniques could result in important improvements of the system accuracy, maintaining the interpretability to an acceptable level.

Key words: Fuzzy rule-based systems, linguistic modeling, fuzzy rule cooperation, weighted fuzzy rules, genetic algorithms.

1 Introduction

The use of linguistic Fuzzy Rule-Based Systems (FRBSs) allows us to deal with the modeling of systems building a linguistic model clearly interpretable by human beings. This area is known as Linguistic Modeling. In this framework, the linguistic model consists of a set of linguistic descriptions obtained by means of different automatic system identification techniques from input-output data pairs representing the behavior of the system being modeled.

One of the most interesting features of an FRBS is the interpolative reasoning that it develops. This characteristic plays a key role in the high performance of FRBSs and is a consequence of the cooperation among the linguistic rules composing the knowledge base. As it is known, the output obtained from an FRBS is not usually due to a simple linguistic rule but to the cooperative action of several linguistic rules that have been fired because they match the system input to any degree.

There are different ways to induce rule cooperation in the learning process [2, 6, 7]. In [2], a new learning methodology to induce a better cooperation among the fuzzy rules was proposed: the Cooperative Rules (COR) methodology. The learning philosophy was based on the use of *ad hoc data-driven methods*[3] to determine the

---

[3] A family of efficient and simple methods guided by covering criteria of the data in the example set

fuzzy input subspaces where a rule should exist and a set of candidate consequents assigned to each rule. After that, a combinatorial search was carried out in the set of candidate consequents to obtain a set of rules with good cooperation among them. In [1, 3], different combinatorial search techniques were considered with this aim.

On the other hand, other technique to improve the rule cooperation is the use of weighted fuzzy rules [4, 8, 9], in which modifying the linguistic model structure an importance factor (weight) is considered for each rule. By means of this technique, the way in which these rules interact with their neighbor ones could be indicated.

In this work, we propose the hybridization of both techniques to obtain weighted cooperative fuzzy rules. Thus, the system accuracy is increased while the interpretability is maintained to an acceptable level. To do that, we present the Weighted COR (WCOR) methodology, which includes the weight learning within the original COR methodology.

To learn the subset of rules with the best cooperation and the weights associated to them, different search techniques could be considered [10]. In this contribution, we will consider a Genetic Algorithm (GA) for this purpose.

The paper is organized as follows. In the next section the COR methodology is introduced. In Section 3, the weighted fuzzy rule structure is presented. In Section 4, the WCOR methodology to obtain weighted cooperative rules is proposed. Experimental results are shown in Section 5. Finally, some concluding remarks are pointed out in Section 6.

2 The COR Methodology

The COR methodology is guided by example covering criteria to obtain antecedents (fuzzy input subspaces) and candidate consequents [2]. Depending on the combination of this technique with different ad hoc data-driven methods, different learning approaches arise. In this work, we will consider the Wang and Mendel's method [11] (WM) for this purpose —approach guided by examples—. The COR methodology following this approach presents the following learning scheme:

Let $E = \{e_1, \ldots, e_l, \ldots, e_N\}$ be an input-output data set representing the behavior of the problem being solved —with $e_l = (x_1^l, \ldots, x_n^l, y^l)$, $l \in \{1, \ldots, N\}$, N being the data set size, and n being the number of input variables—. And let $\mathcal{A}_j$ be the set of linguistic terms of the i-th input variable —with $j \in \{1, \ldots, n\}$— and $\mathcal{B}$ be the one of the output variable.

1. *Generate a candidate linguistic rule set.* This set will be formed by the rule best covering each example (input-output data pair) contained in the input-output data set. The structure of each rule, RC^l, is obtained by taking a specific example, e_l, and setting each one of the rule variables to the linguistic label associated to the fuzzy set best covering every example component, i.e.,

$$RC_l = \text{IF } X_1 \text{ is } A_1^l \text{ and } \ldots \text{ and } X_n \text{ is } A_n^l$$
$$\text{THEN } Y \text{ is } B^l,$$

   with

$$A^l_j = arg \max_{A' \in \mathcal{A}_j} \mu_{A'}(x^l_j) \text{ and } B^l = arg \max_{B' \in \mathcal{B}} \mu_{B'}(y^l).$$

2. *Obtain the antecedents R^{ant}_i of the rules composing the FRBS and a set of candidate consequents $C_{R^{ant}_i}$ associated to them.* Firstly, the rules are grouped according to their antecedents. Let $R^{ant}_i = $ IF X_1 is A^i_1 and ... and X_n is A^i_n be the antecedents of the rules of the *i*-th group, where $i \in \{1, \ldots, M\}$ (with M being the number of groups, i.e., the number of rules finally obtained). The set of candidate consequents for the R^{ant}_i antecedent combination is defined as:

$$C_{R^{ant}_i} = \{B_k \in \mathcal{B} \mid \exists e_l \text{ where}$$
$$\forall j \in \{1, \ldots, n\}, \forall A'_j \in \mathcal{A}_j, \mu_{A^i_j}(x^l_j) \geq \mu_{A'_j}(x^l_j)\}$$
$$\text{and } \forall B' \in \mathcal{B}, \mu_{B_k}(y^l) \geq \mu_{B'}(y^{l^i})\} .$$

3. *Perform a combinatorial search among the sets $C_{R^{ant}_i}$ looking for the combination of consequents with the best cooperation.* An improvement in the learning process consists of adding a new term to the candidate consequent set corresponding to each rule, the *null consequent* $\mathcal{N}$, such that $C_{R^{ant}_i} = C_{R^{ant}_i} \cup \mathcal{N}, i = 1, \ldots, M$. If this consequent is selected for a specific rule, such rule does not take part in the FRBS finally learned.

Since the search space tackled in step 3 of the algorithm is usually large, it is necessary to use approximate search techniques. In [3] four different well-known techniques were proposed for this purpose. In this work we will consider a GA as search technique.

3 The Use of Weighed Linguistic Rules

Using rule weights [4, 8, 9] has been usually considered to improve the way in which the rules interacts, improving the accuracy of the learned model. In this way, rule weights suppose an effective extension of the conventional fuzzy reasoning system that allow the tuning of the system to be developed at the rule level [4, 9].

When weights are applied to complete rules, the corresponding weight is used to modulate the firing strength of a rule in the process of computing the defuzzified value. From human beings, it is very near to consider this weight as an importance degree associated to the rule, determining how this rule interacts with its neighbor ones. We will follow this approach, since the interpretability of the system is appropriately maintained. In addition, we will only consider weight values in $[0,1]$ since it preserves the model readability. In this way, the use of rule weights represents an ideal framework for extended LM when we search for a trade-off between accuracy and interpretability.

In order to do so, we will follow the weighted rule structure and the inference system proposed in [9]:

IF X_1 is A_1 and ... and X_n is A_n
THEN Y is B with $[w]$,

where X_i (Y) are the linguistic input (output) variables, A_i (B) are the linguistic labels used in the input (output) variables, w is the real-valued rule weight, and *with* is the operator modeling the weighting of a rule.

With this structure, the fuzzy reasoning must be extended. The classical approach is to infer with the FITA (First Infer, Then Aggregate) scheme and compute the defuzzified output as the following *weighted sum*:

$$y_0 = \frac{\sum_i m_i \cdot w_i \cdot P_i}{\sum_i m_i \cdot w_i},$$

with m_i being the matching degree of the i-th rule, w_i being the weight associated to the i-th rule, and P_i being the characteristic value of the output fuzzy set corresponding to that rule. In this contribution, the center of gravity will be considered as characteristic value and the *minimum t-norm* will play the role of the implication and conjunctive operators.

A simple approximation for weighted rule learning would consist in considering an optimization technique to derive the associated weights of the previously obtained rules (e.g., by means of ad hoc data-driven methods as WM, or even COR). However, due to the strong dependency between the consequent selection and the learning of the associated weights, this two step-based technique is not the most useful to obtain weighted rules with good cooperation. Therefore, we need to include the learning of rule weights in the combinatorial search process of cooperative rules within the COR methodology.

4 The WCOR Methodology

In this section, we present the WCOR methodology to obtain weighted cooperative rules. With this aim, we include the weight derivation within the cooperative rule learning process.

4.1 Operation Mode

This methodology involves an extension of the original COR methodology. Therefore, WCOR consists of the following steps:

1. *Obtain the antecedents R_i^{ant} of the rules composing the FRBS and a set of candidate consequents $C_{R_i^{ant}}$ associated to them.*
2. *Problem representation. For each rule R_i we have:*

$$R_i^{ant}, C_{R_i^{ant}}, \text{ and } w_i \in [0, 1].$$

Since R_i^{ant} is kept fixed, the problem will consist of determining the consequent and the weight associated to each rule. Two vectors of size M (number of rules finally obtained) are defined to represent this information, c_1 and c_2, where,

$$c_1[i] = k_i \mid B_{k_i} \in C_{R_i^{ant}}, \; and$$
$$c_2[i] = w_i, \; \forall i \in \{1,\ldots,M\},$$

except in the case of considering rule simplification, in which $B_{k_i} \in C_{R_i^{ant}} \cup \mathcal{N}$. In this way, the c_1 part is an integer-valued vector in which each cell represents the index of the consequent used to build the corresponding rule. The c_2 part is a real-valued vector in which each cell represents the weight associated to this rule. Finally, a problem solution is represented as follows:

$$c = c_1 \, c_2$$

3. *Perform a search on the c vector, looking for the combination of consequents and weights with the best cooperation.* The main objective will be to minimize the *mean square error:*

$$\mathrm{MSE} = \frac{1}{2 \cdot N} \sum_{l=1}^{N} (F(x_1^l, \ldots, x_n^l) - y^l)^2,$$

with $F(x_1^l, \ldots, x_n^l)$ being the output inferred from the FRBS when the example e_l is used and y^l being the known desired output.

4.2 Genetic Algorithm Applied to the WCOR Methodology

The proposed GA performs an approximate search among the candidate consequents with the main aim of selecting the set of consequents with the best cooperation and simultaneously learning the weights associated to the obtained rules. The main characteristics of the said algorithm are presented in the following:

- *Genetic Approach* — An elitist generational GA with the Baker's stochastic universal sampling procedure.
- *Initial Pool* — The initial pool is obtained by generating a possible combination at random for the c_1 part of each individual in the population. And for the c_2 part, it is obtained with an individual having all the genes with value '1', and the remaining individuals generated at random in $[0,1]$.
- *Crossover* — The standard two-point crossover in the c_1 part combined with the max-min-arithmetical crossover in the c_2 part. By using the max-min-arithmetical crossover, if $c_2^v = (c[1], \ldots, c[k], \ldots, c[n])$ and $c_2^w = (c'[1], \ldots, c'[k], \ldots, c'[n])$ are crossed, the next four offspring are obtained:

$$c_2^1 = ac_2^w + (1-a)c_2^v,$$
$$c_2^2 = ac_2^v + (1-a)c_2^w,$$
$$c_2^3 \text{ with } c_3[k] = \min\{c[k], c'[k]\},$$
$$c_2^4 \text{ with } c_4[k] = \max\{c[k], c'[k]\},$$

with $a \in [0, 1]$ being a parameter chosen by the GA designer.

In this case, eight offspring are generated by combining the two ones from the c_1 part (two-point crossover) with the four ones from the c_2 part (max-min-arithmetical crossover). The two best offspring so obtained replace the two corresponding parents in the population.

- *Mutation* — The operator considered in the c_1 part randomly selects a specific fuzzy subspace ($i \in \{1, \ldots, M\}$) almost containing two candidate consequents, and changes at random the current consequent k_i by other consequent k_i' such that $B_{k_i'} \in C_{R_i^{ant}}$ and $k_i' \neq k_i$. On the other hand, the selected gene in the C_2 part takes a value at random within the interval $[0, 1]$.

5 Experiments

To analyze the behavior of the proposed method, we have chosen a real-world problem to estimate the length of low voltage lines for an electric company [5].

5.1 Problem Description

Sometimes, there is a need to measure the amount of electricity lines that an electric company owns. This measurement may be useful for several aspects such as the estimation of the maintenance costs of the network, which was the main goal in this application [5]. Since a direct measure is very difficult to obtain, the consideration of models becomes useful. In this way, the problem involves finding a model that relates the *total length of low voltage line* installed in a rural town with the *number of inhabitants* in the town and the *mean of the distances from the center of the town to the three furthest clients* in it. This model will be used to estimate the total length of line being maintained.

To do so, a sample of 495 rural nuclei has been randomly divided into two subsets, the training set with 396 elements and the test set with 99 elements, the 80% and the 20% respectively. Both data sets considered are available at *http://decsai.ugr.es/~casillas/fmlib/*.

Finally, the linguistic partitions considered are comprised by *five linguistic terms* with triangular-shaped fuzzy sets giving meaning to them (see Figure 1). The corresponding labels, $\{L_1, L_2, L_3, L_4, L_5\}$, stand for very small, small, medium, large and very large, respectively.

5.2 Methods

We will compare the accuracy of different linguistic models generated from our algorithm, named WCORWM [4], to the ones generated from the following methods: the well-known ad hoc data-driven WM method [11], a *method looking for the cooperation among rules* named CORWM [2, 3] and a *GA for weighted rule learning* over WM and CORWM named WRL. Table 1 presents a short description of each of them.

---

[4] With and without rule simplification

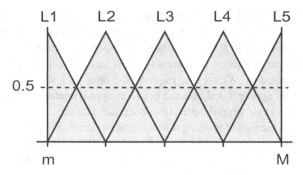

Fig. 1. Linguistic fuzzy partition representation.

Table 1. Methods considered for comparison.

Ref.	Method	Description
[11]	WM	A well-known ad hoc data-driven method
[3]	CORWM	GA applied to the COR methodology (c_1 part of WCOR)
—	WRL	GA for weighted rule learning (c_2 part of WCOR)
—	WCORWM	The proposed algorithm following the WCOR methodology

The values of the parameters used in all of these experiments are presented as follows [5]: 61 individuals, 1,000 generations, 0.6 as crossover probability, 0.2 as mutation probability per chromosome, and 0.35 for the a factor in the max-min-arithmetical crossover.

5.3 Results and Analysis

The results obtained by the analyzed methods are shown in Table 2, where #R stands for the number of rules, and MSE$_{tra}$ and MSE$_{tst}$ respectively for the error obtained over the training and test data. The best results are in boldface.

Table 2. Results obtained in the low voltage line problem.

Method	#R	MSE$_{tra}$	MSE$_{tst}$	2nd stage: WRL MSE$_{tra}$	MSE$_{tst}$
WM	13	298,450	282,029	242,680	252,483
CORWM	13	221,569	196,808	199,128	175,358
WCORWM	13	**160,736**	161,800		
Considering rule simplification					
CORWM	11	218,675	196,399	198,630	176,495
WCORWM	12	161,414	**161,511**		

[5] With these values we have tried easy the comparisons selecting standard common parameters that work well in most cases instead of searching very specific values for each method

Notice that, adding weights (WRL) to the rule sets previously learned with other methods is not sufficient. It is due to the strong dependency among the learned rules and the weights associated to them. Therefore, we need to include the learning of rule weights within the rule learning process to allow an optimal behavior.

The results obtained by WCORWM improve the ones with the remaining techniques. Moreover, an appropriated balance between approximation and generalization (with and without rule simplification) has been maintained.

In the case of the simplified models, it seems that the original COR methodology removes more rules than the desired ones, achieving slight improvements in the results. The use of rule weights takes advantage of rules that at first should be removed improving the way in which they interact.

The decision tables of the models obtained by COR and WCOR are presented in Figure 2. Each cell of the tables represents a fuzzy subspace and contains its associated output consequent, i.e., the correspondent label together with its respective rounded rule weight in the case of WCOR. These weights have been graphically showed by means of the grey colour scale, from black (1.0) to white (0.0).

In these tables we can observe as the use of weighted rules provokes slight changes in the consequents, improving the cooperation among the rules so obtained. Moreover, we can see as the rule in the subspace L1-L4 is maintained when an appropriate interaction level is considered.

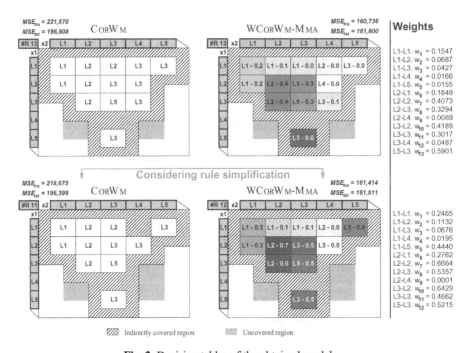

Fig. 2. Decision tables of the obtained models.

6 Concluding Remarks

In this work, we present a methodology to obtain weighted cooperative rules based on the rule cooperation paradigm presented in [2]. To do that, the learning of rule weights has been included within the combinatorial search of cooperative rules. A GA to learn cooperative rules and their associated weights has been developed for this purpose.

The proposed method has been tested in a real-world problem, improving the behavior of the basic linguistic models and the ones considering cooperative rules. Moreover, an appropriated balance between approximation and generalization has been maintained by the proposed methodology.

References

1. Alcalá R, Casillas J, Cordón O, Herrera F (2001) Improvement to the cooperative rules methodology by using the ant colony system algorithm. Mathware & Soft Computing 8:3:321–335
2. Casillas J, Cordón O, Herrera F (2002) COR: A methodology to improve ad hoc data-driven linguistic rule learning methods by inducing cooperation among rules. IEEE Transactions on Systems, Man, and Cybernetics—Part B: Cybernetics 32:4:526–537
3. Casillas J, Cordón O, Herrera F (2002) Different approaches to induce cooperation in fuzzy linguistic models under the COR methodology. In: Bouchon-Meunier B, Gutiérrez-Ríos J, Magdalena L, Yager RR (eds) Techniques for Constructing Intelligent Systems. Springer-Verlag, Heidelberg, 321–334
4. Cho JS, Park DJ (2000) Novel fuzzy logic control based on weighting of partially inconsistent rules using neural network. Journal of Intelligent Fuzzy Systems 8:99–110
5. Cordón O, Herrera F, Sánchez L (1999) Solving electrical distribution problems using hybrid evolutionary data analysis techniques. Applied Intelligence 10:5–24
6. Cordón O, Herrera F (2000) A proposal for improving the accuracy of linguistic modeling. IEEE Transactions on Fuzzy Systems 8:4:335–344
7. Ishibuchi H, Nozaki K, Yamamoto N, Tanaka H (1995) Selecting fuzzy if-then rules for classification problems using genetic algorithms. IEEE Transactions on Fuzzy Systems 9:3:260–270
8. Ishibuchi H, Takashima T (2001) Effect of rule weights in fuzzy rule-based classification systems. IEEE Transactions on Fuzzy Systems 3:3:260–270
9. Pal NR, Pal K (1999) Handling of inconsistent rules with an extended model of fuzzy reasoning. Journal of Intelligent Fuzzy Systems 7:55–73
10. Pardalos PM, Resende MGC (2002) Handbook of applied optimization. Oxford University Press, New York
11. Wang LX, Mendel JM (1992) Generating fuzzy rules by learning from examples. IEEE Transactions on Systems, Man, and Cybernetics 22:1414–1427

XML based Modelling of Soft Computing Methods

Christian Veenhuis and Mario Köppen

Fraunhofer Institute for Production Systems and Design Technology
Department Pattern Recognition
Pascalstr. 8-9, 10587 Berlin, Germany
Email: {christian.veenhuis|mario.koeppen}@ipk.fhg.de

Key words: soft computing, modelling language, algorithm design, XML

Summary. This paper proposes a document-oriented modelling concept for Soft Computing Methods. This modelling concept realizes domain-specific modelling languages derived from XML (Extensible Markup Language). XML is in general considered as the future for internet documents and data exchange. The main concept behind XML is to separate the content of a document from its layout (its appearance). The presented modelling concept uses a document for describing Soft Computing models. Like the content of a document is separated from its layout, the abstract Soft Computing model is separated from a concrete implementation and programming language. The paper presents a way in which modelling concepts for Soft Computing methods can be realized as domain-specific languages based on XML notation. The abstract Soft Computing models can be interpreted or translated by software generators into ready to use source-code (covering the adequate Soft-Computing-functionality) as well as for documentation and exchange of the realized Soft Computing model.

1 Introduction

At present a variety of Soft Computing methods is in existence. Soft Computing is meant as generic term for methods like Evolutionary Algorithms, Neural Networks, Artificial Chemistry, fuzzy algorithms, Cellular Automata, and further bio-inspired methods. L. Zadeh, father of Fuzzy Logic and founder of Soft Computing, defined Soft Computing as follows: *"Soft computing differs from conventional (hard) computing in that, unlike hard computing, it is tolerant of imprecision, uncertainty, partial truth, and approximation. In effect, the role model for soft computing is the human mind. The guiding principle of soft computing is: Exploit the tolerance for imprecision, uncertainty, partial truth, and approximation to achieve tractability, robustness and low solution cost"*.

Every single one of these methods can be called Soft Computing Method (SCM) and represents a complete class of algorithms. The SCM 'Evolutionary Algorithm' (EA) e.g. represents all methods inspired by natural evolution processes like Genetic Algorithms (GA)[5][4], Genetic Programming (GP)[6], Evolutionary Strategies (ES)[8] and so on. Each of these classes represents a huge variety of concrete algorithms, whereby the concrete algorithms differ in their configuration and the used operators.

The number of different classes and concrete algorithms is growing constantly and everyone uses its prefered programming language and operating system. No standardized programming language or modelling concept for SCMs is in existence. This makes the exchange of the concrete algorithms more difficult. In addition, if one wants to use an implemented or a new published algorithm, one has to reprogram or translate the whole algorithm in the worst case (especially, if one wants to use it on another platform or in another programming language). Some software systems are avaible for the simulation and exploration of SCMs. Unfortunately, the created and defined SCMs can rarely be exchanged between these software systems. If one wants to use a created Neural Network (NN) with another simulator e.g. the NN often has to be reprogrammed or redefined.

To overcome the mentioned problems and to start up efforts in exchange and standardization, a concept for creating document-oriented SCM-specific modelling languages is proposed which supports the modelling, description, and documentation of the described concrete algorithm at once. Algorithms described with this modelling languages should be system-independent and separated from an explicit implementation. The concept should also be easy to understand and to learn.

Looking for a suitable concept, different approaches have been studied. Among them are the Extensible Markup Language (XML) [1] [14] and Evolutionary Algorithm Modeling Language (EAML) [12], whereby EAML is derived from XML. Both languages are not traditional programmer-languages that specify a computation flow. They are developed to specify and characterize elements of an environment. These languages abstract several elements by adapted constructs and support their hierarchical order within a document. EAML for example is used to model Evolutionary Algorithms like Genetic Algorithm, Evolutionary Strategy and so on. Every EA is built up by a hierarchical order of elements which represent the different (genetic) operators and methods. This way a semantic model is constructed: the kind of order and the nesting of the elements represent the specific EA.

The paper is organized as follows. Section 2 presents XML and shows first activities in using XML for Soft Computing modelling. Section 3 explains how to create a modelling language for a SCM with XML. For this, the modelling language EAML is used as reference.

2 XML documents and related work

2.1 XML documents

A language for describing a SCM needs a concept which is able to combine different methods, operators, and parameters within a specification. As mentioned before, EAML is an XML-application (i.e. derived from XML). In this section XML is briefly introduced.

The Extensible Markup Language (XML) was derived from SGML. SGML (Standard Generalized Markup Language) was developed by Charles Goldfarb in 1986 as a meta language for all markup languages and will therefore be considered as the mother of all markup languages [1]. A meta language possesses elements for describing markup languages. A markup language possesses a set of markups to structure the content and the layout (appearance) of a document separately. A very popular markup language is HTML (Hypertext Markup Language).

The ability to split the document into its content and layout is the great advantage of XML. For this, so-called semantic markups are used. With a semantic markup one can assign a special meaning to a particular content:

```
<PARTNUMBER> 77465GH7 </PARTNUMBER>
```

A separated file (a so-called 'stylesheet') contains the definition of the layout. This file defines the appearance for all contents, e.g., with the semantic **PARTNUMBER**. For different applications and systems this can be defined in another way. In addition, a system can recognize and process the content between semantic markups more effectively.

The notation of a single XML element is described in the following by using the EBNF (Extended Backus-Naur Form) notation. Thereby it is presented with some modified snippets from the original XML reference [14]:

```
[1] element      ::= EmptyElement | StartTag content EndTag
[2] EmptyElement ::= '<' NameOfElement (S Attribute)* S? '/>'
[3] StartTag     ::= '<' NameOfElement (S Attribute)* S? '>'
[4] Attribute    ::= AttName Eq AttValue
[5] Eq           ::= S? '=' S?
[6] EndTag       ::= '</' NameOfElement S? '>'
[7] S            ::= (#x20 | #x9 | #xd | #xa)+
```

XML allows empty elements and elements which are able to nest other elements as their content (symbol: content). Beside its name an empty element (rule 2) possesses only its attributes [7]. Analogous to empty elements, elements with further elements as content (rules 3 to 6) can also possess attributes. The symbol S of rule 7 represents white-spaces (newline, tabulator, blank etc.) between the single symbols.

The valid structure of an XML-document is defined within a Document Type Definition (DTD). A DTD represents a context-free grammar [7]. XML provides several elements for the definition of DTDs, among others the **ELEMENT** element, as depicted in the following example:

```
<!ELEMENT PERSONS (PERSON)+ >           <!-- 1+ person(s) -->
<!ELEMENT PERSON  (FIRSTNAME|NAME)* > <!-- various data -->
<!ELEMENT FIRSTNAME #PCDATA>
<!ELEMENT NAME       #PCDATA>
```

This example describes the grammar of documents which are able to manage various persons with their names. An example of such a document is given at the end of this section.

Every XML document starts with the following two lines (containing the version and DTD):

```
<?xml version="1.0"?>
<!DOCTYPE PERSONEN SYSTEM "persons.dtd">
```

With the first line the document is identified as XML document of the appropriate version. In the second line the expected structure of the document is defined by a DTD. This DTD is externally loaded from a file called persons.dtd.

After these two lines a hierarchy of the elements can be defined. In the following a complete example:

```
<?xml version="1.0"?>
<!DOCTYPE PERSONEN SYSTEM "persons.dtd">

<PERSONS>
  <PERSON>
    <FIRSTNAME> Douglas </FIRSTNAME> <NAME> Adams </NAME>
  </PERSON>
  <PERSON>
    <FIRSTNAME> Gregory </FIRSTNAME> <NAME> Benford </NAME>
  </PERSON>
</PERSONS>
```

This example is a complete XML document with version specification, DTD, and elements containing the personal data in a structured manner according to the DTD.

Soft Computing models described by a language like XML have obviously the following advantages: They are platform independent and readable for humans (it's text-based) and for software systems (it's structured). XML is very convenient for representing structural data and is the de-facto standard for representation, distribution, and exchange of data. As it is shown in section 3 XML allows also the creation of semantic models.

2.2 Related work

This section presents first efforts in XML based modelling concepts for SCMs. They are still in the beginning but reveal some possibilities.

Veenhuis, Franke, and Köppen have developed EAML (**E**volutionary **A**lgorithm **M**odelling **L**anguage) [12] [13] for the modelling of Evolutionary Algorithms like Genetic Algorithm, Evolutionary Strategy and so on. Every EA is built up by a hierarchical order of elements which represent the different (genetic) operators and methods. The different XML elements representing different operators can be used like building blocks to build an element-hierarchy. This way a semantic model is constructed: the kind of order and the nesting of the elements represent the specific EA. In figure 1 a simple example is given. It depicts a complete EA for finding a predefined bitstring (in this case a bitstring consisting only of the symbol 1). An

EA described this way can be translated (compiled) into ready to use source-code covering the desired EA functionality.

Rubtsov and Butakov have developed NNML (Neural Network Markup Language) [9] [10] for describing Neural Network models. It allows the description of complete Neural Network models with all necessary informations for the full reconstruction of a Neural Network, including data dictionary, pre- and postprocessing, details of structure and parameters, trained weights, and the description of any topology consisting of heterogeneous elements.

CAML (Cellular Automata Modelling Language) [11] was developed for the modelling of different types of Cellular Automata. It works very similar to EAML: Several elements are ordered and nested to describe any set of rules and global properties on an abstract level. Finally, the built element-hierarchy realizes a complete Cellular Automaton. In figure 2 a one dimensional CA for the well-known Sierpinsky triangle is shown. Rule sets with any complexity are defined in CAML documents by using XML notation.

3 Derivation of a Modelling Language

This section gives an impression on how an XML based modelling concept for a Soft Computing Method (SCM) or a part of a SCM can be developed. Thereby EAML is used as an example for a better understanding.

If one wants to develop a modelling language for a SCM, one wants to develop a domain-specific language which is able to describe the whole software system family for this special domain. Here, a concrete algorithm of the domain represents a single software system of this software system family. EAML for example is a domain-specific language for the family of all Evolutionary Algorithms. An EAML description (as XML document) represents a concrete algorithm, i.e. one software system of the domain of EAs.

A good way to develop a domain-specific language for a SCM is by using domain engineering methods. Domain engineering comprises the development of a common model, concrete components, software generators or interpreters, and a framework for a family of software systems. According to [2] this can be done with domain analysis, domain design and domain implementation. With the domain analysis the domain of focus is selected and defined. All domain-specific informations are collected to build a domain model. The domain model represents all properties of a system in the domain. With the domain design the framework (architecture) is designed which is able to describe the main abstract algorithm with its used components of the modeled SCM. The implementation of generators, interpreters, reusable components etc. is done within the domain implementation.

If one would like to do a domain analysis for Soft Computing Methods in general, one would collect at least the following properties:

1. A data representation (genomes for EAs, weights for Neural Networks, cell-states for Cellular Automata, molecules for Artificial Chemistry etc.)

```xml
<?xml version="1.0"?>
<!DOCTYPE EAML SYSTEM "eaml.dtd">

<EAML  standalone="true"  project="bits">

<!-- ================================================================================ -->
<Code name = "objfunc"> <![CDATA[
                                // this is the objective-function
  double obj = ECPARAM(CBits,gene,size); // the worst fitness is the number of bits

  // Compare whole genome bitwise with desired target.
  // Every equal bit-pair causes a decrementation.
  // Thus the best fitness is 0 and the worst is the number of bits within genome.

  for (int i = 0 ; i < ECPARAM(CBits,gene,size) ; i++ )
     if ( GETDATA( i ) == 1 )  obj -= 1;
  return obj;

]]> </Code>

<!-- ================================================================================ -->
<Algorithm  name = "CBits"              <!-- the main-algo (we use only one) -->

    size = "20"                         <!-- size of population = number of individuals -->
    direction = "minimize"              <!-- we want to minimize the objective-function -->
    generations = "100"                 <!-- run max. 100 generations -->
    optimum = "0.0"

    elitistRate  = "10%"                <!-- we use 2 elitists (20 * 0.1 = 2) -->
    operatorRate = "90%"                <!-- create remaining rest of new population with operators -->
>
    <objective><Use ref="objfunc"/></objective>       <!-- reference to objective-function -->

    <genome><BitString size="30" group="1"/></genome>  <!-- use every bit alone -->

    <selection><RouletteWheel/></selection>            <!-- fitness-proportional selection -->

    <operator> <Group> <operators>                     <!-- This is the operator-structur. -->
                                                       <!-- We create two parallel operators: -->
        <Binary rate = "90%" succRate = "10%" >        <!-- a binary (cross-over p=0.9) and -->
                                                       <!-- an unary (mutation p=0.1). -->
          <OnePoint/>                                  <!-- After creating offspring with the binary -->
                                                       <!-- operator, we additionally mutate the -->
          <succUnary>
            <PointIncrement                            <!-- offspring with p = 0.1. -->
              min="0"
              max="1"
              step="1"
            />
          </succUnary>
        </Binary>

        <Unary rate="10%">
          <PointIncrement min="0" max="1" step="1"/>
        </Unary>
    </operators> </Group> </operator>                  <!-- end of operator-structur -->

    <initial> <RandomInitial min="0" max="1"/> </initial> <!-- initialize bitstrings randomly -->

</Algorithm>
</EAML>
```

Fig. 1. A sample element-hierarchy written in EAML

```
<?xml version="1.0" encoding="UTF-8"?>
<!DOCTYPE CAML SYSTEM "caml.dtd">

<!-- Sierpinsky triangle (cell-values: 0 = dead, 1 = alive) -->

<CA  project="sierpinsky"  dimensions="1"  min="0"  max="1">
  <Rule>
    <Cond>
      <OR>
        <Pattern dimx="3" centerx="1" > 0 0 0 </Pattern>
        <Pattern dimx="3" centerx="1" > 1 1 1 </Pattern>
      </OR>
    </Cond>
    <Action>    <Current> <Literal value="0"/> </Current> </Action>
    <Otherwise> <Current> <Literal value="1"/> </Current> </Otherwise>
  </Rule>
</CA>
```

Fig. 2. A simple CA for the Sierpinsky triangle written in CAML

2. Multiple iterations or generations to converge to a solution or state
3. One or more solutions (individuals) per iteration
4. A set of operators, rules, reactions etc.
5. Constraints (ranges for values, thresholds etc.)
6. And often an evaluation (objective) function

But to be more concrete, EAML is taken as example in the following. EAML was briefly introduced in section 2.2. Its focus is the domain of Evolutionary Algorithms. A domain analysis would reveal at least the following properties: *encoding of solutions, initialising functions and states, objective functions, fitness functions, termination criteria, selection methods, genetic operators, probabilities (operators, selection methods etc.), replacement schemes, populations of solutions (individuals), and constraints (gene value range, thresholds etc.)*.

Each of these properties have a variety of attributes. On the basis of the collected properties a domain model or a hierarchical specification can be created. It should be able to organize all SCMs in a hierarchically manner. Each algorithm which can be described by a programming language, can also be represented as tree (every parser transforms the source code into a syntax-tree). E.g., Fiesler shows in [3] that Neural Networks can also be modeled with a tree structure.

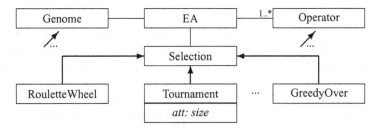

Fig. 3. Object model for a subset of the EA domain

The hierarchical specification can be used to design an object-oriented model for the domain. In figure 3 a subset of the EA domain is modelled. On the one hand this object-oriented model could represent e.g. C++ classes. On the other hand it represents the relationship between the different objects of the domain. These relationships can be directly translated into a structure for XML documents to build a semantic model. For this, the Document Type Definition in which the hierarchical order and semantic is fixed is defined:

```
<!ELEMENT EA (Genome | Operators | Selection ...further elements...)*>
```

```
<!ELEMENT Genome        (...further appropriate elements...)   >
<!ELEMENT Operators     (...further appropriate elements...)+ >
<!ELEMENT Selection     ( RouletteWheel |
                          Tournament     |
                          GreedyOver ...further elements...)  >
```

```
<!ELEMENT RouletteWheel EMPTY >
```

```
<!ELEMENT Tournament EMPTY >
<!ATTLIST Tournament size   CDATA   "2"  >
```

```
<!ELEMENT GreedyOver EMPTY >
```

An instance of this DTD (an XML document) could look like:

```
<?xml version="1.0"?>
<!DOCTYPE EA SYSTEM "EA.dtd">

<EA>
 <Genome>   ...a nested element...   </Genome>
 <Selection>  <Tournament size = "10" />  </Selection>
 <Operators> ...further nested elements... </Operators>

 ...further nested elements...
</EA>
```

The objects of the domain are ordered hierarchically. Each object (element) can be considered as building block. In figure 1, an operator element also exists. It encloses the operator-hierarchy where all genetic operators are arranged in a nested order configured with their attributes. This hierarchy of elements builds a semantic model. The Document Type Definition (DTD) describes the general software system family and an instance of this DTD describes a concrete software system of this software system family.

Within the domain implementation phase, e.g., software generators or interpreters for the DTD are developed. In an object-oriented programming language the

XML elements can easily be realized by defining a class for each element and class-attributes for every attribute of an element. Thereby the object-oriented model is directly realizable into a software system. The different building blocks e.g. for the `Selection` element can be realized with polymorphic mechanisms, as depicted in figure 3.

Having a set of building blocks for the different methods of the domain is not enough. During the domain design phase the framework of the software system family has to be developed as well. Here, the framework means a very general main algorithm of the modelled domain. For EAML e.g. a full configurable and adaptable EA algorithm was designed. It possesses polymorphic interfaces to all the categories of building blocks (selection methods, operators, fitness functions etc.). The root element `EA` represents this framework. All nested elements are the components (building blocks) for the different tasks. The program which reads the XML document (the parser) would set the appropriate instances of classes to the appropriate interfaces and configures the EA this way with the building blocks out of the XML document.

4 Conclusions

First efforts are done in using XML documents for the modelling of SCMs. This paper showed that Soft Computing Methods could be specified in a hierarchical order. One way for creating an XML based modelling concept for a SCM was presented. After employing Domain Engineering methods a domain model is obtained which can be transformed into a Document Type Definition (DTD) this way defining a modelling language. XML documents for this DTD contain complete models of a SCM by arranging the different components (building blocks) for the possible methods and procedures of that domain in an element-hierarchy. Because XML elements possess attributes, the models are configurable.

At least the introduced modelling languages EAML (Evolutionary Algorithm Modelling Language) and CAML (Cellular Automata Modelling Language) were developed with the presented approach. Using the XML notation turns out to be of great advantage: The XML documents can be used as model descriptions and as file formats. They can be exchanged between different software systems which can process documents of the DTDs. Up to now, a variety of XML editors are in existence which allow the convenient creation and editing of the models within the XML documents. The possibility of describing abstract models with their semantics and their parameters could make XML a sufficient candidate for standardized modelling languages of SCMs in the future.

Last but not least, the authors would like to motivate further discussions on the presented method as well as for standardization and exchange within the Soft Computing community.

5 Acknowledgements

The authors would like to thank Stephanie Wenzel for proof-reading this paper.

References

1. Henning Behme and Stefan Mintert, *XML in der Praxis: Professionelles Web-Publishing mit der Extensible Markup Language*, Addison-Wesley, Bonn, Deutschland, 1998
2. Krzysztof Czarnecki and Ulrich W. Eisenecker, *Generative Programming: Methods, Tools, and Applications*, Addison-Wesley, 2000
3. E. Fiesler and H.J. Caulfield, *Neural Network Formalization*, Computer Standards and Interfaces, 1994 vol. 16(3), pp. 231-239
4. D. E. Goldberg, *Genetic Algorithms ins Search, Optimization and Machine Learning*, Addison Wesley, Reading, MA, 1989
5. John H. Holland, *Adaption in Natural and Artificial Systems*, MIT Press, 1992
6. J.R. Koza, *Genetic Programming: On the Programming of Computers by Natural Selection*, MIT Press, Cambridge, MA, 1992
7. Oliver Pott and Gunter Wielage, *XML: Praxis und Referenz*, Markt+Technik Verlag, Kempten, Germany, 2000
8. Ingo Rechenberg, *Evolutionsstrategie '94*, Friedrich Frommann Verlag, Stuttgart, Germany, 1994
9. Rubtsov D., Butakov S., *A Unified format for trained neural network description*, Proc. of the 2001 Int. Joint Conf. on Neural Networks, Washington, 2001, paper #476
10. Rubtsov D., Butakov S., *Neural Network Markup Language*, URL: http://www.nnml.alt.ru, April 16, 2002
11. C. Veenhuis, *Cellular Automata Modelling Language*, URL: http://vision.fhg.de/ veenhuis/CAML, February 12, 2002
12. C. Veenhuis, K. Franke, M. Köppen, *A Semantic Model for Evolutionary Computation*, Proc. 6th International Conference on Soft Computing (IIZUKA2000), Iizuka, Japan, 2000
13. C. Veenhuis, K. Franke, M. Köppen, *Evolutionary Algorithm Modelling Language*, URL: http://vision.fhg.de/ veenhuis/EAML, February 12, 2002
14. W3C (World Wide Web Consortium), *Extensible Markup Language (XML) 1.0 (Second Edition)*, URL: http://www.w3.org/TR/2000/REC-xml-20001006, Feb 24, 2002

Intelligent information retrieval: some research trends

Gabriella Pasi

ITC-National Council of Research, Via Ampère, 56, 20131 Milano, Italy
gabriella.pasi@itc.cnr.it

Summary. In this paper some research trends in the field of Information Retrieval are presented. The focus is on the definition of "intelligent" systems, i.e. systems that can represent and manage the vagueness and uncertainty which is characteristic of the process of information searching and retrieval.

1 Introduction

In recent years, the increasing quantity of information available on the World Wide Web has raised the need for effective systems that allow an easy and flexible access to information relevant to users' needs. As a consequence there has been a strong resurgence of interest in the research area of Information Retrieval [2, 19, 38, 41, 45]. Information Retrieval (IR) aims at modelling, designing and implementing systems able to provide a content-based access to a large amount of information [2]. Information can be of any kind: textual, visual, or auditory, although most actual IR systems store and enable the retrieval of only textual information organised in documents. Search engines are a recent outgrowth of the research in IR, where documents are Web pages. The ultimate aim of an IRS is to estimate the relevance of documents to a user's information need explicitly expressed in a query. This is a very difficult task as relevance is a subjective notion: the same document may be fully relevant to a user and fully not relevant to another user, although both posing the same query to the same IR system. An IRS simulates then a decision-making activity: what the users expect from an IRS is a list of the information items relevant to their needs, ordered according to their preferences [36]. In order to do so an IRS is based on a representation of both information items and users' needs, necessary to compare them. The representation process introduces uncertainty and loss of information: the user's expression of information needs is often uncertain and vague, and a formal representation of the documents' content (typically based on keywords extraction) introduces a loss of the informative content of a document, expressed in natural language, thus generating an uncertainty about the real semantics of the information carried by the document. Moreover the representation process should depend on the users' interests, while IRSs usually assume that a document has the same representation for all users. The effectiveness of an IRS (usually measured as its ability to select relevant information and to reject irrelevant one) is crucially related to the system's flexibility, intended as its capability both to deal with the vagueness and uncertainty of the retrieval process, and to learn the user's concept of relevance through an adaptive behaviour. In recent years some retrieval techniques based on IR research have become common in most IRSs; among such techniques we recall term weighting, "natural language" queries, ranked retrieval results, "query-by-example", and query formulation assistance [15]. Moreover some advanced systems attempt to tackle the uncertainty inherent in the IR task based on probabilistic models. Some examples of such models can be found in [10]. However, despite of these applications, we are still far from having systems which are highly satisfactory for users: search engines are a clear example of this situation: many irrelevant documents are retrieved in response to a user query. To make systems more effective a big deal of research in IR is aimed at trying to add some kind of intelligence to IRSs. A component of an intelligent behaviour is flexibility, intended as the capability of learning a context and adapting to it. A possible direction to add flexibility to Information Retrieval Systems is to make them tolerant to uncertainty and imprecision in the user system interaction; another aspect that ensures flexibility is the ability to learn the user's notion of relevance [3, 7, 9, 11, 12, 13, 21]. These are both means to adapt to the users' needs: the former is a way to allow a more natural expression of users needs, the latter is the capability of elicitating from the user her/his actual information preferences. The term "intelligent information retrieval" is increasingly used in the literature, and refers to techniques aimed at the definition of systems that offer a flexible access to the huge availability of documents in digital form. Personalized indexing, relevance feedback, text categorization, text mining, cross-lingual information retrieval, question-answering tools, flexible user interfaces are examples of such techniques [1, 4, 9, 11, 13, 17, 18, 20, 22, 24, 37, 44]. In this paper a synthetic introduction to some of these techniques is presented. A special emphasis is given to the intelligent retrieval approaches defined by means of "Soft Computing" methodologies. Soft Computing is an expression used to indicate a synergy of methodologies useful for solving problems requiring some form of intelligence that diverts from traditional computing. The principal constituents of Soft Computing are: fuzzy logic, neural networks, probabilistic reasoning, evolutionary computing, chaotic computing and parts of machine learning theory. Soft Computing differs from conventional (hard) computing in that, unlike hard computing, it is tolerant to imprecision, vagueness, partial truth, and approximation. Because of these properties Soft Computing can provide very powerful tools for modelling flexible systems for the Information Access. In section 2 the basic definition of an IRS is introduced, and in section 3 some main techniques are described by which the basic scheme of an IRS may be enriched to the aim of making it more flexible. In section 4 the important problem of document indexing is presented, and some research works aimed at defining more flexible and personalized indexing procedures of semi-structured documents are briefly sketched. In our personal view this is an hot research topic in the field of IR. Finally, in section 5 some approaches to better the user system interaction trough flexible query languages tolerant to an approximate expression of users' needs is described.

2 Information Retrieval Systems

Information Retrieval (IR) aims at defining systems able to provide a fast and effective content-based access to a large amount of stored information [2, 41, 45]. Information can be of any kind: textual, visual, or auditory, although most actual IR systems store and enable the retrieval of only textual information organized in documents. A user accesses the IRS by formulating a query, which the IRS evaluates to the aim of retrieving all documents that it estimates relevant to the user's needs expressed by the query. The problem of identifying the information relevant to specific needs is a decision-making problem, based on the assessment of the subjective notion of relevance. An Information Retrieval System simulates this decision process, by automatically "comparing" the formal representations of both documents and user queries. The main components of an IRS are: a collection of information items, an indexing mechanism (which automatically constructs the representation of documents), a query language, and a matching mechanism (which estimates the relevance - or probability of relevance - of information items to queries). The input of these systems is constituted by a user query; their output is usually an ordered list of selected items, which have been estimated relevant to the information needs expressed in the user query. The choice of the formal background to define both the document and query representations characterises the model of an IRS. In the IR literature different models have been proposed. The ultimate aim of an IRS is to estimate the relevance of documents to users' information needs. This is a very hard and complex task, since it is pervaded with imprecision and uncertainty. The effectiveness of an IRS is usually measured as its ability to find relevant documents and to reject irrelevat ones. The basic scheme showed of IR is usually enriched with a number of techniques which allows to improve the retrieval capabilities, with the consequence of improving the system's effectiveness. Among such techniques there are Relevance Feedback, Text Categorization, Use of Thesauri etc. In section 3 some of these techniques are shortly described. A promising direction to increase the effectiveness of IRSs is to model the concept of partiality intrinsic in the IR process and to make the systems adaptive, i.e.able to "learn" the users' concept of relevance. In recent years the research has explored many different directions to the aim of modelling the vagueness and uncertainty that invariably characterize the management of information [9, 12, 13]. A big deal of research in IR is devoted to the definition of retrieval models that account for this imprecision and uncertainty. In section 2.1 some research trends concerning the definition of flexible IR models are shortly presented. An important aspect which affects the effectiveness of IRSs is related to the way in which the information items are formally represented; the document representation is extremely simple, usually based on keywords extraction and weighting; moreover the IRSs generally produce a unique representation of documents for all users, not taking into account the each user looks at a document content in a personalized way, which should be taken into account in the indexing phase. This adaptive view of the document is not modelled. Another important aspect is related to the fact that on the WWW some standard for the representation of semi-structured information are becoming more and more employed (such as XML). For this reason it is important to exploit the structure of documents in order to represent the information they contain. In section 4 some approaches to personalized document indexing are synthesized.

3 Making IRSs flexible: some "intelligent" techniques

In this section some techniques are synthetically presented, which constitute important directions in the IR research field, and that also make use of soft computing methodologies to model a behaviour adaptive and tolerant to uncertainty and vagueness. The expression Soft computing denotes a synergy of methodologies useful to solve problems using some form of intelligence that divert from traditional computing. The principal constituents of SC are: fuzzy logic, neural networks, probabilistic reasoning, and evolutionary computing, which in turn subsume belief networks, genetic algorithms, parts of learning theory, multivalued logics. SC differs from conventional (hard) computing in that, unlike hard computing, it is tolerant to imprecision, uncertainty, partial truth, and approximation. Because of these properties, SC can provide very powerful tools for IR. In [12] some techniques and applications of Soft Computing in Information Retrieval are presented.

3.1 IR models that deal with uncertainty and vagueness

The most long standing set of approaches that try to model the uncertainty intrinsic in IR goes under the name of Probabilistic IR [11, 19, 45]. The Probabilistic model [45] ranks documents in decreasing order of their evaluated probability of relevance to a user's information need. Past and present research has made much use of formal theories of probability and of statistics in order to estimate the probability of relevance. Without going into the details of any of the large number of probabilistic models of IR that have been proposed in the literature (for a survey see [10], if we assume that a document is either relevant or not relevant to a query, the task of a probabilistic IR system is to rank documents according to their estimated probability of being relevant. Probabilistic relevance models base this estimation on evidence about which documents are relevant to a given query. The problem of estimating the probability of relevance for every document in the collection is difficult because of the large number of variables involved in the representation of documents in comparison to the small amount of document relevance information available. The various proposed models differ, primarily, in the way they estimate this or related probabilities. Probabilistic inference models apply concepts and techniques originating from areas such as logic and artificial intelligence. Another class of approaches which try to model at some extent the approximate nature of the retrieval activity goes under the name of Logical IR models. The main motivation advocated in the literature to model IR in the logical framework is the need for a more general formal discipline, as logic, to reason about the foundational principles of IR [27, 42]. A common basis of the logical models proposed in the literature is to represent both documents and queries as formulae of the language of the adopted logic. The relationship between a document and a query is expressed by an implication of a query by a document. The estimate of the relevance of a document with respect to a query consists in determining the "logical status" of the implication. To overcome some limitations encountered in adopting classical logic, non-classical logics have been considered and employed to model IR, such as modal logic, logical imaging, terminological logic, and fuzzy logic [14, 29, 33, 35]. In particular, by the logical interpretation of information retrieval introduced by van Rijsbergen [46], the implication of a query by a document is estimated with a given degree of certainty or strenght. In order to increase the flexibility of IRSs some approaches based on the application of fuzzy set theory have been defined. A fuzzy set allows the characterisation of its elements by means of the concept of "graduality"; this concept supports a more accurate description of a class of elements when a sharp boundary of membership cannot be naturally devised. The entities involved in an IRS are well suited to be formalised within this formal framework to the aim of capturing their inherent vagueness; the main levels of application of fuzzy set theory to IR have concerned the definition of extensions of the Boolean model (concerning both the representation of documents and the query language), and the definition of associative mechanisms, such as fuzzy thesauri and fuzzy clustering [9, 25, 30]. Fuzzy generalizations of the Boolean model have been defined to the aim of defining IRSs able to produce discriminated answers in response to users' queries. A survey of fuzzy extensions of IRSs and of fuzzy generalizations of the Boolean IR model can be found in [9, 25]. In section 3 a fuzzy model of personalized document indexing is described, while in section 4 some fuzzy approaches for defining flexible query languages are presented. A different approach is based on the application of the connectionist theory to IR. Neural networks have been used in this context to design and implement IRSs that are able to adapt to the characteristics of the IR environment, and in particular to the user's interpretation of relevance. The application of connectionist models to Information Retrieval (IR) is not a recent phenomenon. Indeed, a number of papers have appeared on this subject, and much research is in progress. In [13] a review is presented of some neural approaches to IR based on the two most important paradigms of learning used in the NN field: supervised learning and unsupervised learning. Some approaches have applied Rough Set Theory and Multivalued Logics to define IR models [12]. Genetic Algorithms have been mainly applied to IR for improving document representation and indexing, and for defining relevance feedback mechanisms [12, 26].

3.2 Relevance Feedback

Relevance feedback is a process aimed at improving the IRS effectiveness, by means of an iterative process of query refinement directed by the user [16, 26, 28, 34, 40]. Users identify relevant documents in an initial list of retrieved documents, and the relevance feedback mechanism creates a new query based on those sample relevant documents. The improvement of the original query is then based on the user direct evaluation of the relevance of the retrieved documents. A relevance feedback mechanism is a module which can be incorporated in the IRS. Such a mechanism performs a decisional task based on the assumption that the representations of relevant documents have some degree of similarity: if a retrieved document is estimated relevant with respect to the initial query some information can be derived from its representation and then used to formulate a new refined query able to retrieve additional relevant documents. Relevance feedback mechanisms first identify in the relevant documents those new terms which might represent potential concepts of interest to the user, and then reformulate the query by refining it with those terms. Meaningful terms are generally identified on the basis of statistical criteria, essentially by evaluating their frequency within the set of relevant documents selected by the user. The feedback mechanism reformulates then a new and more refined query including the terms present in the original one. Algorithms for automatic relevance feedback have been studied in IR for more than thirty years, but there are some practical difficulties that have delayed the general adoption of this technique. As outlined in [15] the central problems in relevance feedback are both the selection of "features" (words, phrases) from relevant documents, and the computation of weights for these features in the context of a new query. Feedback techniques are generally based on the assumption that a few relevant documents (the top ten, for example) would be provided by the user. As Croft outlines, however, in many real interactions users specify only a single relevant document. Sometimes the user is, in effect, browsing using feedback with the consequence that the selected relevant document may not even be strongly related to the initial query. Research aimed at correcting this problem is underway and more operational systems using relevance feedback can be expected in the near future.

3.3 Automated text categorization

In the last 10 years the problem of text categorization has emerged as an important problem connected with IR. It is aimed at the automated categorization (classification) of texts into predefined categories, thus organizing them and making retrieval more flexible and consequently more effective. TC is applied in many domains, such as for example document indexing based on a controlled vocabulary, document filtering, document sense disambiguation etc. As well illustred in [43] the dominant approach to text categorization is based on machine learning techniques: a general inductive process automatically builds a classifier by learning from a set of preclassified documents the characteristics of the categories.

3.4 Vocabulary expansion and intelligent users' interfaces

The query representation in IRSs is commonly based on keywords (or strings) specification. The retrieval mechanism performs in this case a lexical match of words. One of the main problems of IR systems is vocabulary mismatch. This means that the information need is often described using different words than are found in relevant documents. Moreover, distinct users employ distinct words to describe the same documents. An important research direction in IR is aimed at defining methods for vocabulary expansion. Vocabulary expansion can result from transforming the document and query representations, as with Latent Semantic Indexing [4], or it can be done as a form of automatic thesaurus built by corpus analysis. Latent Semantic Indexing (LSI) tries to overcome the limitations of lexical matching by using for retrieval some statistically derived conceptual indices instead of individual words [4]. The basic assumption of LSI is that in the word usage there is an underlying or latent structure. LSI is based on the use of Singular Value Decomposition (SVD) to make explicit this latent structure among terms contained in the archived documents. Fuzzy associative mechanisms based on thesauri or clustering techniques [30, 31] have been defined in order to cope with the incompleteness characterizing either the representation of documents or the users' queries. In [30] a wide range of methods for generating fuzzy associative mechanisms is illustred. Fuzzy thesauri and pseudothesauri can be used to expand the set of index terms of documents with new terms by taking into account their varying significance in representing the topics dealt with in the documents; the degree of significance of the associated terms depends on the strength of the associations with the documents' descriptors. An alternative use of fuzzy thesauri and pseudothesauri is to expand each of the search terms in the query with associated terms, by taking into account their distinct importance in representing the concepts of interest; the varying importance is dependent on the associations' strength with the search terms. Fuzzy clustering can be used to expand the set of the documents retrieved by a query with associated documents; their degrees of association with respect to the documents originally retrieved influence their estimated degree of relevance. In [18] a survey of approaches to intelligent user interfaces is summarized. The aim of this research is to provide users with conceptual retrieval, rather than on the usual string matching. This means that the query terms specified by the users are considered as representatives of the concepts in which the users are interested. Qyery expansion provides additional terms , based on the use of a knowledge base or thesaurus of related words.

4 Personalized indexing in IR

The production of effective retrieval results depends on both subjective factors, such as the users' ability to express their information needs in a query, and the characteristics of the Information Retrieval System. A component of IRSs which plays a crucial role in determining their effectiveness is the indexing mechanism, which has the aim of generating a formal representation of the contents of the information items (documents' surrogates). The most used automatic indexing procedures are based on term extraction and weighting: the documents are represented by means of a collection of index terms with associated weights (the index term weights); an index term weight expresses the degree of significance of the index term as a descriptor of the document information content [39, 41]. The vector space model, the probabilistic models and fuzzy models adopt a weighted document representation [41, 45]. The automatic computation of the index term weights is based on the occurrences count of a term in the document and in the whole archive. The adoption of weighted indexes allows for an estimate of the relevance or of a probability of relevance of the documents to the considered query [41, 45]. The weighted representation of documents has the limitation of not taking into account that a term can play a different role within a text, according to the distribution of its occurrences. Let us think for example at scientific papers organised into the sections title, authors, abstract, introduction, references, etc. (this kind of structure can be explicitly defined by means of the XML language). An occurrence of a term in the title has a distinct informative role than an occurrence in the references. This problem is particular important if considering that XML is becoming a standard for the representation of semi-structured information [50]. Moreover, usual indexing procedures behave as a black box producing the same document representation for all users; this enhances the system's efficiency but implies a severe loss of effectiveness. In fact, when examining a document structured in logical sections the users have their personal views of the document's information content; according to this retrieval phase they would naturally privilege the search in some subparts of the view in the documents' structure, depending on their preferences. This last consideration outlines the fact that relevance judgments should be driven by a user's interpretation of the document's structure, and supports the idea of an adaptive indexing [3, 7, 8]. By adaptive indexing we intend personalized indexing procedures which take into account the users' indications to interpret the document contents and to "build" their synthesis on the basis of this interpretation. It follows that if an archive of semi-structured documents is considered (e.g. XML documents), flexible indexing procedures should be defined by means of which the users are allowed to direct the indexing process by explicitly specifying some constraints on the document structure (preference elicitation on the structure of a document). This preference specification should be exploited by the matching mechanism to the aim of privileging the search within the most preferred sections of the document, according to the users' indications. The user/system interaction can then generates a personalized document representation, which is distinct for distinct users [3, 7, 8]. In [7, 8] a user adaptive indexing model has been proposed, based on a weighted representation of semi-structured documents that can be tuned by users according to their search interests to generate their personal document representation in the retrieval phase. A document is represented as an entity composed of sections (such as title, authors, introduction, references, in the case of a scientific paper). The model is constituted by a static component and by an adaptive query-evaluation component; the static component provides an a priori computation of an index term weight for each logical section of the document. The adaptive component is activated by the user in the phase of query formulation and provides an aggregation strategy of the n index term weights (where n is the number of sections) into an overall index term weight. The aggregation function is defined on the basis of a two level interaction between the system and the user. At the first level the user expresses preferences on the document sections, outlining those that the system should more heavily take into account in evaluating the relevance of a document to a user query. This user preference on the document structure is exploited to enhance the computation of index term weights: the importance of index terms is strictly related to the importance for the user of the logical sections in which they appear. At the second level, the user can decide which aggregation function has to be applied for producing the overall significance degree. This is done by the specification of a linguistic quantifier such as at least k and most [48]. By adopting this document representation the same query can select documents in different relevance orders depending on the user indications. It is very important to notice that the elicitation of users' preferences on the structure of a document is a quite new and recent research approach, which can remarkably improve the effectiveness of IRSs. In [32] another representation of structured documents is proposed, which produces a weighted representation of documents written in HyperText Markup Language. An HTML document has a syntactic structure, in which its subparts have a given format specified by the delimiting tags. In this context tags are seen as syntactic elements carrying an indication of the importance of the associated text: when writing a document in HTML, one associates a distinct importance with distinct documents' subparts, by delimiting them by means of appropriate tags. On the basis of these considerations, an indexing function has been proposed, which computes the significance of a term in a document by taking into account the distinct role of term occurences according to the importance of tags in which they appear.

5 Flexible query languages

Vagueness and imprecision are present mostly in the users' expression of their information needs, both in its initial form in the query and in later forms of relevance assessments. It has been shown by much research that the user often does not have a clear picture of what he is looking for and can only represent her/his information need in vague and imprecise terms [23]. A crucial aspect affecting the effectiveness of the system is related to the characteristics of the query language, which should represent in the more accurate and faithful way the user's information needs. The available query languages are mostly based on keyword specifications, and do not allow to express uncertainty and vagueness in the specification of constraints that the relevant information items must satisfy [38, 45]. A flexible query language is a language that makes possible a simple and approximate expression of subjective information needs. By means of fuzzy set theory some flexible query languages have been defined as generalizations of the Boolean query language. In this context a flexible query may consists of either both of the following two soft components or just one: the first component is constituted by selection conditions that are interpreted as soft constraints on the significance of the index terms in each document representation. The second component is constituted by soft aggregation operators which can be applied to the soft constraints in order to define compound selection conditions. The selection conditions are expressed by weighted terms, in which a weight can be either a numeric value in [0,1] or a linguistic expression such as very important; the compound conditions are expressed by means of linguistic quantifiers used as aggregation operators. When soft constraints are specified, the query evaluation mechanism is regarded as performing a fuzzy decision process that evaluates the degree of satisfaction of the query constraints by each document representation by applying a partial matching function. This degree (the Retrieval Status Value) is interpreted as the degree of relevance of the document to the query and is used to rank the documents. The definition of the partial matching function is strictly dependent on the query language definition and specifically on the semantics of the soft constraints. In [5] a linguistic extension of the Boolean language is defined, based on the concept of linguistic variable [47]. By this language the user can associate with query terms either the primary term "important", or some compound terms, such as "very important" or "fairly important" to qualify the desired importance of the search terms in the query. A second level of softening of the Boolean query language concerns the specification of aggregation operators. In the Boolean query language, the AND and OR connectives allow for crisp aggregations which do not capture any vagueness. For example, the AND used for aggregating M selection conditions does not tolerate the unsatisfaction of a single condition; this may cause the rejection of useful items. Within the framework of fuzzy set theory a generalization of the Boolean query language has been defined, based on the concept of linguistic quantifiers: they are employed to specify both crisp and vague aggregation criteria of the selection conditions [6]. New aggregation operators can be specified by linguistic expressions, with a self-expressive meaning such as at least k and most of. They are defined with a behavior between the two extremes corresponding to the AND and the OR connectives, which allow, respectively, requests for all and at least one of the selection conditions. The linguistic quantifiers used as aggregation operators, are defined by Ordered Weighted Averaging (OWA) operators [49]. By adopting linguistic quantifiers, the requirements of a complex Boolean query are more easily and intuitively formulated. For example when desiring that at least 2 out of the three selection conditions "politics", "economy", "inflation" be satisfied, one should formulate the following Boolean query: (politics AND economy) OR (politics AND inflation) OR(economy AND inflation), which can be replaced by the simpler one at least 2(politics, economy, inflation). In [8] a generalisation of the Boolean query language that allows to personalize the search in structured documents (as illustred in section 4) is proposed; both content-based selection constraint, and soft constraints on the document structure can be expressed. The atomic component of the query (basic selection criterion) is defined as follows:

$$aq = t \text{ in } Q \text{ preferred sections}$$

in which t is a search term expressing a content-based selection constraint, and Q is a linguistic quantifier such as all, most, or at least kselection constraint. It is assumed that the quantification refers to the sections that are semantically meaningful to the user. Q is used to aggregate the significance degrees of t in the desired sections and then to compute the global Retrieval Status Value of the document d with respect to the atomic query condition aq.

6 Conclusions

In this paper some approaches to the definition of flexible Information Retrieval Systems have been presented. In particular some promising research directions that could guarantee the development of more effective IRSs have been outlined. Among these, the research efforts aimed at defining new indexing techniques of semi-structured documents (such as XML documents) are very important: the possibility of creating in a user-driven way the documents' surrogates would ensure a modeling of the users' interests also at the indexing level (usually this is limited to the query formulation level).

References

1. Agosti M, Crestani F and Pasi G eds. (2001). Lectures on Information Retrieval. Lecture Notes in Computer Science, Springer Verlag.
2. Baeza-Yates R and Ribeiro-Nieto B (1999). Modern Information Retrieval. Addison-Wesley, Harlow, UK.
3. Berrut C, Chiaramella Y (1986). Indexing medical reports in a multimedia environment: the RIME experimental approach. In proc. ACM-SIGIR 89, Boston, USA, 187–197.
4. Berry MW, Dumais ST, and O'Brien GW (1995). Using linear algebra for intelligent information retrieval. SIAM Review, 37(4): 573–595.
5. Bordogna G and Pasi G (1993). A fuzzy linguistic approach generalizing Boolean information retrieval: a model and its evaluation. Journal of the American Society for Information Science 44(2): 70–82.
6. Bordogna G and Pasi G. and G. Pasi (1995) Linguistic aggregation operators in fuzzy information retrieval. International Journal of Intelligent systems 10(2): 233–248.
7. Bordogna G and Pasi G (1995) Controlling retrieval trough a user-adaptive representation of documents. International Journal of Approximate Reasoning 12: 317–339.
8. Bordogna G and Pasi G (2000) Flexible Representation and Querying of Heterogeneous Structured Documents. Kibernetika: 36(6): 617–633.
9. Bordogna G and Pasi G (2001) Modelling Vagueness in Information Retrieval. In: Agosti M, Crestani F and Pasi G eds. Lectures in Information Retrieval. Springer Verlag.
10. Crestani F, Lalmas M, van Rijsbergen CJ, and Campbell I (1998) Is this document relevant? Probably. ACM Computing Surveys 30(4):528-552.
11. Crestani F, Lalmas M, van Rijsbergen CJ (eds) (1998) Information Retrieval: Uncertainty and Logics. Kluwer Academic Publisher, Norwell, MA, USA.
12. Crestani F and Pasi G (eds) (2000). Soft Computing in Information Retrieval: Techniques and Applications. Physica Verlag. Series Studies in Fuzziness.
13. Crestani F and Pasi G (1999). Soft Information Retrieval: Applications of Fuzzy Set Theory and Neural Networks. In: Kasabov N and Kozma R (eds). Neuro-fuzzy Techniques for Intelligent Information Systems. Physica-Verlag. Springer-Verlag Group.
14. Crestani F and van Rijsbergen CJ (1995). Information retrieval by logical imaging,. Journal of Documentation 51(1): 293–331.
15. Croft B (1995). The Top 10 Research Issues for Companies that Use and Sell IR Systems. D-Lib Magazine November 1995.
16. Dillon M and Desper J (1980). The Use of Automatic Relevance Feedback in Boolean Retrieval Systems. Journal of Documentation 36(3): 197–208.

17. Dumais ST, Landauer TK, and Littman LM (1996). Automatic cross-linguistic information retrieval using Latent Semantic Indexing. In proc. ACM SIGIR 96 - Workshop on Cross-Linguistic Information Retrieval: 16–23, August 1996.
18. Gauch S (1992). Intelligent Information Retrieval: An Introduction.Journal of the American Society of Information Science 43(2): 175–182.
19. Glover EJ, Lawrence S, Gordon MD, Birmingham WP, and Lee Giles C (1999). Web Search – YourWay. Communications of the ACM.
20. Goker A (1989). Machine learning for "intelligent" information retrieval. In: proc. of the 11th BCS IRSG Research Colloquium on Information Retrieval. Huddersfield Polytechnic: 211-27.
21. Goker A (1999). Capturing Information Need by Learning User Context. In proc. Sixteenth International Joint Conference in Artificial Intelligence: Learning About Users Workshop: 21-27.
22. Haverkam D and Gauch S (1998). Intelligent Information Agents: Review and Challenges for Distributed Information Sources. Journal of the Society for Information Science 49(4): 304-311.
23. Ingwersen P (1992). Information Retrieval Interaction. Taylor Graham. London, UK.
24. Kao B, Lee J, Ng CY, and Cheung D (2000). Anchor Point Indexing in Web Document Retrieval. IEEE Transactions on Systems, Man, and Cybernetics Part C: Applications and Reviews 30(3): 364–373.
25. Kraft D, Bordogna G, Pasi G (1999). Fuzzy Set Techniques in Information Retrieval. In: Bezdek JC, Dubois D and Prade H (eds). Fuzzy Sets in Approximate Reasoning and Information Systems. The Handbooks of Fuzzy Sets Series. Kluwer Academic Publishers 469–510.
26. Kraft DH, Petry FE, Buckles BP, and Sadasivan T (1997). Genetic Algorithms for Query Optimization in Information Retrieval: Relevance Feedback. In: Sanchez E, Shibata T, and Zadeh LA (eds). Genetic Algorithms and Fuzzy Logic Systems Singapore: World Scientific.
27. Lalmas M (1998). Logical models in Information Retrieval: introduction and overview, Information Processing and Management 34:19–33.
28. Lenk PJ and Floyd BD (1988). Dynamically updating relevance judgements in probabilistic information systems via users' feedback. Management Science, 34(12): 1450–1459.
29. Meghini C, Sebastiani F, Straccia U, Thanos C (1995). A model of information retrieval based on terminological logic. In Proc. ACM Sigir Conference on Research and Development in Information Retrieval, Pittsburgh, U.S.A., 298-307.
30. Miyamoto S (1990). Fuzzy sets in Information Retrieval and Cluster Analysis. Kluwer Academic Publishers.
31. S. Miyamoto (1990). Information retrieval based on fuzzy associations. Fuzzy Sets and Systems 38(2): 191–205.
32. Molinari A and Pasi G (1996). A Fuzzy Representation of HTML Documents for Information Retrieval Systems. In: Prodcings of the IEEE International Conference on Fuzzy Systems, 8-12 September, New Orleans, U.S.A., Vol 1, 107–112.
33. Nie YJ (1989). An information retrieval model based on modal logic. Information Processing and Management 25(5): 477–491.
34. Pasi G, Marques Pereira RA (1999). A decision making approach to relevance feedback in information retrieval: a model based on a soft consensus dynamics. International Journal of Intelligent Systems 14(1): 105–122.

35. Pasi G (1999). A logical formulation of the Boolean model and of weighted Boolean models. In: Proceedings Workshop on Logical and Uncertainty Models for Information Systems (LUMIS 99). University College London, 5-6 July 1999.
36. Pasi G (2002). Modelling the notion of preference in Information Systems. International Journal of Intelligent Systems. To appear,
37. Rhodes J, Maes P (2000). Just-in-time information retrieval agents. IBM Systems Journal 39(3-4): 685–704.
38. Salton G (1989). Automatic Text Processing - The Transformation, Analysis and Retrieval of Information by Computer. Addison Wesley Publishing Company.
39. Salton G and Buckley C (1988). Term weighting approaches in automatic text retrieval. Information Processing and Management 24(5): 513–523.
40. Salton G and Buckley C (1990). Improving retrieval performance by relevance feedback. J. Of the American Society for Information Science 41(4): 288–297.
41. Salton G and McGill MJ (1983). Introduction to modern information retrieval. New York, NY: McGraw-Hill.
42. Sebastiani F (2002). On the role of logic in Information retrieval. Information Processing and Management 34: 1-18.
43. Sebastiani F (2002). Machine Learning in automated text categorization. ACM Computing Surveys 34(1), March 2002.
44. Smeaton A (1992). Progress in the application of Natural Language Processing to Information Retrieval tasks. The Computer Journal 35(3): 268–278.
45. van Rijsbergen CJ (1979). Information Retrieval. London, England, Butterworths and Co., Ltd.
46. Van Rijsbergen CJ (1986). A non-classical logic for information retrieval. The Computer Journal 29(6).
47. Zadeh LA (1975). The concept of a linguistic variable and its application to approximate reasoning, parts I, II. Information Science 8: 199–249, 301–357.
48. Zadeh LA (1983). A computational Approach to Fuzzy Quantifiers in Natural Languages, Computing and Mathematics with Applications 9: 149–184.
49. Yager RR (1988). On Ordered Weighted Averaging Aggregation Operators in Multi-criteria Decision Making. IEEE Transactions on Systems Man and Cybernetics 18(1): 183–190.
50. Special Topic Issue: XML. (2002). Journal of the American Society for information Science 53(6). BaezaYates R, Carmel D, Maarek Y, Soffer A eds.

Representation of Concept Specialization Distance through Resemblance Relations

Miguel-Ángel Sicilia[1], Elena García[2], Paloma Díaz[1], and Ignacio Aedo[1]

[1] DEI Laboratory, Computer Science Department, Carlos III University,
Avd. Universidad 30 – 28911 Leganés, Madrid (Spain)
{msicilia, pdp}@inf.uc3m.es, aedo@ia.uc3m.es
[2] Computer Science Department, University of Alcalá
Ctra. Barcelona km. 33.600 – 28871, Alcalá de Henares, Madrid (Spain)
elena.garciab@uah.es

Summary. Generalization-specialization (*gen-spec*) relationships between pairs of classifiers can be assigned a grade of strength or relative distance, somewhat representing the level of similarity between the class and its subclass. In many cases, these distances can not be adequately computed from the structural features or properties of the classifiers, since class-subclass discrimination semantics are often not represented explicitly, and distances can only be properly assessed subjectively by humans. In this work, we describe a novel approach that uses resemblance relations to model graded specializations, both from a specific classifier (locally) and also along a subset of the generalization hierarchy (globally). We also show how that approach can be combined with current Web-enabled ontology description languages to carry out adaptive behaviors that involve crawling the *gen-spec* hierarchy.

Key words: Generalization/specialization relationship, subtyping, ontology, resemblance relations, RDF

1 Introduction

The concept of generalization (and its inverse specialization) plays a central role in current approaches to knowledge representation using ontologies, in general-purpose object-oriented modelling notations (like the U*nified Modeling Language*, UML [12]), and also in other related fields like object-oriented databases [11] or programming languages. In addition, the resulting taxonomic relations have been integrated within reasoning in approaches like many-sorted logic [10], order-sorted logic [2] and description logic [6]. Generalization (often called '*is-a*' or generalization/specialization – *gen-spec* –) is a relation between classes (or classifiers, in a more generic sense) that implies a taxonomic relation, and its subsequent inheritance semantics. This notion has been studied in the field of object-oriented programming and design under the name of *type/subtype* relation, and the essential property has been considered to be the *subtype requirement* [8], according to which if $f(x)$ is a property provable about objects x of type T, then $f(y)$ should be true for objects y of type S, where S is an specialization of T (this is essentially the same interpretation underlying *subsumption* [2]). This basic assumption appears in one form or another in all the above mentioned fields. Without breaking that requirement, some approaches allow a specialization to have an empty set of extensions. For example, the DAML+OIL ontology markup language [7] allows the subclass relation between classes to be acyclic (while the RDF [14] language do not), providing a way to assert class equality, but this can be considered an extreme case.

The common understanding of the '*is-a*' relation considers it as '*all-or-nothing*', in the sense that the relation is equally strong between a class and any of its sub-classes, and also at every level of the hierarchy. This assumption is in many cases an oversimplification of the psychological account of the real-world relations we are modelling. In other words, some sub-classes can be considered to be closer to a given super-class than others. As a somewhat extreme example, let's suppose we have a hierarchy rooted in the mammal class, with sub-classes domestic-cat, and primate, and siamese-cat as a subclass of domestic-cat. We can (subjectively) consider that the first specialization level represents a bigger step than the second, and that the distance from the abstract mammal category to primate is somewhat shorter than its distance to domestic-cat[1] (in the sense that the latter is a more specific category, while the former is still rather abstract). Most current *gen-spec* semantics simply neglect this fact, resulting in a subtle problem of *epistemological adequacy* (using the term in the sense given in [9]).

As a second example, let's suppose a Web shopping recommender agent is operating on the UNSPSC[2] product ontology to recommend items to the user according to a set of products the user is believed to like. If a user has a preference on the class Pianos, the agent may crawl to its direct superclass Keyboard-instruments, and then try to show related products by going down in the hierarchy to Musical-organs and Accordions. The *distance* at that level can be considered short, in comparison to crawling the generalization level from Keyboard-instruments to its superclass Musical-instruments, and in turn, this latter level is somewhat at a shorter distance than the one from Musical-instrument to the general category represented by the class Musical Instruments, Recreational Equipment, Supplies and Accessories (which include disparate products categories like fitness equipment and toys). In consequence, the agent would tend to crawl shorter levels more than larger ones, since the recommendation proximity of the products decrease as distance from the superclass increases.

A third example is the class Bulldozer in the *Cyc Transportation Ontology*[3], which is a subclass of the classes named RoadWork-Vehicle and Transportation-

---

[1] Obviously, this example is an oversimplification of the *mammals* taxonomy, but analogous cases are very often found in other contexts.

[2] See http://www.unspsc.org

[3] A modified version of the Cyc's taxonomy of transportation devices has been used as a case study for the approach described in this paper. It's available at http://opencyc.sourceforge.net/daml/cyc-transportation.daml

`Device-Vehicle`. Intuitively, it's clear that the first superclass is closer to the class than the latter, and thus, an agent would decide to take the first path – before the second – in a shallow reasoning process.

The just presented notion of 'graded' specializations has been somewhat addressed *at the instance level* in fuzzy conceptual modelling, by constraining the membership grades of a *fuzzy subclass* [5], with a threshold that represent the minimum distance from the superclass. But it would be more convenient to separate it clearly from that notion of fuzzy subclass and deal with it *at the class-level*, to enrich conceptual definitions with semantics that can be used with no regard to the instance level.

In this work, we describe the semantics of an approach to generalization in ontologies that allows the definition of graded specializations at the class (or term) level, measuring *how much* a classifier specializes a more general concept (that is, a *distance* notion). In Section 2, the problem is described, and resemblance relations are proposed as a measure for closeness between classes in classifier hierarchies. Section 3 describes how these fuzzy resemblance measures can be integrated in a modern ontology definition language and sketches a simple scenario that takes advantage from them. Finally, conclusions and future work are provided in Section 4.

2 Resemblance as a Metric for Specialization Distance

We'll denote a generalization relationship (or link) between two classifiers in the universal set C of classifiers as (informally) defined in (1).

$$a \succ^d b \quad a,b \in C \tag{1}$$

$$d = \{\phi_i(a,y) \mid a \succ^d y \land y \in C\} \tag{2}$$

The discriminator d determines the taxonomic criterion that justifies the relationship, and can be represented in the most general case by a set of predicates (2) – one for each direct specialization – that determines the specific properties of the instances of each subclass. Each of the predicates ϕ_i characterize one of the subclasses discriminated, and these characterizations can be expressed (for example, discriminating by ranges, 'a viola is a *alto* string-instrument', while 'a violin is a *treble* string instrument', and so on with the cello - *tenor* - and the bass). Note that on a single specific (super-)class, an arbitrary number of discriminators $d_1, d_2, \ldots, d_n$ can be defined, corresponding to different specialization criteria.

Discriminators are considered to be predicates on the structure of the involved classifiers, but in practice, they are often denoted simply as a set of constants or labels. That is, an specific discriminator d_j is not described as a set of predicates, but as the simple enumeration of the subclasses that participate in the discrimination (3). We'll use this definition from here on, for simplicity's sake.

$$d_j = \{c_1, c_2, \ldots c_k\}, \; c_i \in C \tag{3}$$

This approach correspond to the way in which they are represented in the UML by the discriminator *meta-attribute* in *meta-class Generalization* [12].

For example, the *Cyc Transportation Ontology* includes the concepts of RoadVehicle-Electric, RoadVehicle-ICE (*internal-combustion-engine*), Bus-RoadVehicle, Automobile and Motorcycle, as subclasses of the class RoadVehicle. At least two discriminators originate the taxonomy at that level. The first and second subclasses can are clearly discriminated by *motor-type* (4), while the third, fourth and fifth are chiefly distinguished by the room they provide for passengers (5) (according to the ontology documentation), so that we have two discriminators that originate from RoadVehicle).

$$d_{motor-type} = \{\text{RoadVehicle} - \text{Electric}, \text{RoadVehicle} - \text{ICE}\} \qquad (4)$$

$$d_{room} = \{\text{Bus} - \text{RoadVehicle}, \text{Automobile}, \text{Motorcycle}\} \qquad (5)$$

Given a classifier, its specialization links to its direct subclasses are divided in disjoint sets (partitions), according to their discriminators. Each partition represents an orthogonal dimension of specialization, and as such, should be handled separately. P denotes the set of (local) partitions of a model or ontology defined on a set C of classifiers (6). For example, $p_{(motor-type, \text{RoadVehicle})} = \{\text{RoadVehicle} - \text{Electric}, \text{RoadVehicle} - \text{ICE}\}$.

$$P = \{p_{(d,a)} \mid a \in C\} \ \ where \ p_{(d,a)} = \{c \mid c \in C \wedge a \succ^d c\} \qquad (6)$$

Mechanisms can be devised for which an estimation of resemblance measures can be automatically obtained [13]. For example, the number of varying structural features from class to subclass could be used as a metric [1]. But all these mechanisms are inherently flawed in practical settings since common conceptual models have incomplete contextual information items that often are not encoded in the model itself (i.e. hierarchy modelers abstract many details that would be needed for the measure of specialization distance, since many specializations seems obvious for them). Therefore, for practical reasons, it would ultimately be required that a human (the modeler or even the users of the model) gives an assessment of specialization measures. But humans find it difficult to give such measures in a global way, since distance is a relative concept.

We have designed a number of small experiments (using the above mentioned ontologies) to gather some evidence about how people tend to assess the relative distance between classes and subclasses. Results have leaded us to sketch an approach for the task, which involves human assessments at two levels:

- At a *micro-level*, in which the distance between a class and its subclasses for a specific discriminator (i.e. for a specific partition $p \in P$) is assessed.
- At a *sub-tree level*, in which distances (obtained at the micro-level) inside a hierarchy tree including descendants of a given classifier are somehow '*harmonized*' (we do not cover this level in detail in this paper).

Due to the subjective nature of such assessments, a large population would be required to come up with a statistically reliable measure, but the acquisition process is outside of the scope of the present work. In order to represent assessments, we have used *resemblance relations* to model specialization distance (note that resemblance or similarity relations can be used also as general semantic relationships that are not related with the subtype requirement, but they're not considered here). A resemblance relation R on a crisp domain D is a binary fuzzy relation (7).

$$R : D \times D \rightarrow [0 \ldots 1] \tag{7}$$

which satisfies reflexive (8) and symmetric (9) properties.

$$R(x,x) = 1 \ \forall x \in D \tag{8}$$

$$R(x,y) = R(y,x) \ \forall x,y \in D \tag{9}$$

Given this definition, a separate partial resemblance relation R can be obtained (from micro-level assessments) locally for each partition of subclasses, so that we operate on a set of relations (10) in the form (11).

$$\Pi_D = \bigcup_{x \in P} R_x \tag{10}$$

$$R_x : p_{(d,c)} \cup \{c\} \times p_{(d,c)} \cup \{c\} \rightarrow [0 \ldots 1] \tag{11}$$

Relations are labelled partial since they only contain class-subclass relationships, that is, relations are really defined in the form $R_x : \{c\} \times p_{(d,c)} \rightarrow [0 \ldots 1]$, i.e. from a specified super-class to all its subclasses that are discriminated by an specific d (although this can easily be extended to siblings: a complete resemblance relation could be derived for each local partition by the properties of the resemblance relation). This enables a form of stepwise simple reasoning in which concepts at hierarchy level i can be substituted with the closest concept in the $i \pm 1$ level traversing *gen-spec* relations through the different discriminators.

The partial resemblance relations are obtained by directly taking the resemblance grades obtained experimentally from *micro-level* assessments with a sample population knowledgeable in the domain (which requires the conversion of human relative distance assessments – like '*subclass B is rather closer from superclass A than subclass C*' – to numerical values in the unit interval[4]). Once the local resemblance structured, subtree-level assessments can be used to adjust resemblance values entire subtrees of the generalization hierarchy, using only a discriminator for each super-class. This approach only works for limited-size subtrees, since it's necessary to set comparison between all the elements in a set where at least one generalization link is included for each of the partitions.

---

[4] We have not already mathematically formalized such a procedure, but it'd possible to do so.

3 Extending Semantic Markup Languages for Specialization Distance

We have specified an extension to DAML+OIL to encode resemblance relations within RDF files. The concept of discriminator was added to the language, along with a way to encode local resemblance relations. The following example RDF fragment sketches the essentials of this extension, which uses a higher-order statement [14] called withDiscriminator and withResemblance about subClassOf statements.

```
<daml:Class rdf:ID="A">
  <rdfs:subClassOf rdf:resource="#B">
      <ext:withDiscriminator rdf:ID="d1"/>
      <ext:withResemblance grade="0.8"/>
  </rdfs:subClassOf>
</daml:Class>

<daml:Class rdf:ID="K">
  <rdfs:subClassOf rdf:resource="#B">
      <ext:withDiscriminator rdf:ID="#d1"/>
      <ext:withResemblance grade="0.7"/>
  </rdfs:subClassOf>
  <rdfs:subClassOf rdf:resource="#D">
      <ext:withDiscriminator rdf:ID="#d2"/>
      <ext:withResemblance grade="very low"/>
  </rdfs:subClassOf>
</daml:Class>

<ext:HarmonizedHierarchy rdf:ID="h1">
<ext:fromClass rdf:ID="#B">
<ext:discriminants>
  <rdf:Bag>
    <rdf:li resource="#d1"/>
    <rdf:li resource="#d3"/>
  </rdf:Bag>
</ext:discriminants>
<ext:HarmonizedHierarchy>
```

The example shows that either numeric values or labels in ordered label sets (that could be defined as *XML Schema* dataypes [15]) could be used for the resemblance values. Note also that harmonized hierarchies are explicitly encoded by specifying the root class, and the set of discriminators that have been assessed.

As a case study, a simple personalized recommender agent has been developed, that uses resemblance relations in the filtering process inside a Web application about products. More specifically, the agent takes content items from a database of existing ones for user U that has previously demonstrated interest in a set of classifiers C_U.

The items are annotated by terms in the agent's internal ontology. The agent searches for items that match C_U (and that have not been previously visited by U), taking into account the extended annotations by using resemblance to direct subclass or superclasses as a partial match in the absence of full matches. Figure 1 shows an example of the prototype, where the left frame shows changing links about UNSPSC categories, ordered by relevance for the specific user (this is commonly referred to as *adaptive sorting* in adaptive hypermedia research [3]), and taking into account resemblance in filtering related items.

Fig. 1. Overall layout of the Web prototype that uses resemblance relations to navigate product categories.

For example, if *piano* $\in C_U$ then the agent has two options to generate recommendations from the *gen-spec* hierarchy:

- Going 'up' to the more abstract term Keyboard-instruments
- Or going 'down' to specializations of piano (like spinet, console or grand pianos).

The agent would choose first the shorter distance (i.e. the larger resemblance). In the example, it would choose going 'down' (and sort the piano subclasses by descending resemblance), since the way up represents a bigger step, due to the highly abstract nature of the term Keyboard-instruments.

This behavior may prevent the agent to jump to "excessively" abstract categories (e.g. reaching the awkward Musical Instruments, - Recreational Equipment, Supplies and Accessories class mentioned above).

Other resemblance-filtering schemes can be implemented and studied, and existing ontologies can be easily annotated to support this approach, but we have found that ontology-level overall relations are difficult to acquire from users.

4 Conclusions and Future Work

Current approaches to generalization in conceptual modelling and ontology engineering lack the notion of distance between classes and subclasses. This notion, in many cases, cannot be computed from the structural characteristics of the ontology (or model) due, for example, to incompleteness or ill-defined hierarchies.

We have described the notion of distance from a class to its subclasses, and how it can be represented through partial local resemblance (fuzzy) relations, and added to current RDF-based ontology description languages. This distance notion can be assessed and applied easily to filtering processes, but further empirical user testing is needed to come up with a measure of its impact from the user-interaction perspective. Our approach can be considered as complementary to fuzzy subtyping schemes [4], in which uncertainty and/or partial truth about types of objects is considered.

Further research should address additional properties of *gen-spec* hierarchies, for example, the relationship between discriminators at various levels of the hierarchy, or the implications of the notion of distance in '*sibling*' classes inside the ontology.

References

1. AlGhamdi J, Elish M, Ahmed M (2002) A tool for measuring inheritance coupling in object-oriented systems. Information Sciences, 140(3-4):217–227
2. Beierle C (1995) Type inferencing for polymorphic order-sorted logic programs. In: Sterling L (editor) Proceedings of the Twelfth International Conference on Logic Programming. Springer-Verlag, Lecture Notes in Computer Science 2401:765–780
3. Brusilovsky P (2001) Adaptive hypermedia. User Modeling and User Adapted Interaction 11(1–2):87–110
4. Cao T H, Creasy P N (2000) Fuzzy types: a framework for handling uncertainty about types of objects. International Journal of Approximate Reasoning, Elsevier Science, 25(3):217–253
5. Chen G (1998) Fuzzy logic in data modeling: semantics, constraints, and database design. Kluwer Academic Publishers
6. Guarino N, Welty C (2000) Ontological analysis of taxonomic relationships. In: Laender A, Storey V (editors) Proceedings of the 19th International Conference on Conceptual Modeling. Springer-Verlag, Lecture Notes in Computer Science 1920: 210–224
7. Horrocks I (2002) DAML+OIL: a reason-able Web ontology language. In: Jensen, C S et al. (eds.) Proceedings of the 8th International Conference on Extending Database Technology. Springer-Verlag Lecture Notes in Computer Science 2287: 2–13
8. Liskov B, Wing J M (1994) A behavioral notion of subtyping. ACM Transactions on Programming Languages and Systems 16(6):1811–1841
9. McCarthy J, Hayes P (1969) Some philosophical problems from the standpoint of artificial intelligence. Machine Intelligence 4:463–502

10. Meinke K, Tucker J V (1993) Many-sorted logic and its applications. Wiley, Chichester
11. Norrie M C, Reimer U, Lippuner P, Rys M, Schek H J (1994) Frames, objects and relations: three semantic levels for knowledge-based systems. In: Baader F et al. (eds.) Proceedings of the Workshop on Reasoning about Structured Objects: Knowledge Bases meets Databases, CEUR Workshop Proceedings
12. Object Management Group (OMG) (2001) The unified modeling language specification, version 1.4, available at http://www.uml.org
13. Spanoudakis G, Constantopoulos P (1994) Similarity for analogical software reuse: a computational model. In: Proceedings of the 11th European Conference on Artificial Intelligence(ECAI '94),John Wiley and Sons: 18–22
14. World Wide Web Consortium (W3C) (1999) Resource Description Framework (RDF) Model and Syntax. W3C Recommendation, 22 February 1999, available at http://www.w3.org/TR/1999/REC-rdf-syntax-19990222
15. World Wide Web Consortium (W3C) (2001) XML Schema Part 2: Datatypes. W3C Recommendation, 2 May 2001, available at http://www.w3.org/TR/xmlschema-2/

Kernel-based Face Recognition by a Reformulation of Kernel Machines

Pablo Navarrete and Javier Ruiz del Solar

Department of Electrical Engineering,
Center for Web Research - Department of Computer Science Universidad de Chile.
Av. Tupper 2007, Santiago - CHILE
{pnavarre, jruizd}@cec.uchile.cl

Summary. Until now, linear methods like PCA and FLD have been widely tested for solving the problem of face recognition, showing good recognition rates. In this paper a general framework is going to be introduced in order to solve these problems using kernels, i.e. applying the same linear methods in high dimensional spaces. For this purpose a general solution of kernel machines is obtained in a different way to the current approaches. As a result of this process the problem of KFD, originally solved for two-class problems, is solved for an arbitrary number of classes, so that it becomes applicable for face recognition. Simulations are performed using a small face database (Yale Face Database) and a large face database (FERET).

Key words: Face Recognition, Kernel-PCA, KFD, Multiclass-KFD

1 Introduction

Since the development of Support Vector Machine (SVM) [13] as an optimal classifier for two-class problems, a great interest has been generated in the method used to generalize the linear decision rule to non-linear ones, i.e. using kernel functions. Linear methods like Principal Components Analysis (PCA) [8] and Fisher Linear Discriminant (FLD) [3] has been formulated using kernels. The extension of linear methods to non-linear ones, using the so-called kernel trick, is what we call kernel machines.

The generalization to non-linear methods using kernels works as follows: if the algorithm to be generalized uses the training vectors only in the form of Euclidean dot-products, then all the dot-products like $\mathbf{x}^T\mathbf{y}$, can be replaced by a so-called kernel function $K(\mathbf{x},\mathbf{y})$. If $K(\mathbf{x},\mathbf{y})$ fulfills the Mercer's condition, i.e. the operator K is semi-positive definite [1], then the kernel can be expanded into a series $K(\mathbf{x},\mathbf{y}) = \sum_{i=1}^{M} \phi_i(\mathbf{x})\phi_i(\mathbf{y})$. In this way the kernel represents the Euclidean dot-product on a different space, called feature space $\mathcal{F}$, in which the original vectors are mapped using the eigenfunctions: $\Phi : \mathbb{R}^N \rightarrow \mathcal{F}$. Depending on the kernel function, the feature space $\mathcal{F}$ can be even of infinite dimension, as the case of Radial Basis

Function (RBF) kernel, but we are never working in such space. If the kernel function does not fulfills the Mercer's condition the problem probably can still be solved, but the geometrical interpretation and its associated properties will not apply in such case.

The article is structured as follows. In section 2, the solution of KPCA and multiclass-KFD kernel machines is shown, using a general methodology. In section 3, a toy experiment is shown in order to test the novel result of multiclass-KFD, and then KPCA and multiclass-KFD are used to solve the problem of face recognition. Finally, in section 4 some conclusions are given.

2 General Solution of Kernel Machines

2.1 Kernelization Strategy

Fundamental Correlation Problem - FCP

The most common way in which kernel machines has been obtained [8] [3] is based on the results of Reproducing Kernel Hilbert Spaces (RKHS) [6], which establishes that any vector in $\mathcal{F}$ that have been obtained from a linear system using the training vectors, must lie in the span of all the training samples mapped in $\mathcal{F}$. Although this step can be used to obtain many kernel machines, it does not give a general solution for them. For instance, in the case of Fisher Discriminant there is a restriction for working only with two classes.

We are going to show that a general solution of kernel machines can be obtained using the so-called Fundamental Correlation Problem (FCP). For this purpose, first we must consider the mapped vectors $\Phi^i = \Phi(\mathbf{x}^i) \in \mathcal{F}$ $i = 1, \ldots, NV$, were the components of Φ corresponds to the M eigenfunctions of a given kernel (i.e. $M = \dim(\mathcal{F})$), and $\Phi = \left[\Phi(\mathbf{x}^1) \cdots \Phi(\mathbf{x}^{NV}) \right]$ is the matrix of mapped vectors. Thereafter, the FCP consists in solving the eigensystem for the correlation matrix $\mathbf{R} = \frac{1}{NV-1} \Phi\Phi^T \in \mathcal{M}^{M \times M}$, or any matrix that can be written in this form that we called the correlation decomposition. Specifically, we do not need the whole matrix of eigenvalues $\Lambda_R \in \mathcal{M}^{M \times M}$ but only the matrix of non-zeros eigenvalues $\widetilde{\Lambda}_R \in \mathcal{M}^{q \times q}$ (where $q \leqslant NV - 1$ is the number of non-zeros eigenvalues), neither the whole matrix of eigenvectors $\mathbf{W}_R = \left[\mathbf{w}_R^1 \cdots \mathbf{w}_R^M \right]$ but only the matrix of eigenvectors associated with non-zeros eigenvalues $\widetilde{\mathbf{W}}_R = \left[\mathbf{w}_R^1 \cdots \mathbf{w}_R^q \right]$.

As $\mathcal{F}$ has a high dimensionality, the FCP cannot be directly solved, i.e. solved in its primal form. This problematic is well known in Principal Component Analysis (PCA) when the dimensionality of the feature vectors is higher than the number of vectors [2]. Moreover, the solution of this problem is well-known in applications like Face Recognition [12] [4], and is also the key of the formulation of Kernel-PCA [8] [11]. In summary we change the primal problem for the dual problem, in which we solved the eigensystem of the inner-product matrix $\mathbf{K}_R = \frac{1}{NV-1} \Phi_R^T \Phi_R \in \mathcal{M}^{NV \times NV}$. The NV eigenvalues of $\mathbf{K}_R$ are equal to a subset of NV eigenvalues of $\mathbf{R}$, including all its non-zeros eigenvalues. Then, $\widetilde{\mathbf{W}}_R$ can be obtained from the eigenvectors of

$\mathbf{K}_R$ associated with its non-zeros eigenvalues $\widetilde{\mathbf{V}}_R = \left[\mathbf{v}_R^1 \cdots \mathbf{v}_R^q \right]$, using the following equation [2]:

$$\widetilde{\mathbf{W}}_R = \frac{1}{\sqrt{NV-1}} \Phi \widetilde{\mathbf{V}}_R \widetilde{\Lambda}_R^{-1/2}. \tag{1}$$

Expression (1) explicitly shows that the set of vectors $\widetilde{\mathbf{W}}_R$ lie in the span of the training vectors Φ, in accordance with the theory of reproducing kernels [6]. Even if $\widetilde{\mathbf{W}}_R$ cannot be computed, the projection of a given mapped vector $\Phi(\mathbf{x}) \in \mathcal{F}$ onto the subspace spanned by $\widetilde{\mathbf{W}}_R$, i.e. $\widetilde{\mathbf{W}}_R^T \Phi(\mathbf{x})$, can be computed using kernels. However, it must be noted that this requires a sum of NV inner products in the feature space $\mathcal{F}$ that could be computationally very expensive if NV is a large number.

The Statistical Representation Matrix - SRM

The methods in which we are focusing our study, can be formulated as the maximization or minimization (or a mix of both) of positive objective functions that have the following general form:

$$f(\mathbf{w}) = \frac{1}{NV-1} \sum_{n=1}^{NS} \left\{ (\Phi, \mathbf{b}_E^n)^T \mathbf{w} \right\}^2 = \mathbf{w}^T \frac{\Phi \mathbf{B}_E \mathbf{B}_E^T \Phi^T}{NV-1} \mathbf{w} \tag{2}$$

$$= \mathbf{w}^T \frac{\Phi_E \Phi_E^T}{NV-1} \mathbf{w} = \mathbf{w}^T \mathbf{E} \mathbf{w} ,$$

where $\mathbf{B}_E = \left[\mathbf{b}_E^1 \cdots \mathbf{b}_E^{NS} \right] \in \mathcal{M}^{NV \times NS}$ is the so-called Statistical Representation Matrix (SRM) of the estimation matrix $\mathbf{E}$, NS is the number of statistical measures (e.g. $NS = NV$ for the correlation matrix), and $\Phi_E = \Phi \mathbf{B}_E \in \mathcal{M}^{M \times NS}$ forms the correlation decomposition of $\mathbf{E}$.

Note that $f(\mathbf{w})$ represents the magnitude of a certain statistical property in the projections on the $\mathbf{w}$ axis. This statistical property is estimated by NS linear combinations of the mapped vectors $\Phi \mathbf{b}_E^n$, $n = 1, \ldots, NS$. Then, it is the matrix $\mathbf{B}_E$ that defines the statistical property, and is independent from the mapped vectors Φ. Therefore, the SRM is going to be useful in order to separate the dependence on mapped vectors from estimation matrices. Examples of SRMs are going to be shown in the following sub-sections.

2.2 Kernel Principal Component Analysis - KPCA

In this problem, the objective function, to be maximized, represents the projection variance:

$$\sigma^2(\mathbf{w}) = \frac{1}{NV-1} \sum_{i=1}^{NV} \left\{ (\Phi^i - \mathbf{m})^T \mathbf{w} \right\}^2 = \mathbf{w}^T \frac{\Phi_C \Phi_C^T}{NV-1} \mathbf{w} = \mathbf{w}^T \mathbf{C} \mathbf{w} , \tag{3}$$

where $\mathbf{m} = \frac{1}{NV} \sum_{n=1}^{NV} \Phi^n$ is the mean mapped vector, and $\mathbf{C}$ is the covariance matrix. Then it is simple to obtain:

$$\Phi_C = \left[\left(\Phi^1 - \mathbf{m} \right) \cdots \left(\Phi^{NV} - \mathbf{m} \right) \right] \in \mathcal{M}^{M \times NV} , \tag{4}$$

$$(\mathbf{B}_C)_{ij} = \delta_{ij} - \frac{1}{NV} , \quad i = 1, \ldots, NV; \; j = 1, \ldots, NV , \tag{5}$$

where δ_{ij} is the Kronecker delta.

It is well-known that the maximization of (3) is obtained by solving the FCP of $\mathbf{C}$ for non-zeros eigenvalues. Then, we can directly write the solution by using expression (1):

$$\widetilde{\mathbf{W}}_C = \frac{1}{\sqrt{NV-1}} \, \Phi \, \mathbf{B}_C \, \widetilde{\mathbf{V}}_C \, \widetilde{\Lambda}_C^{-1/2} . \tag{6}$$

As in (1), (6) shows that the set of vectors $\widetilde{\mathbf{W}}_C$ lies in the span of the training vectors Φ, but in this case this is due to the presence of the SRM $\mathbf{B}_C$.

2.3 Kernel Fisher Discriminant - KFD

In this problem the input vectors (and mapped vectors) are distributed in NC classes, in which the class number i has n_i associated vectors. We denote $\Phi^{(i,j)}$ the mapped vector number j ($1 \leqslant j \leqslant n_i$) in the class number i ($1 \leqslant i \leqslant NC$). Then we have two objective functions:

$$s_b(\mathbf{w}) = \frac{1}{NV-1} \sum_{i=1}^{NC} n_i \left\{ \left(\mathbf{m}^i - \mathbf{m} \right)^{\mathrm{T}} \mathbf{w} \right\}^2 , \tag{7}$$

$$s_w(\mathbf{w}) = \frac{1}{NV-1} \sum_{i=1}^{NC} \sum_{j=1}^{n_i} \left\{ \left(\Phi^{(i,j)} - \mathbf{m}^i \right)^{\mathrm{T}} \mathbf{w} \right\}^2 , \tag{8}$$

where $\mathbf{m} = \frac{1}{NV} \sum_{n=1}^{NV} \Phi^n$ is the mean mapped vector, and $\mathbf{m}^i = \frac{1}{n_i} \sum_{j=1}^{n_i} \Phi^{(i,j)}$ is the class mean number i. The problem consists in maximizing $\gamma(\mathbf{w}) = s_b(\mathbf{w})/s_w(\mathbf{w})$, so that the separation between the individual class means respect to the global mean (7) is maximized, and the separation between mapped vectors of each class respect to their own class mean (8) is minimized. As we want to avoid the problem in which $s_w(\mathbf{w})$ becomes zero, a regularization constant μ can be added in (8) without changing the main objective criterion and obtaining the same optimal $\mathbf{w}$ [10]. Then, the estimation matrices of (7) and (8) are:

$$\mathbf{S}_b = \frac{1}{NV-1} \sum_{i=1}^{NC} n_i \left(\mathbf{m}^i - \mathbf{m} \right) \left(\mathbf{m}^i - \mathbf{m} \right)^{\mathrm{T}} = \frac{1}{NV-1} \Phi_b \Phi_b^{\mathrm{T}} , \tag{9}$$

$$\mathbf{S}_w = \frac{1}{NV-1} \sum_{i=1}^{NC} \sum_{j=1}^{n_i} \left(\Phi^{(i,j)} - \mathbf{m}^i \right) \left(\Phi^{(i,j)} - \mathbf{m}^i \right)^{\mathrm{T}} = \frac{1}{NV-1} \Phi_w \Phi_w^{\mathrm{T}} . \tag{10}$$

The correlation decomposition in (9) and (10) is obtained using:

$$\Phi_b = \left[\sqrt{n_1} \left(\mathbf{m}^1 - \mathbf{m} \right) \cdots \sqrt{n_{NC}} \left(\mathbf{m}^{NC} - \mathbf{m} \right) \right] \in \mathcal{M}^{M \times NC}, \tag{11}$$

$$\Phi_w = \left[\left(\Phi^1 - \mathbf{m}^{C_1} \right) \cdots \left(\Phi^{NV} - \mathbf{m}^{C_{NV}} \right) \right] \in \mathcal{M}^{M \times NV}, \tag{12}$$

where we denote as C_i the class of the vector number i ($1 \leqslant i \leqslant NV$). Therefore, the corresponding SRMs so that $\Phi_b = \Phi \mathbf{B}_b$ and $\Phi_w = \Phi \mathbf{B}_w$ are:

$$\begin{aligned}
(\mathbf{B}_b)_{ij} &= \sqrt{n_j} \left(\tfrac{1}{n_j} \delta_{C_i j} - \tfrac{1}{NV} \right), \quad (\mathbf{B}_w)_{ij} = \delta_{ij} - \tfrac{1}{n_{C_i}} \delta_{C_i C_j}, \\
i &= 1, \ldots, NV; \, j = 1, \ldots, NC, \quad i = 1, \ldots, NV; \, j = 1, \ldots, NV.
\end{aligned} \tag{13}$$

where δ_{ij} is the Kronecker delta.

It is well-known that the solution of the Fisher discriminant is obtained by solving the general eigensystem:

$$\begin{aligned}
\mathbf{S}_b \mathbf{w}^k &= \lambda_k \mathbf{S}_w \mathbf{w}^k, \quad \mathbf{w}^k \in \mathbb{R}^N, k = 1, \ldots, N, \\
\|\mathbf{w}^k\| &= 1.
\end{aligned} \tag{14}$$

with $\gamma_k = s_b(\mathbf{w}^k)/s_w(\mathbf{w}^k)$, and $\mathbf{S}_b$ and $\mathbf{S}_w$ the scatter matrices defined in (9) and (10).

Unfortunately, this system cannot be directly solved because of the high dimensionality of $\mathcal{F}$. The problem was originally introduced in [3], where it has been solved using kernels, i.e. not working explicitly in $\mathcal{F}$. The main step of that formulation was to use the fact that the solution can be written as linear combinations of the mapped vectors [3]. In this way a reduced general eigensystem can be obtained, but it was necessary to constrain the problem to two classes only.

Using the concepts introduced in section 2.1, we are going to solve the KFD for an arbitrary number of classes. This method is an adaptation of the solution for Fisher Linear Discriminants (FLDs) shown in [10], using FCPs. The key of this solution is to solve two problems instead of one, using the properties of (14).

The solution of the general eigensystem (14), is formed by the eigenvectors of $\mathbf{S}_w^{-1} \mathbf{S}_b$. But, as the rank of $\mathbf{S}_w \in \mathcal{M}^{M \times M}$ is less or equal to $NV - NC$, $\mathbf{S}_w$ is always singular. To avoid this problem, we force $\mathbf{S}_w$ to be invertible by adding a regularization matrix $\mu \mathbf{I}$ to it. Then, in order to solve the Fisher discriminant in the feature space $\mathcal{F}$, we are going to follow the procedure stated in [10] using FCPs for the solution of each problem. This procedure is motivated by the fact that the solution of (14) diagonalizes both matrices $\mathbf{S}_b$ and $\mathbf{S}_w$. Therefore, the following steps must be executed:

- First, in order to diagonalize $\mathbf{S}_w$, we solve the FCP of $\mathbf{S}_w$ for which we know its correlation decomposition (10). In this way we obtain the non-zeros eigenvalues $\Lambda_w \in \mathcal{M}^{p \times p}$ with $p \leqslant (NV - NC)$, and their associated eigenvectors, by using expression (1):

$$\widetilde{\mathbf{W}}_w = \frac{1}{\sqrt{NV-1}} \Phi \mathbf{B}_w \widetilde{\mathbf{V}}_w \widetilde{\Lambda}_w^{-1/2}. \tag{15}$$

- Next, in order to diagonalize $\mathbf{S}_b$, maintaining the diagonalization of $\mathbf{S}_w$, we are going to diagonalize an "hybrid" matrix:

$$\mathbf{H} = \left(\mathbf{W}_w \check{\Lambda}_w^{-1/2}\right)^{\mathrm{T}} \mathbf{S}_b \left(\mathbf{W}_w \check{\Lambda}_w^{-1/2}\right) = \mathbf{W}_{\mathrm{H}} \, \Lambda_{\mathrm{H}} \, \mathbf{W}_{\mathrm{H}}^{\mathrm{T}} \quad \in \mathcal{M}^{M \times M}, \tag{16}$$

where $\check{\Lambda}_w = \Lambda_w + \mu \mathbf{I}$ is the regularized matrix of eigenvalues of $\mathbf{S}_w$, so that it becomes invertible. Note that all the eigenvalues and eigenvectors of $\mathbf{S}_w$ are used, since the smallest eigenvalues of $\mathbf{S}_w$ are associated with the largest general eigenvalues γ_k. This fact is essential in order of this solution to be valid for kernels functions with infinite eigenfunctions (e.g. RBF kernel).

As we know the correlation decomposition of $\mathbf{S}_b$ (9), we can write the correlation decomposition of $\mathbf{H}$ as:

$$\mathbf{H} = \tfrac{1}{NV-1} \underbrace{\left(\check{\Lambda}_w^{-1/2} \mathbf{W}_w^{\mathrm{T}} \, \Phi \mathbf{B}_b\right)}_{\Phi_{\mathrm{H}} \,\in\, \mathcal{M}^{M \times NC}} \underbrace{\left(\check{\Lambda}_w^{-1/2} \mathbf{W}_w^{\mathrm{T}} \, \Phi \mathbf{B}_b\right)^{\mathrm{T}}}_{\Phi_{\mathrm{H}}^{\mathrm{T}} \,\in\, \mathcal{M}^{NC \times M}} \tag{17}$$

Solving the Dual FCP of $\mathbf{H}$, we obtain its non-zeros eigenvalues $\widetilde{\Lambda}_{\mathrm{H}} \in \mathcal{M}^{q \times q}$, with $q \leqslant (NC - 1)$, and its associated eigenvectors by using expression (1):

$$\widetilde{\mathbf{W}}_{\mathrm{H}} = \tfrac{1}{\sqrt{NV-1}} \, \Phi_{\mathrm{H}} \widetilde{\mathbf{V}}_{\mathrm{H}} \widetilde{\Lambda}_{\mathrm{H}}^{-1/2} = \tfrac{1}{\sqrt{NV-1}} \check{\Lambda}_w^{-1/2} \mathbf{W}_w^{\mathrm{T}} \Phi \mathbf{B}_b \widetilde{\mathbf{V}}_{\mathrm{H}} \widetilde{\Lambda}_{\mathrm{H}}^{-1/2}. \tag{18}$$

In order to solve the Dual FCP of $\mathbf{H}$, we need to compute $\mathbf{K}_{\mathrm{H}} = \tfrac{1}{\sqrt{NV-1}} \Phi_{\mathrm{H}}^{\mathrm{T}} \Phi_{\mathrm{H}}$. As all the eigenvectors and eigenvalues of $\mathbf{S}_w$ are considered, we need to find some way of using only the ones associated with non-zeros eigenvalues. As $\mathbf{W}_w$ is an orthonormal matrix, i.e. $\mathbf{W}_w^{\mathrm{T}} \mathbf{W}_w = \mathbf{W}_w \mathbf{W}_w^{\mathrm{T}} = \mathbf{I}$, the projection matrices $\mathbf{W}_w^0 \left(\mathbf{W}_w^0\right)^{\mathrm{T}}$ (with $\mathbf{W}_w^0$ the matrix of $\mathbf{S}_w$'s eigenvectors associated with zeros eigenvalues) and $\widetilde{\mathbf{W}}_w \widetilde{\mathbf{W}}_w^{\mathrm{T}}$, fulfill the condition $\mathbf{W}_w^0 \left(\mathbf{W}_w^0\right)^{\mathrm{T}} = \mathbf{I} - \widetilde{\mathbf{W}}_w \widetilde{\mathbf{W}}_w^{\mathrm{T}}$. With this consideration, using some straightforward algebra, we obtain the following expression for computing $\mathbf{K}_{\mathrm{H}}$:

$$\mathbf{K}_{\mathrm{H}} = \mathbf{B}_b^{\mathrm{T}} \mathbf{K}_{\mathrm{R}} \mathbf{B}_w \widetilde{\mathbf{V}}_w \widetilde{\Lambda}_w^{-1/2} \left\{ \left(\widetilde{\Lambda} + \mu \mathbf{I}\right)^{-1} - \tfrac{1}{\mu} \mathbf{I} \right\} \widetilde{\Lambda}_w^{-1/2} \widetilde{\mathbf{V}}_w^{\mathrm{T}} \mathbf{B}_w^{\mathrm{T}} \mathbf{K}_{\mathrm{R}} \mathbf{B}_b$$
$$+ \tfrac{1}{\mu} \mathbf{B}_b^{\mathrm{T}} \mathbf{K}_{\mathrm{R}} \mathbf{B}_b \tag{19}$$

with $\mathbf{K}_{\mathrm{R}} = \tfrac{1}{NV-1} \Phi^{\mathrm{T}} \Phi$ as introduced for the FCP of the correlation matrix $\mathbf{R}$.

- Finally, from (16) is easy to see that the matrix

$$\widetilde{\mathbf{W}}_{\mathrm{KFD}} = \mathbf{W}_w \check{\Lambda}_w^{-1/2} \widetilde{\mathbf{W}}_{\mathrm{H}} \tag{20}$$

diagonalizes both matrices $\mathbf{S}_b$ and $\mathbf{S}_w$, as well as $\mathbf{S}_w^{-1} \mathbf{S}_b$ with $\widetilde{\Lambda}_{\mathrm{H}}$ the matrix of non-zeros general eigenvalues, and therefore it solves the problem. Then, replacing previous results, we obtain the complete expression for computing $\widetilde{\mathbf{W}}_{\mathrm{KFD}}$:

$$\widetilde{\mathbf{W}}_{\text{KFD}} = \mathbf{\Phi} \frac{1}{\sqrt{NV-1}} \left(\mathbf{B}_w \widetilde{\mathbf{V}}_w \widetilde{\Lambda}_w^{-1/2} \left\{ \left(\widetilde{\Lambda} + \mu \mathbf{I} \right)^{-1} - \frac{1}{\mu} \mathbf{I} \right\} \cdot \right.$$

$$\left. \widetilde{\Lambda}_w^{-1/2} \widetilde{\mathbf{V}}_w^{\text{T}} \mathbf{B}_w^{\text{T}} \mathbf{K}_{\text{R}} \mathbf{B}_b \widetilde{\mathbf{V}}_{\text{H}} \widetilde{\Lambda}_{\text{H}}^{-1/2} + \frac{1}{\mu} \mathbf{B}_b \widetilde{\mathbf{V}}_{\text{H}} \widetilde{\Lambda}_{\text{H}}^{-1/2} \right). \qquad (21)$$

As (1) and (6), (21) shows that the set of vectors $\widetilde{\mathbf{W}}_{\text{KFD}}$ lies in the span of the training vectors $\mathbf{\Phi}$. Then, due to the solution of the FCP of $\mathbf{S}_w$ and the FCP of $\mathbf{H}$, that are explicitly present in (21), we have solved the KFD for an arbitrary number of classes.

3 Simulations

3.1 Toy Experiment

In order to see how the multiclass-KFD works with more than two classes, in Figure 1(a) it is shown an artificial 2D-problem of 4 classes, that we solved using KFD with a RBF kernel, $K(\mathbf{x}, \mathbf{y}) = e^{-\|\mathbf{x}-\mathbf{y}\|^2/0.1}$, and a regularization parameter $\mu = 0.001$. Figure 1(b) shows the first feature found by KFD, in which we see that the vertical and horizontal classes become well separated from each other. Figure 1(c) shows the second KFD feature, in which the two diagonal classes become well separated from each other. Finally, Figure 1(d) shows the third KFD feature, in which the horizontal and vertical classes become well separated from the two diagonal classes. These three features are sufficient to discriminate between all the 4 classes. It also interesting to see that the RBF solution does not show a clear decision in the shared center area, but it tries to take the decision as close as it can. If we decrease the variance of the RBF kernel, the discrimination in this zone will be improved, but the discrimination far from this zone will worsen. Therefore, the kernel function must be adjusted in order to obtain better results.

3.2 Face Recognition

We have seen that, following the results of RKHS, the solution of kernel machines as KPCA and KFD can be written as $\mathbf{W} = \mathbf{\Phi} \mathbf{A}$, where $\mathbf{A}$ is the matrix of parameters for a given kernel machine. Specifically, form (6) and (21), we obtain the following matrix of parameters for KPCA and KFD:

$$\mathbf{A}_{\text{KPCA}} = \frac{1}{\sqrt{NV-1}} \mathbf{B}_{\text{C}} \widetilde{\mathbf{V}}_{\text{C}} \widetilde{\Lambda}_{\text{C}}^{-1/2}, \qquad (22)$$

$$\mathbf{A}_{\text{KFD}} = \frac{1}{\sqrt{NV-1}} \left(\mathbf{B}_w \widetilde{\mathbf{V}}_w \widetilde{\Lambda}_w^{-1/2} \left\{ \left(\widetilde{\Lambda} + \mu \mathbf{I} \right)^{-1} - \frac{1}{\mu} \mathbf{I} \right\} \cdot \right.$$

$$\left. \widetilde{\Lambda}_w^{-1/2} \widetilde{\mathbf{V}}_w^{\text{T}} \mathbf{B}_w^{\text{T}} \mathbf{K}_{\text{R}} \mathbf{B}_b \widetilde{\mathbf{V}}_{\text{H}} \widetilde{\Lambda}_{\text{H}}^{-1/2} + \frac{1}{\mu} \mathbf{B}_b \widetilde{\mathbf{V}}_{\text{H}} \widetilde{\Lambda}_{\text{H}}^{-1/2} \right). \qquad (23)$$

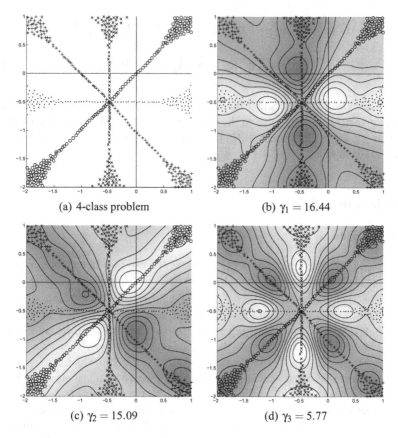

(a) 4-class problem   (b) $\gamma_1 = 16.44$

(c) $\gamma_2 = 15.09$ (d) $\gamma_3 = 5.77$

Fig. 1. (a) Artificial example of a 4-class problem. Features of the KFD using a RBF kernel: (b) first feature, (c) second feature, and (d) third feature.

Then, for a given vector $\mathbf{x}$, its correspondent feature vector using kernel machines is $\mathbf{p} = \Phi^T \Phi(\mathbf{x})$. This can be written as:

$$\mathbf{p} = \mathbf{A}^T \mathbf{k}, \tag{24}$$
$$(\mathbf{k})_i = \Phi(\mathbf{x}_i)^T \Phi(\mathbf{x}) = K(\mathbf{x}_i, \mathbf{x}) \qquad i = 1, \dots, NV. \tag{25}$$

In this notation, the vector $\mathbf{k} \in \mathbb{R}^{NV}$ represents the so-called kernel projection that is equal for any kind of kernel machine, and the vector $\mathbf{p}$ represents the feature vector for a specific kernel machine.

In Figure 2 we show a general framework in order to use the feature vectors of kernel machines to solve the problem of face recognition. A preprocessing module performs changes on input images prior to the face recognition process. In our experiments we used basically the same preprocesses introduced in [9] for face detection problems. The preprocess includes: window resizing (scale the face image, given fixed proportions, to obtain face images of 100×185), masking (to avoid border pixels not corresponding to the face in the image), illumination gradient compensation (substract a best-fit brightness plane to compensate heavy shadows caused by extreme lighting angles), histogram equalization (to spread the energy of all intensity values on the image), and normalization (to make all input images of the same energy). The result of the preprocess, $\mathbf{x} \in \mathbb{R}^N$, mapped onto the feature space, $\Phi(\mathbf{x}) \in \mathcal{F}$, is then projected on the support images, mapped in the feature space, using the kernel function $K : \mathbb{R}^N \times \mathbb{R}^N \to \mathbb{R}$. Therefore, the face database needs to be store as complete images. This is due to the fact that kernel machines need the image vectors in order to reproduce the eigenvectors in the feature space $\mathcal{F}$. The parameters of the given kernel machine, $\mathbf{A}^T \in \mathcal{M}^{NV \times m}$, is then applied on the kernel projection vector, $\mathbf{k} \in \mathbb{R}^{NV}$, in order to obtain the feature vector $\mathbf{p} \in \mathcal{R}^m$. The *Similarity Matching* module compares the similarity of the reduced representation of the query face vector $\mathbf{p}$ with the reduced vectors $\mathbf{p}^k$, $\mathbf{p}^k \in \mathbb{R}^m$, that correspond to reduced vectors of the faces in the database. Therefore, the face database needs also to be store as reduced vectors for the recognition process. By using a given criterion of similarity the *Similarity Matching* module determines the most similar vector $\mathbf{p}^k$ in the database. The class of this vector is the result of the recognition process, i.e. the identity of the face. In addition, a *Rejection System* for unknown faces is used if the similarity matching measure is not good enough.

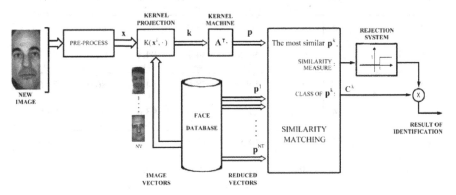

Fig. 2. General framework of a kernel-based face recognition system.

The main objective of similarity measures is to define a value that allows the comparison of feature vectors. With this measure the identification of a new feature vector will be possible by searching the most similar vector into the database. This is the well-known nearest-neighbor method. One way to define similarity is to use a measure of distance, $d(\mathbf{p}, \mathbf{p}^k)$, in which the similarity between vectors, $S(\mathbf{p}, \mathbf{p}^k)$ is inverse to the distance measure. In our experiments we used the Euclidean distance measure and two similarity measures: Cosine, cosine of the angle between $\mathbf{p}$ and $\mathbf{p}^k$, and FFC (Fuzzy Feature Contrast), a fuzzy similarity measure originally proposed in [7]. The same similarity measures are also applied after a so-called Whitening transformation, equivalent to a Mahalanobis distance if the Euclidean distance is used. More details of these similarity measures are given in previous studies of linear eigenspace-based methods [4] [5].

In our experiments we used a small face database (Yale Face Database) and a large face database (FERET). Table 1 and 2 show the recognition rates obtained when using PCA (Principal Component Analysis) and FLD (Fisher Linear Discriminant) features. Table 3 and 4 show the recognition rates obtained when using KPCA and multiclass-KFD features.

Table 1. Mean recognition rates using different numbers of training images per class, and taking the average over 20 different training sets. The small numbers are the standard deviation of each recognition rate. All results consider the top 1 match for recognition. Yale Database.

PROJECTION METHOD	IMAGES PER CLASS	AXES	EUCLIDEAN	COS	FFC	WHITENING EUCLIDEAN	WHITENING COS	WHITENING FFC
PCA	6	49	95.7 2.7	95.8 2.7	81.8 5.4	83.3 5.9	89.3 4.1	81.8 5.4
FLD		14	94.6 2.1	95.2 2.5	85.9 5.2	**97.2** **2.2**	97.0 2.5	90.8 5.5
PCA	5	42	94.0 2.5	94.1 2.5	76.8 10.4	82.2 7.2	87.7 5.6	76.8 10.4
FLD		14	94.0 3.2	94.3 2.6	87.3 5.8	**95.0** **3.9**	94.2 4.7	87.7 6.1
PCA	4	35	93.4 1.9	93.4 2.1	78.7 5.5	85.4 4.0	88.0 4.0	78.7 5.5
FLD		14	92.9 2.4	93.5 2.4	84.7 3.8	**94.4** **2.1**	92.9 3.9	85.1 5.7
PCA	3	28	91.9 2.5	92.4 2.2	78.6 6.8	84.2 4.0	86.0 4.5	78.6 6.8
FLD		14	89.8 4.5	90.9 4.4	81.6 5.5	**93.0** **2.2**	92.0 2.8	83.9 5.0
PCA	2	21	88.9 5.0	88.9 5.0	75.5 6.9	83.4 6.3	85.1 4.0	75.5 6.9
FLD		14	88.5 3.4	88.1 4.2	79.6 6.1	**89.9** **4.1**	88.1 2.8	77.2 7.2

4 Conclusions

The main practical result of this study is the solution of multiclass discriminants by use of the solution of two FCPs. The Fisher Linear Discriminant (FLD) is originally formulated for an arbitrary number of classes, and it has shown a very good performance in many applications including face recognition [4]. Afterwards, KFD has shown important improvements as a non-linear Fisher discriminant, but the limitation for two-class problems impeded its original performance for many problems of

Table 2. Recognition rates using different numbers of training images per class. All results consider the top 1 match for recognition. FERET database.

PROJECTION METHOD	IMAGES PER CLASS	AXES	EUCLIDEAN	COS	FFC	WHITENING EUCLIDEAN	WHITENING COS	WHITENING FFC
PCA	3	316	**94.1**	**94.1**	87.4	77.6	92.5	87.4
FLD		253	92.5	92.1	91.7	79.9	92.9	90.9
PCA	2	252	86.4	**86.8**	81.5	73.2	85.6	81.5
FLD		253	85.2	85.0	82.9	73.4	79.9	83.1

Table 3. Mean recognition rates using different numbers of training images per class, and taking the average over 20 different training sets. The small numbers are the standard deviation of each recognition rate. All results consider the top 1 match for recognition. Yale Database.

PROJECTION METHOD	IMAGES PER CLASS	AXES	EUCLIDEAN	COS	FFC	WHITENING EUCLIDEAN	WHITENING COS	WHITENING FFC
KPCA	6	89	96.1 2.7	96.1 2.7	82.7 8.9	92.6 4.6	90.7 5.9	82.7 8.9
KFD		14	**96.9** **2.2**	96.8 1.9	92.4 4.2	96.3 2.6	93.9 3.6	89.8 6.0
KPCA	5	74	94.5 2.5	94.5 2.5	82.9 9.9	88.9 7.5	87.7 7.6	82.9 9.9
KFD		14	94.9 4.1	**95.4** **2.8**	89.4 5.2	94.5 4.1	92.3 5.4	87.6 6.1
KPCA	4	59	93.7 1.9	93.7 1.9	84.6 6.2	89.9 4.9	88.1 4.6	84.6 6.2
KFD		14	94.1 3.4	**95.7** **2.4**	89.1 4.3	92.9 2.6	91.5 3.7	84.3 4.9
KPCA	3	44	92.5 1.9	92.5 1.9	82.6 5.5	90.3 3.3	88.1 3.6	82.6 5.5
KFD		14	92.7 2.6	**94.0** **1.8**	87.1 5.2	93.0 2.4	91.3 3.0	82.3 4.8
KPCA	2	29	89.9 4.3	89.9 4.3	76.2 7.8	90.2 3.7	87.7 3.2	76.2 7.8
KFD		14	90.4 2.6	**92.3** **3.6**	82.0 4.7	89.1 4.1	87.3 4.1	77.5 5.7

Table 4. Recognition rates using different numbers of training images per class. All results consider the top 1 match for recognition. FERET database.

PROJECTION METHOD	IMAGES PER CLASS	AXES	EUCLIDEAN	COS	FFC	WHITENING EUCLIDEAN	WHITENING COS	WHITENING FFC
KPCA	3	761	94.5	94.5	85.4	79.5	95.3	85.4
KFD		253	95.3	94.5	**95.7**	82.7	75.2	60.2
KPCA	2	507	86.6	86.6	83.3	**89.6**	89.4	83.3
KFD		253	87.8	87.8	88.6	77.2	71.9	62.0

pattern recognition. In the toy example shown in section 3.1, we have seen that the multiclass-KFD can discriminate more than two classes with high accuracy, even in complex situations.

Finally, the formulation of kernel machines here proposed allows us to use the results of this study with other objective functions, written as (2). The only change must be applied on the SRMs that code the desired statistical measures. Therefore, our results are applicable to a general kind of kernel machines. These kernel machines use second order statistics in a high dimensional space, so that in the original space high order statistics are used. As the algorithms here formulated present several difficulties in practice, further work must be focused in the optimization of them.

Acknowledgements

This research was supported by the join "Program of Scientific Cooperation" of CONICYT (Chile) and BMBF (Germany), and by the Millenium Nucleous Center for Web Research, Grant P01-029-F, Mideplan (Chile). Portions of the research in this paper use the FERET database of facial images collected under the FERET program.

References

1. R. Courant and D. Hilbert. *Methods of Mathematical Physics*, volume 1. Wiley Interscience, 1989.
2. M. Kirby and L. Sirovich. Application of the Karhunen-Loève procedure for the characterization of human faces. *IEEE Trans. Pattern Anal. Machine Intell.*, 12:103–108, 1990.
3. S. Mika, G. Rätsch, J. Weston, B. Schölkopf, and K. Müller. Fisher discriminant analysis with kernels. In Y. Hu, J. Larsen, E. Wilson, and S. Douglas, editors, *Neural Networks for Signal Processing IX*, pages 41–48. IEEE, 1999.
4. P. Navarrete and J. Ruiz-del Solar. Comparative study between different eigenspace-based approaches for face recognition. In N.R. Pal and M. Sugen, editors, *Advances in Soft Computing - AFSS 2002*, pages 178–184. Springer, 2002.
5. P. Navarrete and J. Ruiz-del Solar. Towards a generalized eigenspace-based face recognition framework. In T. Caelli, A. Amin, R.P.W. Duin, M. Kamel, and D. de Ridder, editors, *Structural, Syntactic, and Statistical Pattern Recognition*, pages 662–671. Springer, 2002.
6. S. Saitoh. *Theory of reproducing Kernels and its applications*. Longman Scientific & Technical, Harlow, UK, 1988.
7. S. Santini and R. Jain. Similarity measures. *IEEE Trans. Pattern Anal. Machine Intell.*, 21(9):871–883, 1999.
8. B. Schölkopf, A. Smola, and K. Müller. Nonlinear component analysis as a kernel eigenvalue problem. *Neural Computation*, 10:1299–1319, 1998.
9. K. Sung and T. Poggio. Example-based learning for view-based human face detection. *IEEE Trans. Pattern Anal. Machine Intell.*, 20(1):39–51, 1998.
10. D. Swets and J. Weng. Using discriminant eigenfeatures for image retrieval. *IEEE Trans. Pattern Anal. Machine Intell.*, 18(8):831–836, 1996.

Kernel-based Face Recognition by a Reformulation of Kernel Machines 195

11. M. Tipping. Sparse kernel principal component analysis. In T. Leen, T. Dietterich, and V. Tresp, editors, *Advances in Neural Information Processing Systems*, volume 13. MIT Press, 2001.
12. M. Turk and A. Pentland. Eigenfaces for recognition. *J. Cognitive Neuroscience*, 3:71–86, 1991.
13. V. Vapnik. *The Nature of Statistical Learning Theory*. Springer Verlag, New York, 2nd edition, 2000.

Face Recognition Using Multi Log-Polar Images and Gabor Filters

María José Escobar[1], Javier Ruiz-del-Solar[2] and José Rodriguez[1]

[1] Department of Electronics, Universidad Téc. Fed. Santa María, Valparaíso,CHILE
mjescobar@ieee.org,jrp@elo.utfsm.cl
[2] Department of Electrical Engineering, Universidad de Chile, Santiago, CHILE
jruizd@cec.uchile.cl

Summary. The MLPG architecture, a new biologically based architecture for face recognition is here proposed. In this architecture log-polar images and Gabor Filtering are employed for modeling the way in which face images are processed between the retina and the primary visual cortex. Some simulations of the recognition abilities of the MLPG using the Yale Face Database are presented, together with a comparison with the EBMG and LPG architectures.

Key words: Face Recognition, Gabor Filtering, Log-Polar Transformation

1 Introduction

Face Recognition is a very lively and expanding research field. The increasing interest in this technology is mainly driven by applications like access control to buildings, identification for law enforcement, and recently passive recognition of criminals in public places as airports, stadiums, etc.

The capacity of recognize and identify other human beings using the information contained in a face image, is one of the distinctive characteristic of our visual system and where it shows a high specialization compared with the visual systems of other mammals [6]. For this reason and considering that the implementation of computational system for face recognition is a tough task, it seems natural to imitate the mechanisms used in our visual system to implement a face recognition system. Taking this idea into account, the present work describes a face recognition system inspired in the processing models of our visual system.

Many different approaches have been proposed to solve the task of face recognition. The most successful can be divided into the ones that analyze the faces in a holistic sense (Eigenfaces, Fisherfaces, etc.) and the ones that analyze the constitutive parts of the faces (eyes, mouth, nose, etc.) as for example, the so-called dynamic link-architecture. The dynamic link architecture is a general face recognition technique that represents the faces by projecting them onto an elastic grid where a Gabor filter bank response is measured at each grid node. The recognition of the faces is performed by measuring the similarity of the filter response at each node (Gabor-jets) between different face images. The nodes of the grid normally, but not necessary, correspond to so-called fiducial points (center of the eyes, top of the nose, corners of the lips, etc.). One of the most successful dynamic link architectures is the Elastic Bunch Graph Matching - EBGM (see for example [12]).

Gabor analysis, a particular case of joint spatial/ frequency analysis, is biologically based, and their oriented filters (Gabor filters) model the kind of visual processing carried out by the simple and complex cells of the primary visual cortex of higher mammals [7]. The shape of the receptive fields of these cells and their organization are the results of visual unsupervised learning during the development of the visual system in the first few months of life [11]. Based on these facts one can say that the dynamic link approach for face recognition is biologically motivated, at least in the stages where Gabor analysis is employed. The aim of this work is to take a step further into the construction of a face recognition system that is biologically based. To achieve this objective the Log-Polar Transformation - LPT [8], which models the Retinothopic Mapping of the visual information between the retina and the area V1 of the visual cortex [7][11], is used.

The use of LPT in face recognition applications is not new, different research groups have employed it. Tistarelli and Grosso [9][10] have implemented an active face recognition system that uses the LPT together with Principal Analysis Components. Chien and Choi have used the LPT to locate landmarks in faces [1]. Minut et al. have used the LPT together with Hidden Markov Models for the recognition of faces [5]. However, to our knowledge our work is the first where the LPT have been included in a dynamic-link face recognition architecture. In a previous work, we implemented an architecture that mixes EBMG with LPT, named Log-Polar Gabor architecture - LPG [4]. In the LPG architecture a Gabor Filter Bank is applied to each node of a grid in a log-polar image.

The architecture here proposed, called Multi Log-Polar Gabor - MLPG, consists of a biological inspired system that uses LPT and Gabor Filters but in a different way than LPG. In the MLPG architecture a log-polar image for each fiducial point considerate important is obtained, then a Gabor Filter Bank is applied in each of these new images, obtaining as result the Gabor-Jets for each image with less computational cost than EBMG.

The article is structured as follows. The proposed face recognition architecture is described in section II. In section III some simulations of the recognition abilities of the MLPG using the Yale Face Database are presented, together with a comparison with the EBMG and LPG architectures. Finally, in section IV some conclusions of this work are given.

2 Architecture Proposed

A block diagram of the proposed architecture is presented in Figure 1. When a new face image arrives to the system, the Face Alignment block finds the F_p fiducial points to be used for the recognition. In our current implementation this block is not implemented in an automatic way, but manually. Afterwards the *Log-Polar Transformation* is applied over the input face image using as origin of the transformation the coordinates of the fiducial points, obtaining F_p different Log-Polar images. After that, *Gabor Filtering* is applied to each of these F_p images, obtaining the values of the Gabor-Jet for each fiducial point. Finally, to determine the identity of the input face, its Gabor-Jets are compared with the Gabor-Jets of each face image that forms the *Face Graphs Database* (P classes and N log-polar images per class), using the *Similarity Matching block*.

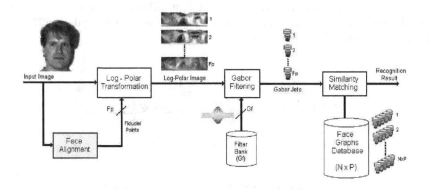

Fig. 1. Block diagram of the architecture proposed.

2.1 Log-Polar Transformation

Studies of optical nerves and visual image projection in the cerebral cortex show that the global retinothopic structure of the cortex can be characterized in terms of the geometrical properties of the so-called retinothopic mapping [7][11]. This mapping can be modeled by the Log-Polar Transformation - LPT [8].

If $I(x,y)$ is a rectangular image in Cartesian coordinates, then the LPT with origin (x_0, y_0) will be given by:

$$I^*(u,v) = L_{x_0,y_0}\{I(x,y)\} \tag{1}$$

where

$$u = Mlog(\rho), \tag{2}$$

$$\rho = \sqrt{(x - x_0)^2 + (y - y_0)^2}, \tag{3}$$

$$v = \phi = tan_\alpha^{-1}\left(\frac{y-y_0}{x-x_0}\right) \tag{4}$$

The magnification parameter M and the angular resolution α, are chosen considering the height and width of the transformed image. An example of LPT applied to an image is shown in Figure 2.

The LPT allows a significant reduction of the visual data to be processed. This data diminution is produced by the logarithmic sampling of the input signal in the radial direction and by the constant sampling (the same number of points is taken) in each angular sector to be transformed. Additionally, this mapping provides an invariant representation of the objects, because rotations and scalings of the input signal are transformed into translations [8], which can be easily compensated.

In the presented architecture, we will use four different Log-Polar images, obtained using as center of the transformation the position of each pupil, the center of the mouth and the middle point between the eyes.

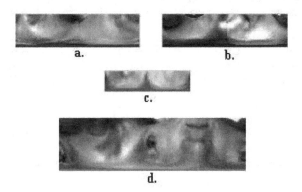

a.   b.

c.

d.

Fig. 2. Log-Polar images obtained using four different fiducial points as center of the log-polar transformation.

2.2 Gabor Filtering

Bidimensional Gabor Filters, originally proposed by Daugman [2][3], correspond to a family of bidimensional Gaussian functions modulated by a cosine function (real part) and a sinus function (imaginary part). These filters are given by ($\mathbf{x} = (x,y)$):

$$\psi_j(\vec{x}) = \frac{1}{\sigma}exp\left(-\frac{\vec{x}^2}{2\sigma^2}\right)\left[exp(i\vec{k}_j\vec{x}) - exp\left(-\frac{\sigma^2\vec{k}_j^2}{2}\right)\right] \tag{5}$$

$$\vec{k}_j = \begin{pmatrix} k_{jx} \\ k_{jy} \end{pmatrix} = \begin{pmatrix} k_v\cos(\phi_\mu) \\ k_v\sin(\phi_\mu) \end{pmatrix}, k_v = 2^{-\frac{v+2}{2}}\pi, \phi_\mu = \mu\frac{\pi}{8}$$

with v and μ the frequency and orientation parameter respectively. The width σ of the Gaussian depends of the frequency and is given by $\sigma = 2\pi/k_j$.

In the implementation of our architecture, 40 filters (5 frequencies and 8 orientations, $j = 0,...,40$) compose the filter bank. The filter parameters are given by $v = 0,...,4$ and $\mu = 0,...,7$.

The Gabor Filtering is applied to our F_p different Log-Polar images, centering the filters in the bottom left pixel of each one. When the whole family of filters is applied to each image, we obtain a Gabor-Jet formed by a vector of 40 components of complex numbers. Each complex number J_j is the result of the convolution of the Log-Polar image I with each of the 40 Gabor Filters (ψ_j) (see equation (6)).

$$J_j(\vec{x}) = J_j(x,y) = \sum_{i=x-2\sigma}^{x+2\sigma} \sum_{j=y-2\sigma}^{y+2\sigma} I(i,j)\psi_j(x-i,y-j) \qquad (6)$$

2.3 Similarity Matching

This block measures the similarity between two groups of Gabor-Jets, no matter if they come from an elastic grid (LPG architecture) or from a simple set of them (MLPG architecture). It compares the Gabor Jet of an input image with the existing ones in the face database.

To compare the group of F_p Gabor-Jets (J_n^T) of the image I^T, with the group of F_p Gabor-Jets (J_n^M) of the image I^M we use the following expression:

$$S_G\left(I^T,I^M\right) = \frac{1}{F_p} \sum_{n=1}^{F_p} S_a\left(J_n^T,J_n^M\right) \qquad (7)$$

This expression is the mean value of the similarity measures $S_a()$, which is given by:

$$S_a(J^T,J^M) = \frac{\sum_{j=1}^{G_f} a_j^T a_j^M}{\sqrt{\sum_{j=1}^{G_f} (a_j^T)^2 \sum_{j=1}^{G_f} (a_j^M)^2}} \qquad (8)$$

where $a_j^i, j = 1,...,G_f$ is the magnitude of the jth complex component of the Gabor Jet J^i.

3 Simulations

We performed a comparison between our proposed MLPG architecture, with the LPG and the EBMG architectures. In order to perform the test, we realized several simulations using Yale University Face Image Database. We employed 165 images of 15 different classes. The size of the images was 128 x 128 pixels. In Table 1 we show the results of several simulations. For each simulation we used a fixed number of training images (N), using the same type of images per class, according with the Yale database specification. In order to obtain representative results we take the average of 20 different set of images for each fixed number of training images. All the images not used for training are used for testing.

In the EBGM implementation 16 fiducial points were used. The points are shown in Figure 3. The Gabor-Jet for each one of these fiducial points was obtained applying the 40 Gabor Filters mentioned in section 2.

Fig. 3. Fiducial Points taken to implement EBMG.

For the LPG implementation, the sizes of the log-polar images were 60x219 pixels. These log-polar images were obtained applying the LPT to the original face image using the middle point between the eyes as origin of the transformation. We used the same Gabor Filters and fiducial points than EBMG. The position of the fiducial points was transformed using the LPT (see [4]).

For the MLPG architecture we just employed three fiducial points (pupils and center of the mouth) to obtain three different Log-Polar images using the coordinates of these points as origin of the transformation. Beside these three Log-Polar images, we also used a Log-Polar image obtained applying TPL to the whole face using the middle point between the eyes as origin of the transformation. The sizes of the Log-Polar images used for the identification were 37x147, 37x147, 23x103 and 60x219. These sizes correspond to the log-polar images centered at the pupils, center of the mouth and middle point between the eyes respectively.

In Table 1 we can see that the recognition rate of MLPG is very similar than LPG, considering that only four Gabor-Jets were used. EBMG shows slightly better results than the ones obtained using LPG or MLPG.

The size of the Gabor Filters used, depends of their frequency. With the Gabor Filters defined in (5), the size of the filters was 16x16, 24x24, 32x32, 44x44 and 64x64 pixels for the five frequencies used. Now if we calculate the number of multiplications needed for EBMG to form the Gabor-Jets, considering 8 orientations for each frequency, we obtain that is $N_M = 1009664$ (see equation (9)), which means a high computational cost. If we realize the same calculation for LPG architecture, in the worst case, assuming that the Gabor Filters are completly contained inside the

Table 1. Mean recognition rates using different numbers of training images per class (N), and taking the average of 20 different training sets (small numbers correspond to the standard deviations).

N	EBGM	LPG	MLPG
2	91,96%	74,11%	70,41%
	(3,58%)	(5,63%)	(4,85%)
3	93,42%	79,38%	82,00%
	(2,01%)	(4,04%)	(4,09%)
4	95,05%	83,33%	83,90%
	(1,87%)	(3,29%)	(3,68%)
5	94,88%	85,55%	87,89%
	(1,97%)	(3,02%)	(2,98%)
6	94,87%	89,13%	88,27%
	(2,12%)	(2,83%)	(2,25%)
7	96,25%	89,75%	90,25%
	(1,82%)	(3,09%)	(3,55%)
8	95,55%	91,22%	91,77%
	(2,11%)	(3,48%)	(3,45%)

Table 2. Total number of multiplications needed to obtain the Gabor-Jets for recognition.

EBGM	LPG	MLPG
1009664	946176	63104

log-polar image, we obtain a smaller number ($N_M = 946176$, also see equation (9)). For the MLPG architecture, by locating the filters in the bottom left pixel of the images, we obtain that the number of multplications needed to get the Gabor-Jets, in the worst case, is just $N_M = 63104$ (also see equation (9)), 16 times smaller than EBMG and 150 times smaller than LPG. The total number of multiplications needed to form the Gabor-Jets for recognition, for the three approaches here analized, is shown in Table 2.

$$N_M = F_p * N_{orient} \left(\sum_{i=1}^{N_{frec}} FilterSize_i \right) \qquad (9)$$

$$EBMG : N_M = 16 * 8 * (16 * 16 + 24 * 24 + 32 * 32 + 44 * 44 + 64 * 64)$$
$$LPG : N_M = 16 * 8 * (16 * 16 + 24 * 24 + 32 * 32 + 44 * 44 + 60 * 60)$$
$$MLPG : N_M = 16 * 8 * \frac{(16*16+24*24+32*32+44*44+60*60)}{4}$$

4 Conclusions

A new biologically based approach for face recognition was here presented. Under this approach, the way in which face images are processed between the retina and the primary visual cortex of our visual system, is modeled using the Log-Polar Transformation and Gabor Filtering.

Some simulations of the recognition abilities of the proposed architecture using the Yale Face Database were presented, together with a comparison with the EBMG and LPG architectures. The recognition rates obtained with the proposed architecture are slightly smaller than EBGM and very similar with LPG. However MLPG has an smaller computational cost than EBMG or LPG. MLPG is 16 times faster than EBMG and 15 times faster than LPG.

5 Acknowledgements

This research was supported by the joint "Program of Scientific Cooperation" of CONYCIT (Chile) and BMBF (Germany).

References

1. Chien S, Choi I (2000) Face and Facial Landmarks location based on Log-Polar Mapping, Lecture Notes in Computer Science LNCS 1811:379–386
2. Daugman J G (1990) An Information-Theoretical View of Analog Representation in Striate Cortex, Computational Neuroscience, MIT Press 403–424
3. Daugman J G (1997) Complete Discrete 2-D Gabor Transforms by Neural Networks for Image Analysis and Compression, IEEE Trans. on Acoustics, Speech, and Signal Proc. vol. 36 7:1169–1179
4. Escobar M J, Ruiz-del-Solar J (2002) Biologically-based Face Recognition using Gabor Filters and Log-Polar Images, Proc. of the IJCNN 2002, Honolulu Hawaii
5. Minut S, Mahadevan S, Henderson J, and Dyer F (2000) Face Recognition using Foveal Vision, Lecture Notes in Computer Science LNCS 1811:424–433
6. Navarrete P, Ruiz-del-Solar J, and Escobar M J (2001) Reconocimiento de Caras mediante Métodos de tipo "Eigenspace", Anales del Instituto de Ingenieros de Chile, Vol. 113, 1:3–25
7. Ruiz-del-Solar J (1998) Biologisch basierte Verfahren zur Objekterkennung und Texturanalyse, Doctoral Degree Thesis, Technical University of Berlin, Germany (ISBN 3-8167-4647-0)

8. Schwartz E L (1980) Computational anatomy and functional architecture of striate cortex: a spatial mapping approach to perceptual coding, Vision Research, 20:645–669
9. Tistarelli M, and Grosso E (2000) Active vision-based face authentication, Image and Vision Computing, 18:299–314
10. Tistarelli M, and Grosso E (1998) Active Vision-based Face Recognition: Issues, Applications and Techniques, in Face Recognition: From Theory to Applications, Springer, 262–286
11. Wilson H R, Levi D, Maffei L, Rovamo J, and DeValois R (1990) The Perception of Form: Retina to Striate Cortex, Visual Perception: The Neurophisiologcal Foundations, Academic Press
12. Wiskott L, Fellous J M, Krüger N, and Von der Malsburg C (1997) Face Recognition by Elastic Bunch Graph Matching, IEEE Trans. on Patt. Analysis and Machine Intell., vol. 19, 7:775–779

Performance Analysis of Statistical Classifier SMO with other Data Mining Classifiers

A B M Shawkat Ali*

School of Computing and Information Technology, Monash University, Victoria 3842, Australia. Shawkat.Ali@infotech.monash.edu.au

Summary. Seven classifiers are compared on sixteen quite different, standard and extensively used datasets in terms of classification error rates and computational times. It is found that the average error rates for a majority of the classifiers are closes with each other but the computational times of the classifiers differ over a wide range. The statistical classifier Sequential Minimal Optimization (SMO) based on Support Vector Machine has the lowest average error rate and computationally it is faster than four classifiers but slightly expensive than other two classifiers.

1 Introduction

Classification of different sizes dataset is one of the important data mining task. Recently researchers in the data mining communities are trying to build a better classifier to overcome the existing limitation [1] compared the accuracy between several decision tree classifiers and non-decision tree classifiers on a large number of datasets. Other studies that are smaller in scale [2, 3, 4, 5, 6, 7].

Our study compared mainly the statistical classifier SMO with others for different size of datasets. Most of datasets were from real life domains. All the data sets are considered with two classes with all the real attributes value. We have organized the present work as follows. In the next section, we describe shortly all the classification algorithms. Following that, we present an experimental setup and comparison of these approaches. Finally we conclude with a discussion of our findings.

2 Algorithms Description

2.1 IBK

IBK is an achievement of the k nearest neighbors classifier. Each case is considered as a point in multi dimensional space and classification is done based on the near-

---

* I would to thanks Ms Suryani Lim for her valuable help to formatting by LaTeX.

est neighbors. The value of 'k' for nearest neighbors can vary. This determines how many cases are to be considered as neighbors to decide how to classify an unknown instance. The time taken to classify a test instance with a nearest neighbor classifier increases linearly with the number of training instances that are kept in the classifier. It needs a large storage requirement [8]. Its performance degrades quickly with increasing noise levels. It also performs badly when different attributes affect the outcome to different extents. One parameter that can be affect the performance of the IBK algorithm is the number of nearest neighbors to be used. By default it uses just one nearest neighbor.

2.2 J48.J48

J48.J48 is a top down decision tree classification algorithm. The algorithm considers all the possible tests that can split the data set and selects a test that gives the best information gain. For each discrete attribute, one test with outcomes as many as the number of distinct values of the attribute is considered. For each continues attribute, binary tests involving every distinct values of the attributes are considered. In order to gather the entropy gain of all these binary tests competently, the training data set belonging to the node in considerations sorted for the values of the continuous attribute and the entropy gains of the binary cut based on each distinct values are calculated in one scan of the sorted data. This process is repeated for each continuous attributes [6].

2.3 J48.PART

The PART algorithm forms rules from pruned partial decision trees built using C4.5's heuristics. The main advantage of PART over C4.5 is that unlike C4.5, the rule learner algorithm does not need to perform global optimization to produce accurate rule sets [9]. To make a single rule, a pruned decision tree is built, the leaf with the largest coverage is made into a rule, and the tree is unneeded. This avoids over fitting by only generalizing once the implications are known. [10] describe the results of an experiment performed on multiple data sets. The result from this experiment showed that PART outperformed the C4.5 algorithm on 9 occasions whereas C4.5 outperformed PART on 6.

2.4 Kernal Density

Kernal Density algorithm works in very simple way as like Naive Bayes. The main difference is that, unlike Naive Bayes, Kernel Density does not assume normal distribution of the data. Kernel Density tries to fit an arrangement of kernel functions. According to [11] Kernel Density estimates are similar to histograms but provide smoother representation of the data. They also illustrate some of the advantage of Kernel Density estimates for data presentation in archaeology. They showed that Kernel Density estimates could be used as a basis for producing contour plots of archeological data, which lead to a useful graphical representation of the data.

2.5 Naive Bayes

The Naive Bayes classification algorithm is created on Bayes rule, which is used to compute the probabilities and it is used to make predictions. Naive Bayes considers that the input attributes are statistically independent. It analyses the relationship between each input attribute and the dependent attribute to derive a conditional probability for each relationship [12]. These conditional probabilities are then combined to classify new cases. An advantage of Naive Bayes algorithm over some other algorithms is that it requires only one pass through the training set to produce a classification model. Naive Bayes works very well when tested on many real world datasets [9]. It can obtain results that are much better than other sophisticated algorithms. However, if a particular attribute value does not occur in the training set in conjunction with every class value, then Naive Bayes may not perform very well. It can also perform poorly on some datasets because attributes were treated, as through if they are not dependent, whereas in reality they are associated.

2.6 OneR

OneR is one of the simplest classification algorithms. As described by [13], OneR produces simple rules based on one attribute only. It generates a one-level decision tree, which are expresses in the form of a set of rules that all test one particular attribute. It is a simple cheap method that often comes up with quit good rules for characterizing the structure in data [8]. It often gets reasonable accuracy on many tasks by simple looking at an attribute.

2.7 SMO

Support Vector Machine (SVM) is an elegant tool for solving pattern recognition and regression problems. Over the past few years, it has attracted a lot of researchers from the neural network and mathematical programming community; the main reason for this being their ability to provide excellent generalization performance. Recently, [14] proposed an iterative algorithm called SMO for solving the regression problem using SVMs. The remarkable feature of the SMO algorithm is that they are fast as well as very easy to implement. SVM requires the solution of a very large quadratic programming (QP) optimization problem during the training time. Unlike previous SVM learning algorithms, which use numerical QP as an inner loop, on the other hand SMO uses an analytic QP step. The reason is SMO spends most of the decision function, rather than performing QP, it can exploit data sets, which contain a substantial number of zero elements. As like divide and conquer, SMO breaks the large QP problem into a series of smallest possible QP problems. These small QP problems are solved analytically, which avoids using a time-consuming numerical QP optimization as an inner loop. The amount of memory required for SMO is linear in the training set size, which permits SMO to handle a huge number of training sets nearly infinity. Because large matrix computation is avoided, SMO scales somewhere between linear and quadratic in the training set size for different test problems, while a standard projected conjugate gradient (PCG) chunking algorithm scales somewhere between linear and cubic in the training set size. The computational cost of SMO is dominated by SVM evaluation, hence SMO is faster for linear SVMs and sparse data sets [15].

We can consider N training data points

$$(x_1,y_1),(x_2,y_2),\ldots,(x_N,y_N)$$

, where $(x_i \varepsilon R^d)$ and $(y_i \varepsilon \pm 1)$.

We would like to construct a linear separating hyperplane:

$$f(x) = sgn(\omega.x - b)$$

We can calculate the value of w and b from the training dataset. Finally the sign will predict the class value either it is positive or negative incase of binary classification.

3 Experimental Setup

In order to evaluate the performance of different classifiers, we have considered all the datasets from the UCI collection [16] and [17]. The system configuration was Pentium III 933 MHz with 256 Mbytes. The name of the all datasets is mentioned in Table 1. Here we considered only the binary classes problems. We choose all the classifiers from Weka data mining tool. Weka is a collection of machine learning algorithms for solving real world data mining problems. It is written in Java and runs on almost any platform. We consider all the classifier with default setting. We also used 10-fold Cross-validation technique. Cross-validation is a method for estimating how good the classifier will perform on new data and is based on "re-sampling" [18]. Cross-validation is good for use especially when the datasets is small. The 10-fold cross validation experiment is conducted as follows:

- The dataset is randomly divided into 10 disjoint subsets, with each containing approximately the same number of cases. The class labels to ensure that the subset class proportions are roughly the same as those in the whole dataset stratify the sampling.
- For each subset, a classifier is constructed using the observations outside that particular subset. The classifier is then tested on the withheld subset to estimate its error rate.
- The 10-fold cross validation error estimates are averaged to provide an estimate for the classifier constructed from all the data.

4 Results

Table 1 shows the percentage of error for each classifier relative to that of the best classifier for each dataset. The last row of the Table 1 gives the average error rates for

Table 1. Results of classifier performance on error rate. The first colume indicates the datasets name and first row indicates the classifiers name

Data Set	IBK	J48.J48	J48.PART	KernelDensity	NaiveBayes	OneR	SMO
pima	29.40	26.26	26.30	28.57	24.27	28.16	22.90
bcw	4.78	5.52	5.23	4.66	3.95	8.20	3.20
echo	44.9	35.55	38.71	36.87	27.88	39.68	28.19
att	42.23	4.46	41.98	41.16	39.54	43.76	38.37
bio	17.1	13.98	14.62	15.83	9.48	20.75	11.54
bld	37.06	37.35	38.97	38.94	49.52	43.42	41.94
crx	18.23	15.14	16.34	17.89	24.00	14.62	14.62
hco	19.24	16.16	18.61	21.36	21.65	18.42	18.66
hea	23.91	21.36	22.12	24.16	15.51	27.98	15.91
hep	18.86	21.79	18.50	19.21	15.00	17.64	12.93
hyp	3.01	0.79	1.17	2.62	2.22	2.46	2.77
pid	33.73	25.84	27.03	31.63	24.43	27.48	22.98
bupa	36.81	34.59	35.36	35.86	44.47	42.95	41.96
german	32.59	27.28	29.93	28.95	24.50	29.83	23.15
h-d	24.73	21.62	21.67	24.46	16.19	30.00	16.49
sonar	13.51	26.67	25.69	14.15	32.03	37.21	22.22
Average Error	25.01	23.15	23.89	24.15	23.42	27.04	21.11

Table 2. The CPU time scale for each classifier. The last row indicates the average time of each classifiers

Data Set	IBK	J48.J48	J48.PART	KernelDensity	NaiveBayes	OneR	SMO
pima	0	0	0.75	0	24.27	0.98	0.64
bcw	0	0	0.75	0	3.95	0.98	0.49
echo	0	0	0.75	0	27.88	0.98	0
att	0.98	0	0	1.30	0.98	0.98	0.28
bio	1.3	0	0	1.3	0.98	0.98	1.17
bld	1.3	0	0	1.3	0.98	0.98	0
crx	0.85	0	0.94	0	0.98	0.98	0
hco	0	0	0	0.9	0.98	0.98	0
hea	0	0	0	0.75	0.98	0.98	0
hep	0	0	0	0.75	0.98	0.98	1.33
hyp	0	1.24	0	0	0.98	0	0.28
pid	1.21	0.85	0	0.94	0.98	0	0
bupa	1.21	0	0	0.94	0.98	0	0.9
german	0.57	0	0	0.85	0.98	0	0
h-d	0	0.75	0.7	0	0.98	0	1.02
sonar	0	0.75	0.7	0.94	0.98	0	0
Average Time	0.46	0.22	0.29	0.62	4.30	0.61	0.38

each classifier. A ranking of the classifiers in terms of average error rates as follows: SMO, J48.J48, Naive Bayes, J48.PART, Kernal Density, IBK, OneR. SMO has the lowest average error rate and OneR is the highest. The easiest datasets to classify are hyp and bcw; the error rates between 0.79 and 8.2. The most difficult dataset are bld, bupa, att and echo, where the minimum error rate is 27.88. Overall SMO shows the maximum number of minimum error rate for the individual dataset.

Table 2 shows the computational times for the classifiers on the basis of each dataset. Overall the fastest classifier is J48.J48 and the slowest classifier is Naive Bayes. J48.J48 classify highest 12 datasets with zero fraction of time. Naive Bayes and OneR require similar time for some datasets.

5 Conclusions

We have tried to find out the best classification algorithms and characteristics of datasets for binary class problem. Our results showed that firstly no single algorithm could outperform any other when the performance measure is the expected generalization accuracy and secondly the average error rates of many classifiers are sufficiently similar that their differences are statistically insignificant. But SMO shows the minimum error rates among others. It is clear that if error rate is the sole criterion, SMO would be the best method of choice. In real world traditional learning techniques for instance neural network is always doing minimization of the empirical risk for pattern recognition problem. On the other hand SMO minimizes the structural risk. Due to this we did not consider neural network classifier. However one disadvantage of SMO is that it can't classify more than two classes problems. We use over the experiment the Weka data mining tools. Weka is more users friendly, much functionality, including classification, clustering, searching for association rules on applied datasets. Therefore, although a single algorithm cannot build the most accurate classifiers in all situations, some algorithms could perform better in specific domains.

References

1. Michie D, Spiegelhalter DJ, Taylor CC (1994) (eds) Machine Learning, neural and statistical classification. Ellis Horwood. London
2. Shavlik JW, Mooney RJ, Towell GG (1991) Symbolic and neural learning algorithms: An empirical comparison. Machine Learning. 6:111–144
3. Brodley CE, Utgoff PE (1992) Multivariate versus univariate decision trees, Technical Report 92-8, Department of computer science, University of Massachusetts, Amherst, MA.
4. Brown DE, Corruble V, Pittard CL, (1993) A comparison of decision tree classifiers with backpropagation neural networks for multimodel classification problems. Pattern Recognition 26:953–961
5. Curram SP, Mingers J (1994) Neural networks, decision tree induction and discriminant analysis: an empirical comparison. Operational Research Society. 45:440–450

6. Joshi KP (1997) Analysis of data mining algorithms. Project Report, University of Missouri-Columbia, USA
7. Ibrahim RS (1999) Data mining of machine learning performance data, MS Thesis, RMIT, Australia
8. Wolpert D, Macready W (1995) No free lunch theorems for search, Technical Report SFI-TR-95-02-010, Santa Fe Institute
9. Witten I, Frank M (2000) Data Mining: practical machine learning tool and technique with java implementation, Morgan Kaufmann, San Francisco, USA
10. Frank E, Witten I (1998) Generating accurate rule sets without global optimization. Machine Learning: Proceedings of the Fifteen International Conference. Morgan Kaufmann Publishers, San Francisco, USA
11. Beardah C, Baxter M, (1996) The archaeological use of kernel density estimates, Internet Archeology, http://www.intarch.ac.uk
12. Brand   D,   Gerritsen   R   (1997)   Naive   bayes   and   nearest   neighbor, http://ww.dbmsmag.com/9807 m07.html
13. Holte R (1993) Very simple classification rules perform well on most commonly used data sets, Machine Learning. 11:63–91
14. Smola J, Scholkopf B (1998) A tutorial on support vector regression, NeuroCOLT Technical Report TR 1998-030, Royal Holloway College, London, UK
15. Platt JC (1999) Fast training of support vector machines using sequential minimal optimization. In Scholkopf B, Burges CJC, Smola AJ (eds), Advances in Kernel Methods: Support Vector Learning, The MIT Press, England
16. Merz CJ, Murphy PM (2002) UCI repository of machine learning data-bases. Department of Information and Computer Science. University of California
17. Lim T S (2001) Knowledge discovery central, http://www.KDCentral.com/
18. Kohavi R (1995) A study of cross-validation and bootstrap for accuracy estimation and model selection, Max-Plank-Institute Proceedings

Two term-layers: an alternative topology for representing term relationships in the Bayesian Network Retrieval Model

Luis M. de Campos[1], Juan M. Fernández-Luna[2], and Juan F. Huete[1]

[1] Dpto. de Ciencias de la Computación e Inteligencia Artificial. E.T.S.I. Informática.
Universidad de Granada. 18071 - Granada, Spain {lci,jhg}@decsai.ugr.es
[2] Dpto. de Informática. E.P.S. Universidad de Jaén. 23071 - Jaén, Spain jmfluna@ujaen.es

Summary. The Bayesian Network Retrieval Model presents the advantage that may capture the main relationships among the terms from a collection by means of a polytree, network that allows the design and application of efficient learning and propagation algorithms. But in some situations where the number of nodes in the graph is very high (the collection represented in the Bayesian network is very large), these propagation methods could be not so fast as needed by an interactive retrieval system. In this paper we present an alternative topology for representing term relationships that avoids the propagation with exact algorithms, very suitable to manage large document collections. This new topology is based on two layers of terms. We compare the retrieval effectiveness of the new model with the old one using several document collections.

1 Introduction

Information Retrieval (IR) is a subfield of Computer Science that deals with the automated storage and retrieval of documents [12]. An *IR system* is a computer program that matches user *queries* (formal statements of information needs) to documents stored in a database (the *document collection*). In our case, the *documents* will always be the textual representations of any data objects. An *IR model* is a specification about how to represent documents and queries, and how to compare them. Many IR models (as well as the IR systems implementing them), as the Vector Space model [12] or Probabilistic models [2], do not use the documents themselves but a kind of document surrogates, usually in the form of vectors of *terms* or *keywords*, which try to characterize the document's information content[3]. Queries are also represented in the same way.

---

[3] In the rest of the paper we will use the word *document* to denote both documents and document surrogates.

When a user formulates a query, this is compared with each document from the collection and a score that represents its relevance (matching degree) is computed. Later, the documents are sorted in decreasing order of relevance and returned to the user.

To evaluate IR systems, in terms of retrieval effectiveness, several measures have been proposed. The most commonly used are *recall (R)* (the proportion of relevant documents retrieved), and *precision (P)* (the proportion of retrieved documents that are relevant, for a given query). The relevance or irrelevance of a document is based, for test collections, on the *relevance judgments* expressed by experts for a fixed set of queries [12]. By computing the precision for a number of values of recall we obtain a recall-precision plot. If a single measure of performance is desired, the average precision for all the recall values considered may be used. Finally, if we are processing together a set of queries, the usual approach is to report mean values of the selected performance measure(s).

Probabilistic IR models use probability theory to deal with the intrinsic uncertainty with which IR is pervaded [7]. Also founded primarily on probabilistic methods, *Bayesian networks* [9] have been proved to be a good model to manage uncertainty, even in the IR environment, where they have already been successfully applied as an extension/modification of probabilistic IR models [13, 11, 6]. The networks are used to compute the posterior probabilities of relevance of the documents in the collection given a query.

In this paper we introduce a modification of the *Bayesian Network Retrieval Model* (BNRM), which aims to improve their efficiency. This model is composed of two subnetworks: the document and term subnetworks. The former stores the documents from the collection and the latter the terms occurring in the documents and their relationships. The fact of capturing term to term relationships within a collection gives a more accurate representation of the collection, improving the effectiveness of the IR system.

The way in which these relationships are represented in the original BNR Model is by means of a polytree that, automatically constructed, captures the main (in)dependence relationships among the terms. Although with this topology a set of efficient learning and propagation algorithms exist, using actual test collection, as TREC, where the number of terms and documents is very large, running a propagation algorithm could be a very time consuming task. If we take into account that in interactive Information Retrieval the user requires the system answer in very few seconds, this kind of structure could not be the most appropriate. This is the problem that lead us to look for an alternative topology for representing these relationships, without having to run a propagation algorithm. The candidate structure is a term subnetwork with two layers, in which the collection terms are duplicated and placed in a second layer, establishing arcs from terms from one layer to terms in the second. This bipartite graph allows to efficiently propagate using a probability function evaluation.

The remainder of the paper is structured as follows: Section 2 introduces the original model, in which the term subnetwork is represented by means of a polytree: general topology, estimation of the probability distributions to be stored in the network and the specific exact probabilities propagation algorithm, designed to allow efficient inference and retrieval. In Section 3 we study the new proposal to learn the relationships between terms and the new topology to store them. The modified inference process necessary to efficiently deal with the new network topology is also described. Section 5 shows the experimental results obtained with our models, using several standard document collections. Finally, Section 5 contains the concluding remarks.

2 The Bayesian Network Retrieval Model

Our retrieval model is based on Bayesian networks, which are graphical models capable of efficiently representing and manipulating n-dimensional probability distributions [9]. A Bayesian network uses two components to codify qualitative and quantitative knowledge: (a) A *Directed Acyclic Graph* (DAG), $G = (V,E)$, where the nodes in V represent the random variables from the problem we want to solve, and the topology of the graph (the arcs in E) encodes conditional (in)dependence relationships among the variables (by means of the presence or absence of direct connections between pairs of variables); (b) a set of conditional probability distributions drawn from the graph structure: for each variable $X_i \in V$ we have a family of conditional probability distributions $P(X_i|pa(X_i))$, where $pa(X_i)$ represents any combination of the values of the variables in $Pa(X_i)$, and $Pa(X_i)$ is the parent set of X_i in G. From these conditional distributions we can recover the joint distribution over V:

$$P(X_1,X_2,\ldots,X_n) = \prod_{i=1}^{n} P(X_i|pa(X_i)) \tag{1}$$

This decomposition of the joint distribution results in important savings in storage requirements. It also allows probabilistic inference (propagation) to be performed (efficiently, in many cases), i.e. computing the posterior probability for any variable given some evidence about the values of other variables in the graph [9].

The set of variables V_B in the Bayesian Network Retrieval Model , G_B, is composed of two different sets of variables, $V_B = \mathcal{T} \cup \mathcal{D}$: the set $\mathcal{T} = \{T_1,\ldots,T_M\}$ of the M terms in the glossary (index) from a given collection and the set $\mathcal{D} = \{D_1,\ldots,D_N\}$ of the N documents that compose the collection[4]. Each term variable, $T_i \in \{\bar{t}_i, t_i\}$, meaning the term in not relevant, and relevant, respectively. Similarly, the domain of each document variable, $D_j \in \{\bar{d}_j, d_j\}$, with the same meaning but now with a document with respect a query.

To determine the topology of the basic Bayesian network, we have taken into account that there is a link joining each term node $T_i \in \mathcal{T}$ and each document node $D_j \in \mathcal{D}$ whenever T_i belongs to D_j. Also, there are not links joining any document nodes D_j and D_k. And finally, any document D_j is conditionally independent of any other document D_k when we know for sure the (ir)relevance values for all the terms

---

[4] We will use the notation T_i (D_j, respectively) to refer to both the term (document, respectively) and its associated variable and node.

indexing D_j. These three assumptions determine the network structure in part: the links joining term and document nodes have to be directed from terms to documents; moreover, the parent set of a document node D_j is the set of term nodes that belong to D_j, i.e., $Pa(D_j) = \{T_i \in \mathcal{T} \mid T_i \in D_j\}$.

The next step is represent term relationships by means of a Bayesian network. To determine these relationships we decided to apply an automatic learning algorithm that has the set of documents as input and generates as the output a polytree of terms (a graph in which there is no more than one undirected path connecting each pair of nodes). The main reason to restrict the structure of the term subnetwork to a polytree is the existence of, firstly, a set of efficient learning algorithms [4, 10], and secondly, a set of exact and efficient inference algorithms, specific for polytrees, that run in a time proportional to the number of nodes [9]. The learning algorithm, completely described in [5], is based on the PA algorithm [4] and Rebane and Pearl's algorithm [10], but with several modifications and new contributions to adapt it to the IR environment. Figure 1 shows the Bayesian network that we have just described.

Term Subnetwork

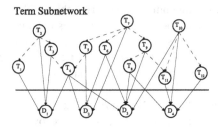

Document subnetwork

Fig. 1. The topology of the Extended Bayesian Network Retrieval Model.

Once the structure has been built, the next step is to estimate the probability distributions stored in each node of the network. Thus, all the root nodes, i.e., those which do not have parents, will store marginal distributions. In our specific case, the only nodes of this type are term nodes. For each root term node, we have to assess $p(t_i)$ and $p(\bar{t}_i)$, task that we put into practice by means of the following estimator: $p(t_i) = \frac{1}{M}$ and $p(\bar{t}_i) = 1 - p(t_i)$ (M is the number of terms in the collection).

The nodes with parents (term and document nodes) will store the conditional probability distributions, one for each of the possible configurations that their parent nodes can take. Starting from the term nodes, these will store the conditional probabilities $p(T_i \mid pa(T_i))$, where $pa(T_i)$ is a configuration of the values associated to the set of parents of T_i. The estimator used in this case is based on the Jaccard coefficient [12]. This measure computes the similarity among two sets by dividing the number of elements in the intersection into the number of elements in the union of both sets. In our context, $p(\bar{t}_i \mid pa(T_i))$ is (later $p(t_i \mid pa(T_i))$ is computed using $p(t_i \mid pa(T_i))$):

$$p(\bar{t}_i \mid pa(T_i)) = n(< \bar{t}_i, pa(T_i) >)/(n(< \bar{t}_i >) + n(pa(T_i)) - n(< \bar{t}_i, pa(T_i) >)), \quad (2)$$

where $n(\Delta)$ is the number of the number of documents that include all the terms that occur as relevant in the configuration and do not include those which are not relevant. The estimation of the conditional probabilities of relevance of a document D_j,

$p(d_j|pa(D_j))$, is not an easy problem. The reason is that the number of conditional probabilities that we need to estimate and store for each D_j grows exponentially with the number of parents of D_j. Instead of explicitly computing and storing these probabilities, we use a *probability function* (also called a canonical model of multicausal interaction [9]). Each time that a given conditional probability is required during the inference process, the probability function will compute and return the appropriate value. We have developed a new general canonical model: for any configuration $pa(D_j)$ of $Pa(D_j)$ (i.e., any assignment of values to all the term variables in D_j), we define the conditional probability of relevance of D_j as follows:

$$p(d_j|pa(D_j)) = \sum_{T_i \in D_j, t_i \in pa(D_j)} w_{ij} \qquad (3)$$

where the weights w_{ij} verify that $0 \le w_{ij} \; \forall i,j$ and $\sum_{T_i \in D_j} w_{ij} \le 1 \; \forall j$. The expression $t_i \in pa(D_j)$ in eq. (3) means that we only include in the sum those weights w_{ij} such that the value assigned to the corresponding term T_i in the configuration $pa(D_j)$ is t_i. So, the more terms are relevant in $pa(D_j)$ the greater is the probability of relevance of D_j. The specific weights, w_{ij}, used by our models, for each document $D_j \in \mathcal{D}$ and each term $T_i \in D_j$, are:

$$w_{ij} = \alpha^{-1} \frac{\mathrm{tf}_{ij} \cdot \mathrm{idf}_i^2}{\sqrt{\sum_{T_k \in D_j} \mathrm{tf}_{kj} \cdot \mathrm{idf}_k^2}} \qquad (4)$$

where α is a normalizing constant (to assure that $\sum_{T_i \in D_j} w_{ij} \le 1 \; \forall D_j \in \mathcal{D}$). Obviously, many other weighting schemes are possible. The weights in eq. (4) have been chosen to resemble the well-known cosine measure [12].

Given a query Q submitted to our system, the retrieval process starts placing the evidences in the term subnetwork: the state of each term T_{iQ} belonging to Q is fixed to t_{iQ} (relevant). Then the inference process is run obtaining, for each document D_j, its probability of relevance given that the terms in the query are also relevant, $p(d_j|Q)$. Finally, the documents are sorted in decreasing order of probability to carry out the evaluation process.

Taking into account the large number of nodes in the Bayesian network and the fact that it contains cycles and nodes with a great number of parents, general purpose inference algorithms cannot be applied due to efficiency considerations, even for small document collections. To solve this problem, we have designed a specific inference method that takes advantage of both the topology of the network and the kind of probability function used for document nodes (eq. 3). This method is called *propagation + evaluation*, is an exact propagation algorithm and is composed of two stages: (1) An exact propagation in the term subnetwork (using Pearl's algorithm for polytrees), which computes $p(t_i|Q), \forall T_i$; (2) An evaluation of the probability function used to estimate the conditional probabilities in document nodes, using the information obtained in the previous propagation. Therefore, the relevance of each document in the collection is computed as follows:

$$p(d_j|Q) = \sum_{T_i \in D_j} w_{ij} \, p(t_i|Q) \qquad (5)$$

3 An alternative representation for term relationships: a topology with two term layers

In large collections, as TREC, the great amount of terms could represent a problem with respect to propagation time, because, although the term subnetwork is represented by a polytree, and the propagation is very fast, if the graph contains a lot of terms and arcs, this process could get slower. Therefore, in this section we present an alternative way of representing term relationships using a Bayesian network, different from a polytree, but trying to assure two very important requirements: on the one hand, the accuracy of the term relationships represented in the graph, and on the other hand, an efficient propagation scheme in the underlying graph to compute the posterior probabilities in each term node.

Let us go to present the new topology that fulfill these requirements. In the new topology we shall include explicit dependence relationships between T_j and each term in $R_p(T_j)$ (the set of those p terms most closely related to T_j, measured in a certain way). Instead of starting from one set of term nodes and including links between the terms in $R_p(T_j)$ and term T_j, as the learning algorithm establishes, we shall use two layers of nodes to represent the term subnetwork: we duplicate each term node T_k in the original layer to obtain another term node T'_k, thus forming a new term layer, $\mathcal{T}'$. The arcs connecting the two layers go from $T'_i \in R_p(T_j)$ to T_j. Therefore, in the new Bayesian network, G_B, the set of variables is $V_B = \mathcal{T} \cup \mathcal{T}' \cup \mathcal{D}$. The parent set of any original term node $T_j \in \mathcal{T}$ is defined as $Pa(T_j) = R_p(T_j)$. We use this topology because conditional probabilities associated to the document nodes in $\mathcal{D}$ (eq. 3) have not to be redefined. Another reason is that dealing with a bipartite graph, which is a particular case of the so called *simple graphs* [3, 8], allows the use of a very fast propagation algorithm, as we shall see later. And finally, the new topology contains three simple layers (see Figure 2), without connections between the nodes in the same layer, and this fact will be essential for the efficiency of the inference process.

Term Subnetwork

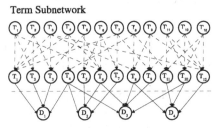

Document subnetwork

Fig. 2. The new extended Bayesian network.

Finally, we have to define the conditional probabilities $p(T_j|Pa(T_j))$ for the terms in the original term layer $\mathcal{T}$. As we put into practice in the document layer of the original model, we use a probability function belonging to the general canonical model defined in eq. (3), where the weights v_{ij} measure the influence of each $T'_i \in Pa(T_j)$ on term T_j:

$$p(t_j|pa(T_j)) = \sum_{T_i' \in Pa(T_j), t_i' \in pa(T_j)} v_{ij} \qquad (6)$$

To learn the graph structure in [5], it was used the Kullback-Leibler's Cross Entropy to determine dependences among terms. Basically, those terms with the highest entropy were connected in the graph. The problem arises because a high value of cross entropy may be due not only to a positive dependence (in the sense that two terms mainly co-occur in the same documents), but also to a negative dependence (when they do not occur in any common document). Taking into account the type of canonical model that we are using to model $p(t_j \mid pa(T_j))$, we need to include in $Pa(T_j)$ only terms that are positively correlated with T_j, hence we can not use cross entropy to measure dependences between terms. Trying to solve this problem, we thought, as a valid solution, to use an approach based on frequencies of co-occurrences of two terms. Before explaining it, let us introduce some notation:

Given a term T_j, if we want to know which terms, from the rest of the collection, are most closely related to it, for each one of these, T_i, counting frequencies, the following values may be computed (t_i means "T_i occurs", and $\bar{t}_i$ stands for "T_i does not occur –and respectively for T_j): $n_{\bar{t}_i\bar{t}_j}$ is the number of times in which neither T_i nor T_j occur in a document; $n_{t_i t_j}$ is the number of times in which both terms occur in the same document and, finally, $n_{\bar{t}_i t_j}$ and $n_{t_i \bar{t}_j}$, the number of documents in which only one of the two terms occur.

The following expression, a maximum likelihood estimator, will be used to measure the strength of their co-occurrence relationship, fixed T_j:

$$strength(T_j, T_i) = n_{t_i t_j}/n_{t_i} \qquad (7)$$

i.e., a coefficient that measures the ratio between the number of documents in which T_j co-occurs with T_i, with respect to the total number of documents in which T_i occurs. When this quotient is close to 1.0, this means all the documents indexed by T_i are also indexed by T_j. But an anomalous behavior is observed when, for instance, T_i occurs only in one document and T_j occurs in that document, the result would be 1.0. At the same time, if we have the case in which T_i and T_j share 5 documents of the 5 in which T_i is in the collection, the ratio is the same, but we would say that T_i and T_j are more closely related in the second example, although the value obtained by eq. (7) is the same in both cases. To solve this problem, we will finally use a Bayesian estimator [1]:

$$strength(T_j, T_i) = (n_{t_i t_j} + 1)/(n_{t_i} + 2) \qquad (8)$$

When this estimator is used, a new problem arises: imagine a pair of terms such that $n_{t_i t_j} = 0$, i.e., they do not co-occur in any document, and also $n_{t_i} = 1$, then $strength(T_j, T_i)$ would be 0.33, value that would be always greater than the one obtained when these terms co-occurs once, for instance, and $n_{t_i} > 4$, situation that is not very logic. To solve this problem, we have adopted a new strength function, $strength'(T_j, T_i)$, that will be 0 when $n_{t_j t_i} = 0$, and $strength(T_j, T_i)$, otherwise.

$$strength'(T_j, T_i) = \{0, \text{ if } n_{t_i t_j} = 0; \text{ and } strength(T_j, T_i), \text{ otherwise}\} \qquad (9)$$

Therefore, to learn the term subnetwork implies to find which are the terms that have the strongest relationships with each one of the terms in the collection, i.e., to determine the sets of parents $R_p(T_j)$, $T_j \in \mathcal{T}$. Thus, for each T_j, the measure $strength(T_j, T_i)$, $\forall T_i \in \mathcal{T}$ is computed. The duplicates of the p terms with the highest values are selected to be elements of $R_p(T_j)$. We have to highlight that the equivalent term to T_j in $\mathcal{T}'$, T_j', is always included in $R_p(T_j)$. In this way, always a term is related with itself, and in case of being instantiated, the posterior probability of that term would be very close to 1.0.

Our proposal for weights v_{ij} in eq. (6) is the following:

$$ v_{ij} = \frac{1-\beta}{S_j} strength'(T_j, T_i), \ \forall T_i' \in Pa(T_j), i \neq j, \text{ and } v_{jj} = \beta \qquad (10) $$

where $S_j = \sum_{T_i' \in Pa(T_j), i \neq j} strength'(T_j, T_i)$ and β is a parameter, $0 < \beta < 1$, that is used to control the importance of the contribution of the term relationships being considered for a term T_j to its final degree of relevance. In this way we are imposing an uniform upper bound for the importance of this combination equal to $1 - \beta$. Using these weights, the posterior probability of relevance for a term T_j given a query Q, $p(t_j \mid Q)$, can be computed (in a way similar as we did in eq. (5)) for $p(d_j \mid Q)$ as follows:

$$ p(t_j | Q) = \sum_{T_i' \in Pa(T_i)} v_{ij} \, p(t_i' | Q) \qquad (11) $$

In this way, what in the original polytree-based model was an instantiation of the terms in the query and the execution of Pearl's propagation algorithm, has been substituted in the new topology by an evaluation of a probability function in the $\mathcal{T}$ layer. The final expression for the calculation of $p(t_j \mid Q)$ is:

$$ p(t_j | Q) = \frac{1-\beta}{S_j} \sum_{T_i' \in Pa(T_j), i \neq j} strength'(T_j, T_i) p(t_i' | Q) + \beta p(t_j' | Q) \qquad (12) $$

where $p(t_i' | Q) = 1.0$ if $t_i' \in Q$, and $1/M$, otherwise (as the terms in the $\mathcal{T}'$ are marginally independent, the posterior probability of the terms which are not in the query coincides with their prior probability, $p(t_i' \mid Q) = p(t_i') = 1/M$).

The posterior probabilities computed using eq. (12) will be used to evaluate the probability function of eq. (11) in the document subnetwork, in order to obtain the final degree of relevance of each document.

4 Experiments and Results

To test the new Bayesian network topology, we have run several retrieval experiments with three medium-size standard collections: ADI, CISI, and CRANFIELD. The main characteristics of these collections with respect to number of documents, terms and queries are (in this ordering): ADI (82, 828, 35), CISI (1460, 4985, 76), and CRANFIELD (1398, 3857, 225).

Our aim is to compare the effectiveness of the two topologies, the original and the new ones. In order to carry out this task, we have tested with a different number of parents, p, for the terms in $\mathcal{T}$, and for several values of the parameter β. To be exact, the first parameter has been set to 5, 10, and 15 parents; the second to 0.6, 0.7 and 0.8. The performance measure considered is the average precision for the *eleven* standard values of recall.

The results of this experimentation are presented in table 1, in which the average precision values for the 11 standard recall points of the original BNR model (Term subnetwork composed of a polytree), for each collection, are shown in the second row (Noted as '*AP-11p (BNRM)*'). The average precisions (*AV-11p*) of the experiments run with the new model for different values of the number of parents and the parameter β (Labels p and β, respectively, in the table) are also shown, as well as the percentage of change with respect to the corresponding average precision of the original model (*%C*).

p	β	*ADI*	*CISI*	*CRANFIELD*	
		0.4130	0.2007	0.4314	*AP-11 (BNRM)*
5	0.6	0.4524	0.216	0.4314	*AV-11p*
		9.54	7.62	0.00	*%C*
5	0.7	0.4547	0.2212	0.4332	*AV-11p*
		10.10	10.21	0.42	*%C*
5	0.8	0.4676	0.2207	0.4316	*AV-11p*
		13.22	9.97	0.05	*%C*
10	0.6	0.4587	0.2182	0.4334	*AV-11p*
		11.07	8.72	0.46	*%C*
10	0.7	0.4681	0.22	0.4347	*AV-11p*
		13.34	9.62	0.76	*%C*
10	0.8	0.4695	0.221	0.4331	*AV-11p*
		13.68	10.11	0.39	*%C*
15	0.6	0.4678	0.2211	0.4332	*AV-11p*
		13.27	10.16	0.42	*%C*
15	0.7	0.4651	0.2203	0.434	*AV-11p*
		12.62	9.77	0.60	*%C*
15	0.8	0.468	0.2208	0.4329	*AV-11p*
		13.32	10.01	0.35	*%C*

Table 1. Results of the experiments with the new topology of the term subnetwork.

Although the results are sensible to the values of the two parameters, p and β, they do not vary greatly. In fact, the means and standard deviations of the percentage of change are, respectively, 12.24 and 1.49 for ADI, 9.58 and 0.81 for CISI, and 0.38 and 0.23 for CRANFIELD. We believe that the number of parents, p, should not be low (because this could prevent the inclusion of useful term relationships).

Analogously, with respect to β, this parameter should not be low (since in this case, the term relationships on a given term could be overload).

We can also observe that the effectiveness of the new model is even better than the performance of the polytree-based model in terms of retrieval success, at least for the three collections considered. This is a good side effect, because our initial goal was to increase the efficiency without degrading the effectiveness.

5 Concluding Remarks

In this paper an new topology for representing term relationships has been presented. Instead of using a polytree as the underlaying structure of the term subnetwork, we have designed a new graph, a bipartite graph (two layers of nodes representing the terms in the collection), that stores the strongest relationships among terms. The main advantage of this graph is that the exact propagation that had to be carried out in the original polytree is substituted by an evaluation of a probability function, being a very efficient method. The main application of this new model will be the retrieval of TREC documents, in which, taking into account its topology and the whole inference method, we think that it will be competitive and efficient.

We have shown with the experiments that this new model in which two term-layers are used to encode term relationships has a better behavior than the original model, although it depends on the collection being tested.

As this work is a first approach to find an alternative structure and substitute the polytree, we have thought several points in which the model may be modified to improve its performance. The first one is the design of more accurate ways of determining the strength of the relationships among terms, reflecting only positive dependences. At the same time, using this previous measure or designing a new one, to develop a method to select the best terms. This selection could be based only on co-occurrences or a combined way between co-occurrences and the cross entropy. To be completely sure that the terms are dependent, we could put into practice an independence test. A second aspect related to this point is the decision about the number of parents of each term. It could be more reasonable that this number was not the same for all the terms, being a term-dependent parameter. Also, the design of a new and more sophisticated probability function to be evaluated in the original layer of terms should be done, in which the β parameter is removed.

References

1. B. Cestnik. Estimating probabilities: A crucial task in Machine Learning. In *Proceedings of ECAI conference*, pages 147–149, 1990.
2. F. Crestani, M. Lalmas, C. J. van Rijsbergen, and L. Campbell. Is this document relevant?... probably. A survey of probabilistic models in information retrieval. *ACM Computing Survey*, 30(4):528–552, 1991.
3. L. M. de Campos and J. F. Huete. On the use of independence relationships for learning simplified belief networks. *Journal of Intelligent Systems*, 12:495–522, 1997.

4. Luis M. de Campos. Independency relationships and learning algorithms for singly connected networks. *Journal of Experimental and Theoretical Artificial Intelligence*, 10(4):511–549, 1998.
5. Luis M. de Campos, Juan M. Fernández, and Juan F. Huete. Query expansion in information retrieval systems using a Bayesian network-based thesaurus. In *Proceedings of the 14th UAI conference*, pages 53–60, 1998.
6. Luis M. de Campos, Juan M. Fernández, and Juan F. Huete. Building Bayesian network-based information retrieval systems. In *2nd LUMIS Workshop*, pages 543–552, 2000.
7. R. Fung and B. Del Favero. Applying Bayesian networks to information retrieval. *Communications of the ACM*, 38(2):42–57, 1995.
8. D. Geiger, A. Paz, and J. Pearl. Learning simple causal structures. *International Journal of Intelligent Systems*, 8:231–247, 1993.
9. J. Pearl. *Probabilistic Reasoning in Intelligent Systems: Networks of Plausible Inference*. Morgan and Kaufmann, 1988.
10. G. Rebane and J. Pearl. The recovery of causal polytrees from statistical data. *Uncertainty in Artificial Intelligence*, pages 175–182, 1989.
11. B. A. Ribeiro-Neto and R. R. Muntz. A belief network model for IR. In H. Frei, D. Harman, P. Schäble, and R. Wilkinson, editors, *19th ACM–SIGIR Conference*, pages 253–260. ACM, 1996.
12. G. Salton and M. J. McGill. *Introduction to Modern Information Retrieval*. McGraw-Hill, 1983.
13. H. R. Turtle and W. B. Croft. Evaluation of an inference network-based retrieval model. *Information Systems*, 9(3):187–222, 1991.

C-FOCUS: A continuous extension of FOCUS

* Antonio Arauzo, Jose Manuel Benítez, and Juan Luis Castro

Department of Computer Science and Artificial Intelligence (DECSAI)
University of Granada (Spain)
arauzo@decsai.ugr.es

Summary. This paper deals with the problem of feature selection. Almuallim and Dieterich [1] developed the FOCUS algorithm which performs optimal feature selection on boolean domains. In this paper an extension of FOCUS is developed to deal with discrete and continuous features. The extension, C-FOCUS, is verified on an artificial geometric figure classification problem and a real world classification problem.

1 Introduction

Feature selection help us to focus the attention of an induction algorithm in those features that are the best to predict a target concept. Although one might think that the more information available to an induction algorithm the better it works, this has revealed to be false for the following two main reasons. First, a large number of features in the input of induction algorithms may turn them very inefficient as memory and time consumers. And second, irrelevant data may confuse algorithms making them to reach false conclusions.

In feature selection, we are interested in finding the minimal set of features which allows us to induce the target concept. John, Kohavi and Pfleger[4] classify the features in three relevance classes: irrelevant, weakly relevant and strongly relevant. The FOCUS algorithm[1] is successful identifying the set with all strongly relevant and the minimal number of weakly relevant features to the target concept. As result of this, FOCUS is an ideal algorithm to use when a minimal set of features is required and noise free samples are available.

FOCUS always finds the optimal set through a complete search on the feature subset space in quasi-polynomial time. In [2] very interesting empirical results of the FOCUS algorithm are presented. It displays good performance even on some datasets with noise..

However FOCUS is limited to boolean domains, while many real problems have discrete and continuous attributes. In order to see if FOCUS good behavior could be

* This research has been supported by project CICYT-TIC2000-1362-C02-01

exported to other problem domains we have extended FOCUS-2[1] (the optimized version of FOCUS) to select features with different data types: nominal, discrete and continuous. The extension to continuous values has been done by defining a concept of what is considered to be distinct in a continuous domain, while the extension to nominal and discrete values is direct since this concept is clear on these domains.

In section 2, we describe FOCUS algorithm and its extension C-FOCUS. In section 3, we create a geometric figure classification problem, which is adequate to apply original FOCUS algorithm but with a mix of continuous and discrete features. Then the results of C-FOCUS application to this problem and a real world problem are shown. Finally, some conclusions and final remarks are collected in section 4.

2 Description of the Algorithm

The main idea of the original FOCUS algorithm is to identify all pairs of examples with a different boolean result. Each of these pairs is called a conflict, and FOCUS goal is to select the minimal set of features that solves all conflicts. A feature is considered to solve a conflict when its value is different between both examples. That is when the feature allow us to distinguish between the two examples.

It is clear when two values are different in a boolean or discrete domain, so it is clear when a conflict is solved by a boolean or discrete variable. But we need to define when two continuous values will be considered different. To this purpose our extension utilizes the absolute difference between the two values in the following simple way. All values in samples of a given feature are normalized to $[0, 1]$. If the difference is greater than a given threshold U the two values will be considered distinct.

FOCUS searches through the space of feature subsets to find the one with a minimal number of features that solves all conflicts.

This search can be done trying sequentially with all sets of $1, 2, 3, \ldots N$ variables until one set that solves all conflicts is found. But if one conflict is solved only by a feature X_i, we know that X_i should belong to the set of features selected. With this idea Almuallim and Dieterich[1] developed an optimized version of FOCUS: FOCUS-2.

Algorithm FOCUS-2(*Sample*)

1. If all the examples in *Sample* have the same class, return $\emptyset$.
2. Let G be the set of all conflicts generated from *Sample*.
3. $Queue = \{M_{\emptyset,\emptyset}\}$.
4. Repeat
   4.1 $M_{A,B}$ = Pop the first element in *Queue*.
   4.2 $OUT = B$.
   4.3 Let a be the conflict in G not covered by any of the features in A, such that $|Z_a - B|$ is minimized, where Z_a is the set of features covering a.
   4.4 For each $x \in Z_a - B$
       4.4.1 If $Sufficient(A \cup \{x\}, Sample)$, return $A \cup \{x\}$.

4.4.2 Insert $M_{A\cup\{x\},OUT}$ at the tail of *Queue*.
4.4.3 $OUT = OUT \cup \{x\}$.

end.

$M_{A,B}$ denotes the space of all feature subsets that include all of the features in the set A and none of the features in the set B.

As the sufficiency test of step 4.4.1, *Sufficient*(*Features*, *Sample*), we have used a simple search through *Sample* of two examples, with values not considered different in selected *Features*, that belong to a different class. If there are no such two examples the *Features* set is sufficient, not being sufficient otherwise.

3 Empirical Study

3.1 Geometric Figure Problem

Problem Description.

To test C-FOCUS we have created a simple geometric figure classification problem. Some examples are drawn from the following classes:

- Equilateral triangle
- Isosceles triangle
- Square
- Rectangle

These are the features considered for every figure:

- Number of sides (NSides)
- Longest side length (LS)
- Shortest side length (SS)
- Perimeter
- Area
- Shortest side length / longest side length (SS/LS)

The formulas and constant values of these features for the considered figures are shown in Fig. 1.

The process used to generate the samples has been the following: Figure samples have been generated by the following procedure:

Repeat N times (where N is the number of examples to generate)

- Choose a figure class (sampled from a uniform random distribution over $\{0,1,2,3\}$)
- Repeat until values satisfy restrictions
    - Generate side lengths (sampled from a uniform random distribution over $[0,1]$)

Fig. 1. Geometric figures and its features

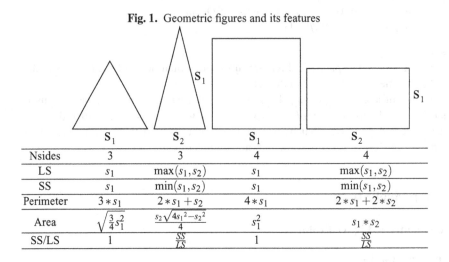

	s_1	s_2	s_1	s_2
Nsides	3	3	4	4
LS	s_1	$\max(s_1,s_2)$	s_1	$\max(s_1,s_2)$
SS	s_1	$\min(s_1,s_2)$	s_1	$\min(s_1,s_2)$
Perimeter	$3*s_1$	$2*s_1+s_2$	$4*s_1$	$2*s_1+2*s_2$
Area	$\sqrt{\frac{3}{4}s_1^2}$	$\frac{s_2\sqrt{4s_1^2-s_2^2}}{4}$	s_1^2	s_1*s_2
SS/LS	1	$\frac{SS}{LS}$	1	$\frac{SS}{LS}$

The restrictions named above are: In isosceles triangles, the sum of the two equal sides should be greater than the other side. And the difference between sides s_1 and s_2 in rectangles and isosceles triangles should be greater than 5%, to avoid them to be almost squares and equilateral triangles, respectively.

All of the above features are related to the classification problem. To test if our extension is able to reject all the irrelevant features, we have introduced other features with random values.

The goal is to select the minimal number of features that allow to classify each example as one of the 4 figure types.

Based on our previous knowledge of the problem, we know that, among the available features, the minimal set that allows to classify the 4 figure types correctly is {Number of sides, Longest side / Shortest side}. While other feature sets like {Longest side, Shortest side, Area} are also good for classification.

Results.

The tests have been made with different sample sets with a varying number of irrelevant features included and different size. We have created three types of samples by adding 1, 10 and 25 irrelevant features added. This way we can study the effect of the number of irrelevant features on the behaviour of the algorithm. We have used samples of 50, 100, 250 and 500 instances, for every of these sample types.

Running on the same datasets the C-FOCUS threshold parameter had been varied in the following values: 0.025, 0.05, 0.1 and 0.2. The results are shown in Tables: 1, 2, 3 and 4, respectively. Some of the feature names are abbreviated as indicated in the feature list at the problem description. Irrelevant variables are referred to as "IrrN" where N is the position of the variable.

C-FOCUS has found a sufficient set of features that allows to classify correctly in 41 cases. It reports that for a given threshold level the problem can not be solved

Table 1. Selected features on each dataset with U=0.025

Examples	Number of irrelevant features		
	1	10	25
50 NSides, SS/LS	NSides, SS/LS	SS, Perimeter
100 NSides, SS/LS	NSides, SS/LS	NSides, SS/LS
250 NSides, SS/LS	NSides, SS/LS	NSides, SS/LS
500 NSides, SS/LS	NSides, SS/LS	NSides, SS/LS

Table 2. Selected features on each dataset with U=0.05

Examples	Number of irrelevant features		
	1	10	25
50 NSides, SS/LS	NSides, SS/LS	NSides, SS/LS
100 NSides, SS/LS	NSides, SS/LS	NSides, SS/LS
250 NSides, SS/LS	NSides, SS/LS	NSides, SS/LS
500 NSides, SS/LS	NSides, SS/LS	NSides, SS/LS

Table 3. Selected features on each dataset with U=0.1

Examples	Number of irrelevant features		
	1	10	25
50 NSides, SS/LS	NSides, SS/LS, Irr0, Irr8	NSides, SS/LS, Irr19
100 NSides, SS/LS, Irr0, Area	NSides, SS/LS, Irr0, Irr2, Irr8	NSides, SS/LS, Irr4
250 NSides, SS/LS	NSides, SS/LS, Irr0, Irr3, Irr4	NSides, SS/LS, Irr0, Irr1, Irr21
500 (Not solved)	(Not solved)	NSides, SS/LS, Irr2, Irr10, Irr12

Table 4. Selected features on each dataset with U=0.2

Examples	Number of irrelevant features		
	1	10	25
50 NSides, SS/LS, Area, Irr0	NSides, SS, Irr0, Irr1, Irr4	NSides, SS/LS, Irr0, Irr2, Irr23
100 (Not solved)	NSides, SS/LS, SS, Irr0, Irr1, Irr6	NSides, SS/LS, SS, Irr2, Irr6, Irr12
250 (Not solved)	NSides, SS/LS, SS, Irr0, Irr1, Irr2, Irr3, Irr4	NSides, SS/LS, Irr0, Irr4, Irr6, Irr7, Irr18
500 (Not solved)	NSides, SS/LS	NSides, SS/LS, SS, Area, Perimeter, Irr2, Irr4, Irr12, Irr17

in 5 cases. And finally, only in 2 cases, which are from the smallest ones (50 example datasets), returns a not sufficient set of feature sets.

The threshold parameter has revealed to be very important to the results, as higher values make C-FOCUS to introduce more features than necessary and sometimes irrelevant.

3.2 Forest CoverType problem

This problem deals with getting the forest cover type for a 30x30 meter cell from a given set of 54 boolean and quantitative features. The dataset for this problem is available at the UCI KDD Archive[3].

We chose randomly 2000 examples from the dataset. C-FOCUS was run on them with different threshold levels, starting with 0.2 and dividing by 2 on each step. The first threshold that gave a feature selection was 0.0125 (previous ones found that the conflict set was unsolvable at that threshold level).

In order to test if the features selected by C-FOCUS are good for this classification problem, we have used a neural network as a classifier. We have compared the results obtained with CFOCUS + NN, NN without using feature selection and Relief-E[6] + NN.

Relief-E has been chosen because it is a very well known feature selection algorithm, and it is common in empirical studies [2]. Besides a similar version of Relief was chosen as a representative of filter feature selection methods to present the wrapper approach [5].

We used four training sets with 4000 examples each one and four disjoint 1000 examples test sets. We employed multilayered perceptrons with one hidden layer, initialized with uniform random weights. Standard back-propagation with a learning rate of 0.05 was used as the training method.

The performance of the neural network is expressed as the percentage of correct classification. Those obtained with a neural net without feature selection, namely, using all of the features are displayed on Table 5.

The features selected by C-FOCUS were: Elevation, Aspect, Slope, Horizontal-Distance-To-Hidrology, Vertical-Distance-To-Hydrology, and Horizontal-Distance-To-Roadways, that is 6 out of 54. Table 6 shows the results obtained with a net using only these features.

The Relief-E algorithm doesn't produce a feature set. Instead it renders a value of the relevance of each feature, which allow us to rank them. We selected the top six features to make it comparable to the C-FOCUS selection. So the features selected have been: Aspect, Horizontal-Distance-To-Roadways, Horizontal-Distance-To-Fire-Points, Horizontal-Distance-To-Hydrology, Slope, and Hillshade-3pm. Table 7 shows the results.

Table 5. Forest problem results without feature selection

Topology	Test set				Max	Mean
	1	2	3	4		
54-4-7	50.4	55.6	57.8	68.4	68.4	58.05
54-5-7	56.8	52.9	54.4	72	72	59.025
54-6-7	43.4	52.1	56.8	70.8	70.8	55.775
54-7-7	48.9	51.5	51.5	70.4	70.4	55.575
54-8-7	41.7	54.8	57.5	69.5	69.5	55.875
54-9-7	44.5	52.7	57.8	68.1	68.1	55.775
54-10-7	53	50.3	52	70.7	70.7	56.5
Max	56.8	55.6	57.8	72	72	60.55
Mean	48.385	52.843	55.400	69.986	69.986	56.654

Table 6. Forest problem results using C-FOCUS selection

Topology	Test set				Max	Mean
	1	2	3	4		
6-4-7	55.6	57.1	64	58.7	64	58.85
6-5-7	61.8	50	64.5	58.2	64.5	58.625
6-6-7	59.5	52.3	69.2	62.2	69.2	60.8
6-7-7	58.7	53	60.3	66.2	66.2	59.55
6-8-7	59.8	52.4	71.7	63.9	71.7	61.95
6-9-7	63.8	54.7	67.7	66.3	67.7	63.125
6-10-7	59.5	52.7	60.8	64.5	64.5	59.375
Max	63.8	57.1	71.7	66.3	71.7	64.725
Mean	59.814	53.171	65.457	62.857	65.457	60.325

Table 7. Forest problem results using RELIEF-E

Topology	Test set				Max	Mean
	1	2	3	4		
6–4-7	21.5	25.5	40.8	41.1	41.1	32.225
6–5-7	25.3	26.6	45	41.3	45	34.55
6–6-7	25.3	28.9	41.4	45.4	45.4	35.25
6–7-7	23.8	25.5	42.5	48.6	48.6	35.1
6–8-7	32	23.8	42	42.8	42.8	35.15
6–9-7	28.5	22.8	43.4	46	46	35.175
6–10-7	25.3	22.3	40.3	47.7	47.7	33.9
Max	32	28.9	45	48.6	48.6	38.625
Mean	25.957	25.057	42.200	44.700	44.700	34.479

4 Summary and Conclusions

We have developed the C-FOCUS algorithm as an extension of the FOCUS [1] algorithm to discrete and continuous domains. In this way it can be used in a wider set of problems.

This algorithm is recommended in classification problems in which we have noise free samples and the main goal is to reduce the number of features. We have created such a problem and found another appropriate real world problem. We have tested C-FOCUS algorithm on both of them and obtained good results with reduced feature sets.

The performance of C-FOCUS improves on the performance of such a high reputed algorithm as Relief-E.

References

1. Hussein Almuallim and Thomas G. Dietterich. Learning boolean concepts in the presence of many irrelevant features. *Artificial Intelligence*, 69(1-2):279–305, 1994.
2. M. Dash and H. Liu. Feature selection for classification. *Intelligent Data Analysis*, 1(1-4):131–156, 1997.
3. S. Hettich and S. D. Bay. The uci kdd archive. http://kdd.ics.uci.edu/, 1999.
4. George H. John, Ron Kohavi, and Karl Pfleger. Irrelevant features and the subset selection problem. In *International Conference on Machine Learning*, pages 121–129, 1994. Journal version in AIJ, available at http://citeseer.nj.nec.com/13663.html.
5. Ron Kohavi and George H. John. Wrappers for feature subset selection. *Artificial Intelligence*, 97(1-2):273–324, 1997.
6. Igor Kononenko. Estimating attributes: Analysis and extensions of RELIEF. In *European Conference on Machine Learning*, pages 171–182, 1994.

2D Image registration with iterated local search

Oscar Cordón[1], Sergio Damas[2], and Eric Bardinet[3]

[1] Department of Computer Science and A. I., University of Granada,
18071 Granada, Spain. ocordon@decsai.ugr.es
[2] Department of Software Engineering, University of Granada,
18071 Granada, Spain. sdamas@ugr.es
[3] INRIA, Epidaure Project, Sophia Antipolis, France. ebard@sophia.inria.fr

Summary. The ability to establish a mapping between the information of two different images and to estimate the geometrical transformation it is supposed it has been applied, are two open problems in computer vision. Indeed, it is a crucial task for a wide range of applications [3] like the integration of information from different sensors, the changes in images taken at different times or under different conditions, the inference of three dimensional information, the model based object recognition, etc. In this work we try to take advantage of the information we can infer from the skeleton of an image. Then, we define a global optimization function for both problems. This is a difficult optimization problem and we apply the well known *Iterated Local Search* (ILS) with great success.

Key words: Image registration, iterated local search, skeleton, medial axis transform.

1 Introduction

Image registration is a fundamental task in image processing used to finding a correspondence (or transformation) among two or more pictures taken under different conditions like different times, using different sensors, from different viewpoints, or a combination of them. Over the years, image registration has been applied to a broad range of situations from remote sensing to medical images or artificial vision and different techniques have been independently studied resulting in a large body of research (in [3], a classification of different registration techniques and applications can be found).

In recent years, a new family of search and optimization algorithms has arised based on extending basic heuristic methods by including them into an iterative framework augmenting its exploration capabilities. This group of advanced approximate algorithms is called *metaheuristics* and an overview on the most prominent ones can be found in [9].

In recent literature, we can find different aproaches to the matching and registration problems ([10], [7]) from the metaheuristics point of view. In this work, we apply the *Iterated Local Search* (ILS) [8] to solve registration problems. Our main contributions are related to the fact of jointly solving matching and registration transformation problems using skeleton derived information.

To do so, in section 2 we present the concept of skeleton and medial axis transformation in the field of shape analysis in computer vision and the way we can use them to generate an object partition. In section 3 we give a brief overview of the concept of image registration and one important method to understand our work. Next, section 4 shows the way we can apply ILS to the registration problem. In section 5 we expose computational results we have achieved applying our method. Finally, in section 6 we review the work we've done and future improvements to be considered.

2 Shape characterization from the medial axis

As we stated in [6] the skeleton of an object is formed by pieces of curves (2D and 3D cases) and surfaces (3D case only) linked together by junctions. The pieces of curves and surfaces which do not contain any junctions are called pure curves and surfaces (therefore the connected components which remain when removing junction elements from a skeleton are pure curves and surfaces). In the following, we will refer to these pieces of skeleton by the expression *skeleton parts*. Finally by frontier points we denote the points which end skeleton parts and are not in contact with junction components (see Figure 1).

As the skeleton $SK(X)$ of an object X is a thin set, it allows us to classify the object topologically, thus allowing the study of the skeleton topology and therefore of the object topology. The topological classification of $SK(X)$, denoted $SK_c(X)$, attaches to each point of the skeleton one of the following labels:

- **Type F**: Frontier Point   - **Type C**: Pure Curve Point
- **Type J**: Junction Point   - **Type S**: Pure Surface Point (3D case only)

depending of which component does the current skeleton point belong to, respectively Frontier, Junction, Pure Curve and Pure Surface.

As an example, Figure 1 illustrates the fact that branches of the skeleton of a 2D object are pure curves linked by junction components, and frontier components are the ends of the skeleton branches which are not in contact with a junction element.

From the **skeleton parts** labeling, we also infer a meaningful partition of the object into regions, each of these regions being associated to one of the skeleton parts. So, now we have skeleton parts and associated object regions. To each skeleton part $P^i_{SK(X)}$ and object region R^i_X we can attach different attributes: i) $P^i_{SK(X)}$ size compared to skeleton size; ii) variation of the distance map along $P^i_{SK(X)}$; iii) R^i_X region size relative to the object size; iv) variation of the curvature sign along $P^i_{SK(X)}$

We have also characterized different skeleton points: **junctions** can be detected and labeled by a connected components extraction of type **J** pixels of $SK_c(X)$. To each skeleton junction $J^i_{SK(X)}$ we can attach different attributes: its order as defined

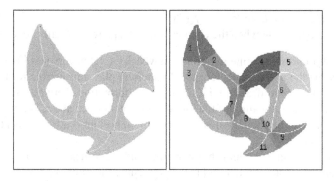

Fig. 1. On the left, topologically characterized skeleton of a 2D object. Skeleton branches (pure curves) are in yellow (or light gray), junction components in red (or black) and frontier components in black. On the right, object partition.

by the number of *skeleton parts* which meet at junction $J^i_{SK(X)}$; the value of ρ at the junction $J^i_{SK(X)}$ compared to the maximal value of ρ in $SK(X)$.

Finally, we have identified **frontier points** in the skeleton. One possible attribute related to these points is the relative value of ρ. The larger the value of ρ, the smoother the curvature change at the boundary will be.

3 Image registration

Image registration can be defined as a mapping between two images (I_1 and I_2) both spatially and with respect to intensity:

$$I_2(x,y,z,t) = g(I_1(f(x,y,z,t))) \tag{1}$$

We can usually find situations where intensity difference is inherent to scene changes, and thus intensity transform estimation given by g is not necessary. In this contribution, we will consider f represents a similarity transformation, i.e. rotation, translation and uniform scaling.

The well known Iterative Closest Point *(ICP)* method was proposed by Besl and McKay [2], and later on extended in different papers ([11], [5]):

- The point set P with N_p points $\mathbf{p_i}$ from the data shape and the model X, with N_x supporting geometric primitives: points, lines, or triangles are given.
- The iteration is initialized by setting $P_0 = P$, the registration transformation by $\mathbf{q_0} = [1,0,0,0,0,0,0]^t$, and $k = 0$. Next four steps are applied until convergence within a tolerance $\tau > 0$.
  1. Compute the matching between the data (scene) and model points by the closest assignment rule: $Y_k = C(P_k, X)$
  2. Compute the registration: $f_k(P_0, Y_k)$

3. Apply the registration: $P_{k+1} = f_k(P_0)$
4. Terminate iteration when the change in mean square error falls below τ

The algorithm presents some important drawbacks: (i) it is very sensitive to outlier presence; (ii) the initial states for global matching play a basic role for the success of the method when dealing with important deformations between model and scene points; (iii) the estimation of the initial states mentioned above is not a trivial task, and (iv) the cost of a local adjustment can be important if a low percentage of oclussion is present.

In view of the later, the algorithm performs bad when dealing with important transformations. As Zhang stated in [11]: "we assume the motion between the two frames is small or approximately known", hence this is a precondition of the algorithm to get reasonable results. Figure 2 shows several examples of that with one of the shapes considered in our experimental study.

ICP Performance							
	Ground truth			ICP estimation			
Tr.	Rot.	Tra.	Scale	Rot.	Tra.		Scale	MSE
1	20°	0	0	1	19.99 °	0	0	0.99	$\simeq 0$
2	90°	0	0	1	2.47 °	0.77	-4.92	-29.92	535.47

Table 1. Ground truth and ICP estimation of different transformations (see figure 2). ICP performance is quite good if small transformations are applied (first row). ICP weak point appears with bigger transformations (second row).Mean Square Error: *MSE* (see section 5) is also presented for both experiments.

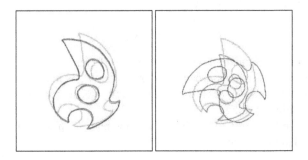

Fig. 2. ICP performance differs when dealing with small or big transforms. On the left: in green (light gray), original shape; in blue (black), original shape with a 20° rotation; in red (dark gray), ICP estimation applied to the original object. On the right: same distribution of colours for a 90° rotation.

4 Iterated local search for the 2D registration problem

Instead of following previous approaches ([1],[4]) based on searching for a good matching and then solving for the registration transformation, in our work we use the ILS metaheuristic for jointly solving our two-fold registration problem finding a good matching between both image points and getting the best similarity registration transformation we supposed it has been applied. To do so, the basics of ILS are first described, and the local search (LS) algorithm considered and the different ILS components are later analyzed.

4.1 The iterated local search metaheuristic

ILS [8] belongs to the group of metaheuristics that extend classical LS methods by adding diversification capabilities. This way, ILS is based on a wrapper around a specific LS algorithm by generating multiple initial solutions to it as follows:

> **procedure** *Iterated Local Search*
> s_0 = GenerateInitialSolution
> s^* = LocalSearch(s_0)
> **repeat**
> s' = Perturbation (s^*, *history*)
> $s^{*'}$ = LocalSearch(s')
> s^* = AcceptanceCriterion(s^*, $s^{*'}$, *history*)
> **until** termination condition met
> **end**

Hence, the algorithm starts by applying LS to an initial solution and iterates a procedure where a strong perturbation is applied to the current solution s^* (in order to move it away from its local neighborhood), and the solution so obtained is then considered as initial starting point for a new LS, from which another locally optimal solution $s^{*'}$ is obtained. Then, a decision is made between s^* and $s^{*'}$ to get the new current solution for the next iteration.

4.2 The local search algorithm considered

As said, this LS procedure allows us to obtain a complete solution to the 2D registration problem: a point matching between the data and the model shapes, and a registration transformation to move the former into the latter. To do so, we only search in the matching space (only the point matching is encoded in the LS solution) and derive the registration by a least squares estimation as done in the ICP based methods (see section 3).

The point matching between both images is represented as a permutation π of size $N = max(N_1, N_2)$, with N_1 and N_2 being the number of points in the data and model shapes, respectively. If $N_1 \geq N_2$, then $\pi(i)$ represents the model point associated to the data point i and viceversa. Notice that this representation has two main pros: (i) it is based on a permutation, a very common structure in the field (used for example to solve the traveling salesman and the quadratic assignment problems), and (ii) it allows us to deal with the case when both images have a different number of points, thus automatically discarding outliers.

Moreover, the other novelty of our method is that the features of the shape are used to guide both the matching and the registration. This way, the objective function will include information regarding both as follows:

$$\min_{M,\theta,t,s} E(M,\theta,t,s) = w_1 \cdot \sum_{i=1}^{N_1} \sum_{j=1}^{N_2} M_{ij} \|X_i - t - sR(\theta)Y_j\|^2 + w_2 \cdot f(M) \qquad (2)$$

where the first term stands for the registration error (M is the binary matrix storing the matching encoded in π and θ,t,s are the similarity transform parameters to be estimated (rotation, translation and scale respectively)), the second one for the matching error, and w_1, w_2 are weighting coefficients defining the relative importance of each.

As regards the second term, there are different ways to define the f function evaluating the goodness of the matching stored in M as a big amount of information can be obtained from the medial axis ([6]). In this contribution, we have chosen the following:

$$f(M) = 0.75 \cdot pointtype + 0.25 \cdot (medaxis + length + izs)$$

where *pointtype* measures the error associated to the assignment of points of different types and the remaining three criteria refer to the variation of the distance map along different branches (see section 2).

Finally, the neighborhood operator is the usual 2-opt exchange, based on selecting two positions in π and exchanging their values. The LS considered is the first improvement variant, where the whole neighboorhod is generated to obtain the best neighbor and the algorithm iterates till the latter is not better than the current solution, i.e., the local search stops in a local optimum.

4.3 The iterated local search components

- *GenerateInitialSolution:* a random permutation is computed.
- *Perturbation:* as a stronger change than the one performed by the 2-opt LS neighborhood operator is needed, we deal with the random exchange of the positions of the values included within a randomly selected sublist of size $\frac{N}{a}$, with $a \in \{2,3,4,5,6\}$ (of course, the smaller the value of a, the stronger the perturbation applied).
- *AcceptanceCriterion:* we select the best of s^* and $s^{*'}$ as current solution, which is shown to yield good performance for a variety of problems.
- *Termination condition:* stop when a fixed number of iterations is reached.

5 Experiments and analysis of results

We have applied the algorithm to two different shapes, over which different rotations, translations and scales have been developed to test the behavior of the technique. In the first row of figure 3 we show the original shapes and the corresponding partitions induced by medial axis transform, the information that will be used during the registration transformation computation to guide the matching among points.

The ILS algorithm has been run during 50 iterations considering $a = 2$ for the perturbation operator. The weights in the objective functions have been defined as $w_1 = 1$ and $w_2 = 2 \cdot \frac{mse(R_{initial})}{f(M(\pi_{initial}))}$, with $R_{initial}$ and $f(M(\pi_{initial}))$ being respectively the registration and matching errors of the initial solution.

The results obtained are shown in rows 2 to 6 of figure 3. The second row corresponds to a simple 60° rotation of the first original object; the next column shows its partition; in the third column we store the representation of the mapping among different points of the different object (we have used several colours to distinguish the mappings of each kind of point). It is important to see in this column the way in which only points of same type match each other, even in the presence of noise in part of the object (third and fourth row) and different projection and junction points differ from the original and the transformed shape. Last column reflects the effect of applying the estimated transformation to the original shape and doing a superposition on the transformed object. Third row corresponds to a 180° rotation with the presence of noise in part of the object. Next row adds a scale transformation of 0.5 to the original object. In the fifth and the sixth rows, $30°$ and $90°$ with a 0.5 scale value transforms have been respectively applied to the top right shape.

An usual error measure in the field of image registration is the *Mean Square Error (MSE)*, tipically given by:

$$MSE = \sum_{i=1}^{N} \|f(x_i) - y_i\|^2$$

where:

- f is the estimated registration function.
- $x_i, i = 1, 2, ..., N$, are the N scene points (the registration function f is applied to every scene point).
- $y_i, i = 1, 2, ..., N$, are the N model points matching the scene ones

Table 2 shows the difference in *MSE* of ICP and ILS when solving the registration problem related to the five transformations of figure 3. First column corresponds to the transformation number. The second one stores ground truth transformation applied to each of the original shapes (solution of the registration problem). Third column includes the ICP estimation of such a transformation. Finally, last column shows the results achieved by ILS.

It can be seen that ILS behaviour is quite good in all the experiments, even in the presence of noise (transformations n. 2 and 3) and this is the reason for the MSE to be bigger than it is in the rest of ILS estimations. Meanwhile, ICP performance is not

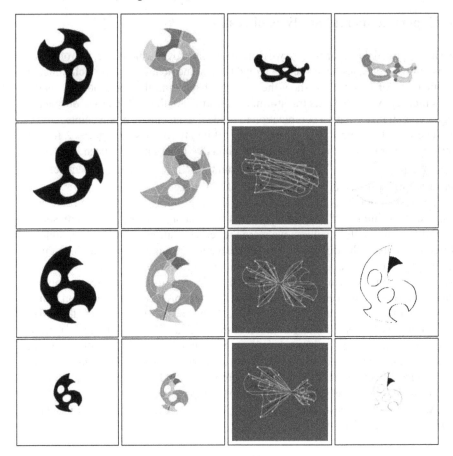

Fig. 3. First row: two original shapes and partition induced by the medial axis. Next rows (from left to right): similarity transform applied to the original shape, object partition, matching results: different point types with different colours (yellow (light gray)→frontiers, blue (dark gray)→junctions, cyan (light gray)→projections).Finally the superposition of the transformed image and the result of applying the estimated transformation to the original shape.

so good but in the fourth experiment. The reason is ICP returns good results with not so strong transformations and this is only the case of the fourth test, a relative small rotation (30°) with neither translation nor scale transformations.

6 Concluding remarks

In this contribution, we have formulated the image registration problem as a two fold one in order to jointly solve for the matching and the search for the similarity

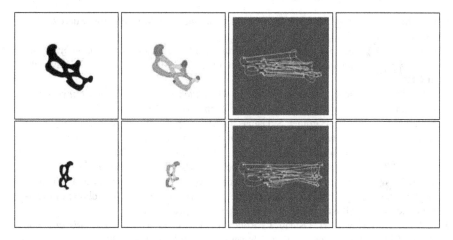

Fig. 4. From left to right: similarity transform applied to the original shape (shown in Figure 3), object partition, matching results: different point types with different colours (yellow (light gray)→frontiers, blue (dark gray)→junctions, cyan (light gray)→projections).Finally the superposition of the transformed image and the result of applying the estimated transformation to the original shape.

ICP vs. ILS Performance												
Tr.	*Ground truth*			*ICP*				*ILS*			
N.	*Rot.*	*Tra.*	*Scale*	*Rot.*	*Tra.*		*Scale*	*MSE*	*Rot.*	*Tra.*		*Scale*	*MSE*
1	60°	0	0	1	4.63 °	43.18	-41.93	0.82	5180.45	59.97 °	0.02	0.01	0.99	2.81
2	180°	0	0	1	14.88 °	19.56	-16.62	0.95	807.69	178.86 °	-3.1	3.7	1.01	58.45*
3	180°	0	0	0.5	4.60 °	-11.37	75.66	0.30	112.58	178.71 °	-1.2	1.1	0.50	18.43*
4	30°	0	0	1	29.99 °	0.18	1.52	1.00	67.77	30.01 °	0.15	0.05	1.00	5.12
5	90°	0	0	0.5	32.07 °	2.65	76.68	0.26	918.74	90.31 °	0.03	-0.02	0.49	1.06

Table 2. ICP and ILS estimation of different transformations shown in figure 3.(*)Noise presence enables a large *MSE* with a good registration estimation.

parameters of the registration transform. To do so, we take advantage of the information infered from the medial axis of the object and we use the ILS metaheuristic to face such a complex optimization task. We have presented results with important transformations applied to different model shapes.

References

1. E. Bardinet, S. Fernández-Vidal, S. Damas, G. Malandain, N. Pérez de la Blanca, *Structural object matching*, In 2nd International Symposium on Advanced Concepts for Intelligent Vision Systems (ACIVS 2000), vol. II, pp. 73-77, Baden-Baden, Germany, 2000.
2. P.J. Besl, N.D. McKay, *A method for registration of 3-D shapes*, IEEE Transactions on Pattern Analysis and Machine Intelligence, vol. 14, pp. 239-256, 1992.

3. L.G. Brown, *A survey of image registration techniques*, ACM Computing Surveys, vol. 24, no. 4, pp. 325-376, 1992.

4. O. Cordón, S. Damas, J. Santamaría, *New Proposals for 3D Image Registration by Means of Evolutionary Computation*, Technical Report DECSAI-02-7-03, DECSAI, University of Granada, 2002

5. J. Feldmar, N. Ayache, *Rigid, affine and locally affine registration of free-form surfaces*, Int. Journal of Computer Vision, vol. 18, no. 2, pp. 99-119, 1996.

6. S. Fernández-Vidal, E. Bardinet, S. Damas, G. Malandain, N. Pérez de la Blanca, *Object representation and comparison inferred from its medial axis*, In International Conference on Pattern Recognition (ICPR 00), vol. 1, pp. 712-715, Barcelona, Spain, 2000.

7. K.P. Han, K.W. Song, E.Y. Chung, S.J. Cho, Y.H. Ha, *Stereo matching using genetic algorithm with adaptive chromosomes* Pattern Recognition, vol. 34, pp. 1729-1740, 2001

8. H. Ramalhinho, O. Martin, O. T. and Stützle, *Iterated Local Search*, Handbook of Meta-heuristics, eds. F. Glover and G. Kochenberger, To appear, 2002.

9. S. Voss, S. Martello, I.H. Osman, C. Roucairol (Eds.) *Advances and Trends in Local Search Paradigms for Optimization*, Kluwer Academic Publishers, 1999.

10. S. M. Yamany, M. N. Ahmed, A. A. Farag, *A new genetic-based technique for matching 3D curves and surfaces*, Pattern Recog., vol. 32, pp. 1817-1820, 1999.

11. Z. Zhang, *Iterative point matching for registration of free-form curves and surfaces*, Int. Journal of Computer Vision, vol. 13, no. 2, pp. 119-152, 1994.

Evolutionary Multiobjective Optimization: Current and Future Challenges

Carlos A. Coello Coello

CINVESTAV-IPN
Evolutionary Computation Group
Departamento de Ingeniería Eléctrica
Sección de Computación
Av. Instituto Politécnico Nacional No. 2508
Col. San Pedro Zacatenco
México, D.F. 07300, MEXICO
ccoello@cs.cinvestav.mx

Summary. In this paper, we will briefly discuss the current state of the research on evolutionary multiobjective optimization, emphasizing the main achievements obtained to date. Achievements in algorithmic design are discussed from its early origins until the current approaches which are considered as the "second generation" in evolutionary multiobjective optimization. Some relevant applications are discussed as well, and we conclude with a list of future challenges for researchers working (or planning to work) in this area in the next few years.

1 Introduction

Several years ago, researchers realized that the principle of "survival of the fittest" used by nature could be simulated to solve problems [11]. This gave rise to a type of heuristics known as Evolutionary Algorithms (EAs). EAs have been very popular in search and optimization tasks in the last few years with a constant development of new algorithms, theoretical achivements and novel applications [15, 1].

One of the emergent research areas in which EAs have become increasingly popular is multiobjective optimization. In multiobjective optimization problems, we have two or more objective functions to be optimized at the same time, instead of having only one. As a consequence, there is no unique solution to multiobjective optimization problems, but instead, we aim to find all of the good trade-off solutions available (the so-called Pareto optimal set).

The first implementation of a multi-objective evolutionary algorithm dates back to the mid-1980s [36]. Since then, a considerable amount of research has been done in this area, now known as evolutionary multi-objective optimization (EMO for short). The growing importance of this field is reflected by a significant increment (mainly during the last eight years) of technical papers in international conferences and peer-reviewed journals, books, special sessions in international conferences and interest groups on the Internet [6].[1]

Evolutionary algorithms seem also particularly desirable for solving multiobjective optimization problems because they deal simultaneously with a set of possible solutions (the so-called population) which allows us to find several members of the Pareto optimal set in a single run of the algorithm, instead of having to perform a series of separate runs as in the case of the traditional mathematical programming techniques. Additionally, evolutionary algorithms are less susceptible to the shape or continuity of the Pareto front (e.g., they can easily deal with discontinuous and concave Pareto fronts), whereas these two issues are a real concern for mathematical programming techniques [6].

This paper deals with some of the current and future research trends in evolutionary multiobjective optimization. The paper is organized as follows. Section 2 presents some basic concepts used in multiobjective optimization. Section 3 briefly describes the origins of evolutionary multiobjective optimization. Section 4 introduces the so-called first generation multiobjective evolutionary algorithms. Second generation multiobjective evolutionary algorithms are discussed in Section 5, emphasizing the role of elitism in evolutionary multiobjective optimization. Finally, Section 7 discusses some of the research trends that are likely to be predominant in the next few years.

2 Basic Concepts

The emphasis of this paper is the solution of multiobjective optimization problems (MOPs) of the form:

$$\text{minimize } [f_1(\mathbf{x}), f_2(\mathbf{x}), \dots, f_k(\mathbf{x})] \tag{1}$$

subject to the m inequality constraints:

$$g_i(\mathbf{x}) \geq 0 \quad i = 1, 2, \dots, m \tag{2}$$

and the p equality constraints:

$$h_i(\mathbf{x}) = 0 \quad i = 1, 2, \dots, p \tag{3}$$

where k is the number of objective functions $f_i : \mathbb{R}^n \to \mathbb{R}$. We call $\mathbf{x} = [x_1, x_2, \dots, x_n]^T$ the vector of decision variables. We wish to determine from among the set $\mathcal{F}$ of all vectors which satisfy (2) and (3) the particular set of values $x_1^*, x_2^*, \dots, x_n^*$ which yield the optimum values of all the objective functions.

---

[1] The author maintains an EMO repository with over 1000 bibliographical entries at: `http://delta.cs.cinvestav.mx/~ccoello/EMOO`, with mirrors at `http://www.lania.mx/~ccoello/EMOO/` and `http://www.jeo.org/emo/`

2.1 Pareto optimality

It is rarely the case that there is a single point that simultaneously optimizes all the objective functions. Therefore, we normally look for "trade-offs", rather than single solutions when dealing with multiobjective optimization problems. The notion of "optimality" is therefore, different in this case. The most commonly adopted notion of optimality is the following:

We say that a vector of decision variables $\mathbf{x}^* \in \mathcal{F}$ is *Pareto optimal* if there does not exist another $\mathbf{x} \in \mathcal{F}$ such that $f_i(\mathbf{x}) \leq f_i(\mathbf{x}^*)$ for all $i = 1, \ldots, k$ and $f_j(\mathbf{x}) < f_j(\mathbf{x}^*)$ for at least one j.

In words, this definition says that $\mathbf{x}^*$ is Pareto optimal if there exists no feasible vector of decision variables $\mathbf{x} \in \mathcal{F}$ which would decrease some criterion without causing a simultaneous increase in at least one other criterion. Unfortunately, this concept almost always gives not a single solution, but rather a set of solutions called the *Pareto optimal set*. The vectors $\mathbf{x}^*$ correspoding to the solutions included in the Pareto optimal set are called *nondominated*. The image of the Pareto optimal set under the objective functions is called *Pareto front*.

3 On the origins of evolutionary multiobjective optimization

The first actual implementation of what it is now called a multi-objective evolutionary algorithm (or MOEA, for short) was Schaffer's *Vector Evaluated Genetic Algorithm* (VEGA), which was introduced in the mid-1980s, mainly aimed for solving problems in machine learning [36]. VEGA basically consisted of a simple genetic algorithm (GA) with a modified selection mechanism. At each generation, a number of sub-populations were generated by performing proportional selection according to each objective function in turn. Thus, for a problem with k objectives, k sub-populations of size M/k each would be generated (assuming a total population size of M). These sub-populations would then be shuffled together to obtain a new population of size M, on which the GA would apply the crossover and mutation operators in the usual way. Schaffer realized that the solutions generated by his system were nondominated in a local sense, because their nondominance was limited to the current population, which was obviously not appropriate. Also, he noted a problem that in genetics is known as "speciation" (i.e., we could have the evolution of "species" within the population which excel on different aspects of performance). This problem arises because this technique selects individuals who excel in one dimension of performance, without looking at the other dimensions. The potential danger doing that is that we could have individuals with what Schaffer called "middling" performance[2] in all dimensions, which could be very useful for compromise solutions, but which will not survive under this selection scheme, since they are not in the extreme for any dimension of performance (i.e., they do not produce the best value for any

---

[2] By "middling", Schaffer meant an individual with acceptable performance, perhaps above average, but not outstanding for any of the objective functions.

objective function, but only moderately good values for all of them). Speciation is undesirable because it is opposed to our goal of finding Pareto optimal solutions. Although VEGA's speciation can be dealt with using heuristics or other additional mechanisms, it remains as the main drawback of VEGA.

From the second half of the 1980s up to the first half of the 1990s, few other researchers developed MOEAs. Most of the work reported back then involves rather simple evolutionary algorithms that use an aggregating function (linear in most cases) [23], lexicographic ordering [14], and target-vector approaches (i.e., nonlinear aggregating functions) [19]. All of these approaches were strongly influenced by the work done in the operations research community and in most cases did not require any major modifications to the evolutionary algorithm adopted.

The algorithms proposed in this initial period are rarely referenced in the current literature except for VEGA (which is still used by some researchers). However, the period is of great importance because it provided the first insights into the possibility of using evolutionary algorithms for multiobjective optimization. The fact that only relatively naive approaches were developed during this stage is natural considering that these were the initial attempts to develop multiobjective extensions of an evolutionary algorithm. Such approaches kept most of the original evolutionary algorithm structure intact (only the fitness function was modified in most cases) to avoid any complex additional coding. The emphasis in incorporating the concept of Pareto dominance into the search mechanism of an evolutionary algorithm would come later.

4 MOEAs: First Generation

The major step towards the first generation of MOEAs was given by David E. Goldberg on pages 199 to 201 of his famous book on genetic algorithms published in 1989 [15]. In his book, Goldberg analyzes VEGA and proposes a selection scheme based on the concept of Pareto optimality. Goldberg not only suggested what would become the standard first generation MOEA, but also indicated that stochastic noise would make such algorithm useless unless some special mechanism was adopted to block convergence. First generation MOEAs typically adopted niching or fitness sharing for that sake. The most representative algorithms from the first generation are the following:

1. **Nondominated Sorting Genetic Algorithm** (NSGA): This algorithm was proposed by Srinivas and Deb [37]. The approach is based on several layers of classifications of the individuals as suggested by Goldberg [15]. Before selection is performed, the population is ranked on the basis of nondomination: all nondominated individuals are classified into one category (with a dummy fitness value, which is proportional to the population size, to provide an equal reproductive potential for these individuals). To maintain the diversity of the population, these classified individuals are shared with their dummy fitness values. Then this group of classified individuals is ignored and another layer of nondominated

individuals is considered. The process continues until all individuals in the population are classified. Since individuals in the first front have the maximum fitness value, they always get more copies than the rest of the population. This allows to search for nondominated regions, and results in convergence of the population toward such regions. Sharing, by its part, helps to distribute the population over this region (i.e., the Pareto front of the problem).

2. **Niched-Pareto Genetic Algorithm** (NPGA): Proposed by Horn et al. [22]. The NPGA uses a tournament selection scheme based on Pareto dominance. The basic idea of the algorithm is the following: Two individuals are randomly chosen and compared against a subset from the entire population (typically, around 10% of the population). If one of them is dominated (by the individuals randomly chosen from the population) and the other is not, then the nondominated individual wins. When both competitors are either dominated or nondominated (i.e., there is a tie), the result of the tournament is decided through fitness sharing [16].

3. **Multi-Objective Genetic Algorithm** (MOGA): Proposed by Fonseca and Fleming [12]. In MOGA, the rank of a certain individual corresponds to the number of chromosomes in the current population by which it is dominated. Consider, for example, an individual x_i at generation t, which is dominated by $p_i^{(t)}$ individuals in the current generation.
   The rank of an individual is given by [12]:

$$\text{rank}(x_i, t) = 1 + p_i^{(t)} \tag{4}$$

All nondominated individuals are assigned rank 1, while dominated ones are penalized according to the population density of the corresponding region of the trade-off surface. Fitness assignment is performed in the following way [12]:
   a) Sort population according to rank.
   b) Assign fitness to individuals by interpolating from the best (rank 1) to the worst (rank $n \leq M$, where M is the total population size) in the way proposed by Goldberg (1989), according to some function, usually linear, but not necessarily.
   c) Average the fitnesses of individuals with the same rank, so that all of them are sampled at the same rate. This procedure keeps the global population fitness constant while maintaining appropriate selective pressure, as defined by the function used.

The main questions raised during the first generation were:

- Are aggregating functions (so common before and even during the golden years of Pareto ranking) really doomed to fail when the Pareto front is non-convex [7]? Are there ways to deal with this problem? Is it worth trying? Some recent work seems to indicate that even linear aggregating functions are not death yet [25].

- Can we find ways to maintain diversity in the population without using niches (or fitness sharing), which requires a process $O(M^2)$ where M refers to the population size?
- If assume that there is no way of reducing the $O(kM^2)$ process required to perform Pareto ranking (k is the number of objectives and M is the population size), how can we design a more efficient MOEA?
- Do we have appropriate test functions and metrics to evaluate quantitatively an MOEA? Not many people worried about this issue until near the end of the first generation. During this first generation, practically all comparisons were done visually (plotting the Pareto fronts produced by different algorithms) or were not provided at all (only the results of the proposed method were reported).
- When will somebody develop theoretical foundations for MOEAs?

Summarizing, the first generation was characterized by the use of selection mechanisms based on Pareto ranking and fitness sharing was the most common approach adopted to maintain diversity. Much work remained to be done, but the first important steps towards a solid research area had been already taken.

5 MOEAs: Second Generation

The second generation of MOEAs was born with the introduction of the notion of elitism. In the context of multiobjective optimization, elitism usually (although not necessarily) refers to the use of an external population (also called secondary population) to retain the nondominated individuals. The use of this external population (or file) raises several questions:

- How does the external file interact with the main population?
- What do we do when the external file is full?
- Do we impose additional criteria to enter the file instead of just using Pareto dominance?

Elitism can also be introduced through the use of a $(\mu + \lambda)$-selection in which parents compete with their children and those which are nondominated (and possibly comply with some additional criterion such as providing a better distribution of solutions) are selected for the following generation.

The previous points bring us to analyze in more detail the true role of elitism in evolutionary multiobjective optimization. For that sake, we will review next the way in which some of the second-generation MOEAs implement elitism:

1. **Strength Pareto Evolutionary Algorithm** (SPEA): This algorithm was introduced by Zitzler and Thiele [43]. This approach was conceived as a way of integrating different MOEAs. SPEA uses an archive containing nondominated solutions previously found (the so-called external nondominated set). At each generation, nondominated individuals are copied to the external nondominated set. For each individual in this external set, a *strength* value is computed. This

strength is similar to the ranking value of MOGA, since it is proportional to the number of solutions to which a certain individual dominates. It should be obvious that the external nondominated set is in this case the elitist mechanism adopted. In SPEA, the fitness of each member of the current population is computed according to the strengths of all external nondominated solutions that dominate it. Additionally, a clustering technique called "average linkage method" [28] is used to keep diversity.

2. **Strength Pareto Evolutionary Algorithm 2** (SPEA2): SPEA2 has three main differences with respect to its predecessor [42]: (1) it incorporates a fine-grained fitness assignment strategy which takes into account for each individual the number of individuals that dominate it and the number of individuals by which it is dominated; (2) it uses a nearest neighbor density estimation technique which guides the search more efficiently, and (3) it has an enhanced archive truncation method that guarantees the preservation of boundary solutions. Thefore, in this case the elitist mechanism is just an improved version of the previous.

3. **Pareto Archived Evolution Strategy** (PAES): This algorithm was introduced by Knowles and Corne [26]. PAES consists of a (1+1) evolution strategy (i.e., a single parent that generates a single offspring) in combination with a historical archive that records some of the nondominated solutions previously found. This archive is used as a reference set against which each mutated individual is being compared. Such a historical archive is the elitist mechanism adopted in PAES. However, an interesting aspect of this algorithm is the mechanism used to maintain diversity which consists of a crowding procedure that divides objective space in a recursive manner. Each solution is placed in a certain grid location based on the values of its objectives (which are used as its "coordinates" or "geographical location"). A map of such grid is maintained, indicating the number of solutions that reside in each grid location. Since the procedure is adaptive, no extra parameters are required (except for the number of divisions of the objective space).

4. **Nondominated Sorting Genetic Algorithm II** (NSGA-II): Deb et al. [8] proposed a revised version of the NSGA [37], called NSGA-II, which is more efficient (computationally speaking), uses elitism and a crowded comparison operator that keeps diversity without specifying any additional parameters. The NSGA-II does not use an external memory as the previous algorithms. Instead, the elitist mechanism consists of combining the best parents with the best offspring obtained (i.e., a $(\mu + \lambda)$-selection).

5. **Niched Pareto Genetic Algorithm 2** (NPGA 2): Erickson et al. [9] proposed a revised version of the NPGA [22] called the NPGA 2. This algorithm uses Pareto ranking but keeps tournament selection (solving ties through fitness sharing as in the original NPGA). In this case, no external memory is used and the elitist mechanism is similar to the one adopted by the NSGA-II. Niche counts in the

NPGA 2 are calculated using individuals in the partially filled next generation, rather than using the current generation.

6. **Micro Genetic Algorithm**: This approach was introduced by Coello Coello & Toscano Pulido [5]. A micro-genetic algorithm is a GA with a small population and a reinitialization process. The micro-GA starts with a random population that feeds the population memory, which is divided in two parts: a replaceable and a non-replaceable portion. The non-replaceable portion of the population memory never changes during the entire run and is meant to provide the required diversity for the algorithm. In contrast, the replaceable portion experiences changes after each cycle of the micro-GA. The population of the micro-GA at the beginning of each of its cycles is taken (with a certain probability) from both portions of the population memory so that there is a mixture of randomly generated individuals (non-replaceable portion) and evolved individuals (replaceable portion). During each cycle, the micro-GA undergoes conventional genetic operators. After the micro-GA finishes one cycle, two nondominated vectors are chosen [3] from the final population and they are compared with the contents of the external memory (this memory is initially empty). If either of them (or both) remains as nondominated after comparing it against the vectors in this external memory, then they are included there (i.e., in the external memory). This is the historical archive of nondominated vectors. All dominated vectors contained in the external memory are eliminated. The micro-GA uses then three forms of elitism: (1) it retains nondominated solutions found within the internal cycle of the micro-GA, (2) it uses a replaceable memory whose contents is partially "refreshed" at certain intervals, and (3) it replaces the population of the micro-GA by the nominal solutions produced (i.e., the best solutions found after a full internal cycle of the micro-GA).

Second generation MOEAs can be characterized by an emphasis on efficiency and by the use of elitism (in the two main forms previously described). During the second generation, some important theoretical work also took place, mainly related to convergence [34, 20]. Also, metrics and standard test functions were developed to validate new MOEAs [41].

The main concerns during the second generation (which we are still living nowadays) are the following:

- Are our metrics reliable? What about our test functions? We have found out that developing good metrics is in itself a multiobjective optimization problem, too. In fact, it is ironic that nowadays we are going back to trusting more visual comparisons than metrics as during the first generation.
- Are we ready to tackle problems with more than two objective functions efficiently? Is Pareto ranking doomed to fail when dealing with too many objec-

---

[3] This is assuming that there are two or more nondominated vectors. If there is only one, then this vector is the only one selected.

tives? If so, then what is the limit up to which Pareto ranking can be used to select individuals reliably?

- What are the most relevant theoretical aspects of evolutionary multiobjective optimization that are worth exploring in the short-term?

6 Applications

An analysis of the evolution of the EMO literature reveals some interesting facts (see [6] for details). From the first MOEA published in 1985 [36] up to the first survey of the area published in 1995 [13], the number of published papers related to EMO is relatively low. However, from 1995 to our days, the increase of EMO-related papers is exponential. Today, the EMO repository registers over 1000 papers, from which a vast majority are applications. The vast number of EMO papers currently available makes it impossible to attempt to produce a detailed review of them in this section. Instead, we will discuss the most popular application fields, indicating some of the specific areas within them in which researchers have focused their main efforts.

Current EMO applications can be roughly classified in three large groups: engineering, industrial and scientific. Some specific areas within each of these groups are indicated next. We will start with the engineering applications, which are, by far, the most popular in the literature. This should not be too surprising, since engineering disciplines normally have problems with better understood mathematical models which facilitates the use of evolutionary algorithms. A representative sample of engineering applications is the following (aeronautical engineering seems to be the most popular subdiscipline within this group):

- Electrical engineering [40]
- Hydraulic engineering [33]
- Structural engineering [27]
- Aeronautical engineering [29]
- Robotics [30]
- Control [39]
- Telecommunications [32]
- Civil engineering [2]

Industrial applications occupy the second place in popularity in the EMO literature. Within this group, scheduling is the most popular subdiscipline. A representative sample of industrial applications is the following:

- Design and manufacture [35]
- Scheduling [38]
- Management [24]

Finally, we have a variety of scientific applications, from which the most popular are (for obvious reasons) those related to computer science:

- Chemistry [21]

- Physics [17]
- Medicine [31]
- Computer science [3]

The above distribution of applications indicates a strong interest for developing real-world applications of EMO algorithms (something not surprising considering that most real-world problems are of a multiobjective nature). Furthermore, the previous sample of EMO applications should give a general idea of the application areas that have not been explored in enough depth yet (e.g., computer vision, coordination of agents, pattern recognition, etc. [6]).

7 Future Challenges

Once we have been able to distinguish between the first and second generations in evolutionary multiobjective optimization, a reasonable question is: where are we heading now? In the last few years, there has been a considerable growth in the number of publications related to evolutionary multiobjective optimization. However, the variety of topics covered is not as rich as the number of publications released each year. The current trend is to either develop new algorithms (validating them with some of the metrics and test functions available) or to develop interesting applications of existing algorithms. We will finish this section with a list of some of the research topics that we believe that will keep researchers busy during the next few years:

- **Incorporation of preferences in MOEAs**: Despite the efforts of some researchers to incorporate user's preferences into MOEAs as to narrow the search, most of the multicriteria decision making techniques developed in Operations Research have not been applied in evolutionary multiobjective optimization [4]. Such incorporation of preferences is very important in real-world applications since the user only needs one Pareto optimal solution and not the whole set as normally assumed by EMO researchers.

- **Highly-Constrained Search Spaces**: There is little work in the current literature regarding the solution of multiobjective problems with highly-constrained search spaces. However, it is rather common to have such problems in real-world applications and it is then necessary to develop novel constraint-handling techniques that can deal with highly-constrained search spaces efficiently.

- **Parallelism**: We should expect more work on parallel MOEAs in the next few years. Currently, there is a noticeable lack of research in this area [6] and it is therefore open to new ideas. It is necessary to have more algorithms, formal models to prove convergence, and more real-world applications that use parallelism.

- **Theoretical Foundations**: It is quite important to develop the theoretical foundations of MOEAs. Although a few steps have been taken regarding proving convergence using Markov Chains (e.g., [34]), and analyzing metrics [41], much more work remains to be done (see [6]).

- **Use of More Efficient Data Structures**: The usage of more efficient data structures to store nondominated vectors is just beginning to be analyzed in evolutionary multiobjective optimization (see for example [10]). Note however, that such data structures have been in use for a relatively long time in Operations Research [18].

8 Conclusions

This paper has provided a general view of the field known as evolutionary multiobjective optimization. We have provided a historical analysis of the development of the area, emphasizing the algorithmic differences between the two main stages that we have undergone so far (the two so-called "generations"). The notion of elitism has been identified as the main responsible of the current generation of algorithms used in this area. Also, some of the most important issues (stated in the form of questions) raised during each of these two generations were briefly indicated.

In the final part of the paper, we have discussed some of the most relevant applications developed in the literature and we identified certain research trends. We have finished this paper with some promising areas of future research in evolutionary multiobjective optimization, hoping that this information may be useful to newcomers who whish to contribute to this emerging research field.

9 Acknowledgments

The author acknowledges support from the mexican Consejo Nacional de Ciencia y Tecnología (CONACyT) through project number 34201-A.

References

1. Thomas Bäck, David B. Fogel, and Zbigniew Michalewicz, editors. *Handbook of Evolutionary Computation*. Institute of Physics Publishing and Oxford University Press, 1997.
2. Richard Balling and Scott Wilson. The maximim fitness function for multi-objective evolutionary computation: Application to city planning. In Lee Spector, Erik D. Goodman, Annie Wu, W.B. Langdon, Hans-Michael Voigt, Mitsuo Gen, Sandip Sen, Marco Dorigo, Shahram Pezeshk, Max H. Garzon, and Edmund Burke, editors, *Proceedings of the Genetic and Evolutionary Computation Conference (GECCO'2001)*, pages 1079–1084, San Francisco, CA, 2001. Morgan Kaufmann Publishers.

3. Stefan Bleuler, Martin Brack, Lothar Thiele, and Eckart Zitzler. Multiobjective Genetic Programming: Reducing Bloat Using SPEA2. In *Proceedings of the Congress on Evolutionary Computation 2001 (CEC'2001)*, volume 1, pages 536–543, Piscataway, New Jersey, May 2001. IEEE Service Center.

4. Carlos A. Coello Coello. Handling Preferences in Evolutionary Multiobjective Optimization: A Survey. In *2000 Congress on Evolutionary Computation*, volume 1, pages 30–37, Piscataway, New Jersey, July 2000. IEEE Service Center.

5. Carlos A. Coello Coello and Gregorio Toscano Pulido. Multiobjective Optimization using a Micro-Genetic Algorithm. In Lee Spector, Erik D. Goodman, Annie Wu, W.B. Langdon, Hans-Michael Voigt, Mitsuo Gen, Sandip Sen, Marco Dorigo, Shahram Pezeshk, Max H. Garzon, and Edmund Burke, editors, *Proceedings of the Genetic and Evolutionary Computation Conference (GECCO'2001)*, pages 274–282, San Francisco, California, 2001. Morgan Kaufmann Publishers.

6. Carlos A. Coello Coello, David A. Van Veldhuizen, and Gary B. Lamont. *Evolutionary Algorithms for Solving Multi-Objective Problems*. Kluwer Academic Publishers, New York, May 2002. ISBN 0-3064-6762-3.

7. Indraneel Das and John Dennis. A closer look at drawbacks of minimizing weighted sums of objectives for pareto set generation in multicriteria optimization problems. *Structural Optimization*, 14(1):63–69, 1997.

8. Kalyanmoy Deb, Amrit Pratap, Sameer Agarwal, and T. Meyarivan. A Fast and Elitist Multiobjective Genetic Algorithm: NSGA–II. *IEEE Transactions on Evolutionary Computation*, 6(2):182–197, April 2002.

9. Mark Erickson, Alex Mayer, and Jeffrey Horn. The Niched Pareto Genetic Algorithm 2 Applied to the Design of Groundwater Remediation Systems. In Eckart Zitzler, Kalyanmoy Deb, Lothar Thiele, Carlos A. Coello Coello, and David Corne, editors, *First International Conference on Evolutionary Multi-Criterion Optimization*, pages 681–695. Springer-Verlag. Lecture Notes in Computer Science No. 1993, 2001.

10. Richard M. Everson, Jonathan E. Fieldsend, and Sameer Singh. Full Elite Sets for Multi-Objective Optimisation. In I.C. Parmee, editor, *Proceedings of the Fifth International Conference on Adaptive Computing Design and Manufacture (ACDM 2002)*, volume 5, pages 343–354, University of Exeter, Devon, UK, April 2002. Springer-Verlag.

11. David B. Fogel, editor. *Evolutionary Computation. The Fossil Record. Selected Readings on the History of Evolutionary Algorithms*. The Institute of Electrical and Electronic Engineers, New York, 1998.

12. Carlos M. Fonseca and Peter J. Fleming. Genetic algorithms for multiobjective optimization: Formulation, discussion and generalization. In Stephanie Forrest, editor, *Proceedings of the Fifth International Conference on Genetic Algorithms*, pages 416–423, San Mateo, CA, 1993. Morgan Kaufmann Publishers.

13. Carlos M. Fonseca and Peter J. Fleming. An overview of evolutionary algorithms in multiobjective optimization. *Evolutionary Computation*, 3(1):1–16, Spring 1995.

14. Michael P. Fourman. Compaction of symbolic layout using genetic algorithms. In *Genetic Algorithms and their Applications: Proceedings of the First International Conference on Genetic Algorithms*, pages 141–153, Hillsdale, NJ, 1985. Lawrence Erlbaum.

15. David E. Goldberg. *Genetic Algorithms in Search, Optimization and Machine Learning*. Addison-Wesley Publishing Company, Reading, MA, 1989.

16. David E. Goldberg and Jon Richardson. Genetic algorithm with sharing for multimodal function optimization. In John J. Grefenstette, editor, *Genetic Algorithms and Their Applications: Proceedings of the Second International Conference on Genetic Algorithms*, pages 41–49, Hillsdale, NJ, 1987. Lawrence Erlbaum.

17. I. Golovkin, R. Mancini, S. Louis, Y. Ochi, K. Fujita, H. Nishimura, H. Shirga, N. Miyanaga, H. Azechi, R. Butzbach, I. Uschmann, E. Förster, J. Delettrez, J. Koch, R.W. Lee, and L. Klein. Spectroscopic Determination of Dynamic Plasma Gradients in Implosion Cores. *Physical Review Letters*, 88(4), January 2002.

18. W. Habenicht. Quad trees: A data structure for discrete vector optimization problems. In *Lecture Notes in Economics and Mathematical Systems*, volume 209, pages 136–145, 1982.

19. P. Hajela and C. Y. Lin. Genetic search strategies in multicriterion optimal design. *Structural Optimization*, 4:99–107, 1992.

20. T. Hanne. On the convergence of multiobjective evolutionary algorithms. *European Journal of Operational Research*, 117(3):553–564, September 2000.

21. Mark Hinchliffe, Mark Willis, and Ming Tham. Chemical process systems modelling using multi-objective genetic programming. In John R. Koza, Wolfgang Banzhaf, Kumar Chellapilla, Kalyanmoy Deb, Marco Dorigo, David B. Fogel, Max H. Garzon, David E. Goldberg, Hitoshi Iba, and Rick L. Riolo, editors, *Proceedings of the Third Annual Conference on Genetic Programming*, pages 134–139, San Mateo, CA, July 1998. Morgan Kaufmann Publishers.

22. Jeffrey Horn, Nicholas Nafpliotis, and David E. Goldberg. A niched pareto genetic algorithm for multiobjective optimization. In *Proceedings of the First IEEE Conference on Evolutionary Computation, IEEE World Congress on Computational Intelligence*, volume 1, pages 82–87, Piscataway, NJ, June 1994. IEEE Service Center.

23. W. Jakob, M. Gorges-Schleuter, and C. Blume. Application of genetic algorithms to task planning and learning. In R. Männer and B. Manderick, editors, *Parallel Problem Solving from Nature, 2nd Workshop*, Lecture Notes in Computer Science, pages 291–300, Amsterdam, 1992. North-Holland Publishing Company.

24. Andrzej Jaszkiewicz, Maciej Hapke, and Pawel Kominek. Performance of Multiple Objective Evolutionary Algorithms on a Distribution System Design Problem—Computational Experiment. In Eckart Zitzler, Kalyanmoy Deb, Lothar Thiele, Carlos A. Coello Coello, and David Corne, editors, *First International Conference on Evolutionary Multi-Criterion Optimization*, pages 241–255. Springer-Verlag. Lecture Notes in Computer Science No. 1993, 2001.

25. Yaochu Jin, Tatsuya Okabe, and Bernhard Sendhoff. Dynamic Weighted Aggregation for Evolutionary Multi-Objective Optimization: Why Does It Work and How? In Lee Spector, Erik D. Goodman, Annie Wu, W.B. Langdon, Hans-Michael Voigt, Mitsuo Gen, Sandip Sen, Marco Dorigo, Shahram Pezeshk, Max H. Garzon, and Edmund Burke, editors, *Proceedings of the Genetic and Evolutionary Computation Conference (GECCO'2001)*, pages 1042–1049, San Francisco, California, 2001. Morgan Kaufmann Publishers.

26. Joshua D. Knowles and David W. Corne. Approximating the nondominated front using the pareto archived evolution strategy. *Evolutionary Computation*, 8(2):149–172, 2000.

27. A. Kurapati and S. Azarm. Immune Network Simulation with Multiobjective Genetic Algorithms for Multidisciplinary Design Optimization. *Engineering Optimization*, 33:245–260, 2000.

28. J.N. Morse. Reducing the size of the nondominated set: Pruning by clustering. *Computers and Operations Research*, 7(1–2):55–66, 1980.

29. Shigeru Obayashi, Takanori Tsukahara, and Takashi Nakamura. Multiobjective genetic algorithm applied to aerodynamic design of cascade airfoils. *IEEE Transactions on Industrial Electronics*, 47(1), February 2000.

30. Andrzej Osyczka, Stanislaw Krenich, and K. Karaś. Optimum design of robot grippers using genetic algorithms. In *Proceedings of the Third World Congress of Structural and Multidisciplinary Optimization (WCSMO)*, Buffalo, New York, May 1999.

31. Andrei Petrovski and John McCall. Multi-objective Optimisation of Cancer Chemotherapy Using Evolutionary Algorithms. In Eckart Zitzler, Kalyanmoy Deb, Lothar Thiele, Carlos A. Coello Coello, and David Corne, editors, *First International Conference on Evolutionary Multi-Criterion Optimization*, pages 531–545. Springer-Verlag. Lecture Notes in Computer Science No. 1993, 2001.

32. W. Pullan. Optimising Multiple Aspects of Network Survivability. In *Congress on Evolutionary Computation (CEC'2002)*, volume 1, pages 115–120, Piscataway, New Jersey, May 2002. IEEE Service Center.

33. Patrick M. Reed, Barbara S. Minsker, and David E. Goldberg. A multiobjective approach to cost effective long-term groundwater monitoring using an elitist nondominated sorted genetic algorithm with historical data. *Journal of Hydroinformatics*, 3(2):71–89, 2001.

34. Günter Rudolph and Alexandru Agapie. Convergence Properties of Some Multi-Objective Evolutionary Algorithms. In *Proceedings of the 2000 Conference on Evolutionary Computation*, volume 2, pages 1010–1016, Piscataway, NJ, July 2000. IEEE Press.

35. Ivo F. Sbalzarini, Sibylle Müller, and Petros Koumoutsakos. Microchannel Optimization Using Multiobjective Evolution Strategies. In Eckart Zitzler, Kalyanmoy Deb, Lothar Thiele, Carlos A. Coello Coello, and David Corne, editors, *First International Conference on Evolutionary Multi-Criterion Optimization*, pages 516–530. Springer-Verlag. Lecture Notes in Computer Science No. 1993, 2001.

36. J. David Schaffer. Multiple objective optimization with vector evaluated genetic algorithms. In *Genetic Algorithms and their Applications: Proceedings of the First International Conference on Genetic Algorithms*, pages 93–100, Hillsdale, NJ, 1985. Lawrence Erlbaum.

37. N. Srinivas and Kalyanmoy Deb. Multiobjective optimization using nondominated sorting in genetic algorithms. *Evolutionary Computation*, 2(3):221–248, Fall 1994.

38. El-Ghazali Talbi, Malek Rahoual, Mohamed Hakim Mabed, and Clarisse Dhaenens. A Hybrid Evolutionary Approach for Multicriteria Optimization Problems: Application to the Flow Shop. In Eckart Zitzler, Kalyanmoy Deb, Lothar Thiele, Carlos A. Coello Coello, and David Corne, editors, *First International Conference on Evolutionary Multi-Criterion Optimization*, pages 416–428. Springer-Verlag. Lecture Notes in Computer Science No. 1993, 2001.

39. K. C. Tan, T. H. Lee, E. F. Khor, and K. Ou. Control system design unification and automation using an incremented multi-objective evolutionary algorithm. In M. H. Hamza, editor, *Proceedings of the 19th IASTED International Conference on Modeling, Identification and Control*. IASTED, Innsbruck, Austria, 2000.

40. Robert Thomson and Tughrul Arslan. An Evolutionary Algorithm for the Multi-Objective Optimisation of VLSI Primitive Operator Filters. In *Congress on Evolutionary Computation (CEC'2002)*, volume 1, pages 37–42, Piscataway, New Jersey, May 2002.

41. Eckart Zitzler, Kalyanmoy Deb, and Lothar Thiele. Comparison of Multiobjective Evolutionary Algorithms: Empirical Results. *Evolutionary Computation*, 8(2):173–195, 2000.

42. Eckart Zitzler, Marco Laumanns, and Lothar Thiele. SPEA2: Improving the Strength Pareto Evolutionary Algorithm. Technical Report 103, Computer Engineering and Networks Laboratory (TIK), Swiss Federal Institute of Technology (ETH) Zurich, Gloriastrasse 35, CH-8092 Zurich, Switzerland, May 2001.

43. Eckart Zitzler and Lothar Thiele. Multiobjective evolutionary algorithms: A comparative case study and the strength pareto approach. *IEEE Transactions on Evolutionary Computation*, 3(4):257–271, November 1999.

Function Finding and the Creation of Numerical Constants in Gene Expression Programming

Cândida Ferreira

Gepsoft, 37 The Ridings, Bristol BS13 8NU, UK candidaf@gepsoft.com

Summary. Gene expression programming is a genotype/phenotype system that evolves computer programs of different sizes and shapes (the phenotype) encoded in linear chromosomes of fixed length (the genotype). The chromosomes are composed of multiple genes, each gene encoding a smaller sub-program. Furthermore, the structural and functional organization of the linear chromosomes allows the unconstrained operation of important genetic operators such as mutation, transposition, and recombination. In this work, three function finding problems, including a high dimensional time series prediction task, are analyzed in an attempt to discuss the question of constant creation in evolutionary computation by comparing two different approaches to the problem of constant creation. The first algorithm involves a facility to manipulate random numerical constants, whereas the second finds the numerical constants on its own or invents new ways of representing them. The results presented here show that evolutionary algorithms perform considerably worse if numerical constants are explicitly used.

Key words: Function finding, Numerical constants, Gene expression programming, Genetic programming, Evolutionary computation

1 Introduction

Genetic programming (GP) evolves computer programs by genetically modifying nonlinear entities with different sizes and shapes [6]. These nonlinear entities can be represented as diagrams or trees. Gene expression programming (GEP) is an extension to GP that also evolves computer programs of different sizes and shapes, but the programs are encoded in a linear chromosome of fixed length [4]. One strength of the GEP approach is that the creation of genetic diversity is extremely simplified as genetic operators work at the chromosome level. Indeed, due to the structural organization of GEP chromosomes, the implementation of high-performing search operators is extremely simplified, as any modification made in the genome always results in valid programs. Another strength of GEP consists of its unique, multigenic nature which allows the evolution of complex programs composed of several simpler sub-programs.

It is assumed that the creation of floating-point constants is necessary to do symbolic regression in general (see, e.g., Ref. 1 and Ref. 6). Genetic programming solved the problem of constant creation by using a special terminal named "ephemeral random constant" [6]. For each ephemeral random constant used in the trees of the initial population, a random number of a special data type in a specified range is generated. Then these random constants are moved around from tree to tree by the crossover operator.

Gene expression programming solves the problem of constant creation differently [4]. GEP uses an extra terminal "?" and an extra domain Dc composed of the symbols chosen to represent the random constants. For each gene, the random constants are generated during the inception of the initial population and kept in an array. The values of each random constant are only assigned during gene expression. Furthermore, a special operator is used to introduce genetic variation in the available pool of random constants by mutating the random constants directly. In addition, the usual operators of GEP plus a Dc specific transposition guarantee the effective circulation of the numerical constants in the population. Indeed, with this scheme of constants manipulation, the appropriate diversity of numerical constants can be generated at the beginning of a run and maintained easily afterwards by the genetic operators.

Notwithstanding, in this work it is shown that evolutionary algorithms do symbolic regression more efficiently if the problem of constant creation is handled by the algorithm itself. In other words, the special facilities for manipulating random constants are indeed unnecessary to solve problems of symbolic regression.

2 Genetic Algorithms with Tree Representations

All genetic algorithms use populations of individuals, select individuals according to fitness, and introduce genetic variation using one or more genetic operators (see, e.g., Ref. 7). In recent years different systems have been developed so that this powerful algorithm inspired in natural evolution could be applied to a wide spectrum of problem domains (see, e.g., Ref. 7 for a review of recent work on genetic algorithms and Ref. 2 for a review of recent work on GP).

Structurally, genetic algorithms can be subdivided in three fundamental groups: i) Genetic algorithms with individuals consisting of linear chromosomes of fixed length devoid of complex expression. In these systems, replicators (chromosomes) survive by virtue of their own properties. The algorithm invented by Holland [5] belongs to this group and is known as genetic algorithm or GA; ii) Genetic algorithms with individuals consisting of ramified structures of different sizes and shapes and, therefore, capable of assuming a richer number of functionalities. In these systems, replicators (ramified structures) also survive by virtue of their own properties. The algorithm invented by Cramer [3] and later developed by Koza [6] belongs to this group and is known as genetic programming or GP; iii) Genetic algorithms with individuals encoded in linear chromosomes of fixed length which are afterwards expressed as ramified structures of different sizes and shapes. In these systems, replicators (chromosomes) survive by virtue of causal effects on the phenotype (ramified structures). The algorithm invented by myself [4] belongs to this group and is known as gene expression programming or GEP.

GEP shares with GP the same kind of ramified structure and, therefore, can be applied to the same problem domains. However, the logistics of both systems differ significantly and the existence of a real genotype in GEP allows the unprecedented manipulation and exploration of more complex systems. Below are briefly highlighted some of the differences between GEP and GP.

2.1 Genetic Programming

As simple replicators, the ramified structures of GP are tied up in their own complexity: on the one hand, bigger, more complex structures are more difficult to handle and, on the other, the introduction of genetic variation can only be done at the tree level and, therefore, must be done carefully so that valid structures are created. A special kind of tree crossover is practically the only source of genetic variation used in GP for it allows the exchange of sub-trees and, therefore, always produces valid structures. Indeed, the implementation of high-performing operators, like the equivalent of natural point mutation, is unproductive as most mutations would have resulted in syntactically invalid structures. Understandingly, the other genetic operators described by Koza [6] – mutation and permutation – also operate at the tree level.

2.2 Gene Expression Programming

The phenotype of GEP individuals consists of the same kind of diagram representation used by GP. However, these complex phenotypes are encoded in simpler, linear structures of fixed length – the chromosomes. Thus, the main players in GEP are the chromosomes and the ramified structures or expression trees (ETs), the latter being the expression of the genetic information encoded in the former. The decoding of GEP genes implies obviously a kind of code and a set of rules. The genetic code is very simple: a one-to-one relationship between the symbols of the chromosome and the functions or terminals they represent. The rules are also very simple: they determine the spatial organization of the functions and terminals in the ETs and the type of interaction between sub-ETs in multigenic systems.

In GEP there are therefore two languages: the language of the genes and the language of ETs. However, thanks to the simple rules that determine the structure of ETs and their interactions, it is possible to infer immediately the phenotype given the sequence of the genotype, and vice versa. This bilingual and unequivocal system is called Karva language. The details of this new language are given in [4].

3 Two Approaches to the Problem of Constant Creation

In this section the problem of constant creation is discussed by comparing the performance of two different algorithms. The first manipulates explicitly the numerical constants and the second solves the problem of constant creation in symbolic regression by creating constants from scratch or by inventing new ways of representing them.

3.1 Setting the System

The comparison between the two approaches (with and without the facility to manipulate random constants) was made on three different problems. The first is a problem of sequence induction requiring integer constants. In this case the following test sequence was chosen:

$$a_n = 5n^4 + 4n^3 + 3n^2 + 2n + 1 \qquad (1)$$

where n consists of the nonnegative integers. This sequence was chosen because it can be exactly solved and therefore can provide an accurate measure of performance in terms of success rate.

The second is a problem of function finding requiring floating-point constants. In this case, the following "V" shaped function was chosen:

$$y = 4.251a^2 + \ln(a^2) + 7.243e^a \qquad (2)$$

where a is the independent variable and e is the irrational number 2.71828183. Problems of this kind cannot be exactly solved by evolutionary algorithms and, therefore, the performance of both approaches is compared in terms of average best-of-run fitness and average best-of-run R-square.

The third is the well-studied benchmark problem of predicting sunspots [8]. In this case, 100 observations of the Wolfer sunspots series were used (Table 1) with an embedding dimension of 10 and a delay time of one. Again, the performance of both approaches is compared in terms of average best-of-run fitness and R-square.

Table 1. Wolfer sunspots series (read by rows)

101	82	66	35	31	7	20	92	154	125	85	68	38	23	10	24	83	132	131	118
90	67	60	47	41	21	16	6	4	7	14	34	45	43	48	42	28	10	8	2
0	1	5	12	14	35	46	41	30	24	16	7	4	2	8	17	36	50	62	67
71	48	28	8	13	57	122	138	103	86	63	37	24	11	15	40	62	98	124	96
66	64	54	39	21	7	4	23	55	94	96	77	59	44	47	30	16	7	37	74

For the sequence induction problem, the first 10 positive integers n and their corresponding term a_n were used as fitness cases. The fitness function was based on the relative error with a selection range of 20% and maximum precision (0% error), giving maximum fitness $f_{max} = 200$ [4].

For the "V" shaped function problem, a set of 20 random fitness cases chosen from the interval [-1, 1] was used. The fitness function used was also based on the relative error but in this case a selection range of 100% was used, giving $f_{max} = 2,000$.

For the time series prediction problem, using an embedding dimension of 10 and a delay time of one, the sunspots series presented in Table 1 result in 90 fitness cases. In this case, a wider selection range of 1,000% was chosen, giving $f_{max} = 90,000$. In all the experiments, the selection was made by roulette-wheel sampling coupled with simple elitism and the performance was evaluated over 100 independent runs. The six experiments are summarized in Table 2.

Table 2. General settings used in the sequence induction (SI), the "V" function, and sunspots (SS) problems. The "*" indicates the explicit use of random constants.

	SI*	SI	V*	V	SS*	SS
Number of runs	100	100	100	100	100	100
Number of generations	100	100	5000	5000	5000	5000
Population size	100	100	100	100	100	100
Number of fitness cases	10	10	20	20	90	90
Function set	+-*/	+-*/	+-*/LE	+-*/LE	4(+-*/)	4(+-*/)
			K~SC	K~SC		
Terminal set	a, ?	a	a, ?	a	a-j, ?	a-j
Rand. const. array length	10	–	10	–	10	–
Rand. const. range	{0,1,2,3}	–	[-1,1]	–	[-1,1]	–
Head length	6	6	6	6	8	8
Number of genes	7	7	5	5	3	3
Linking function	+	+	+	+	+	+
Chromosome length	140	91	100	65	78	51
Mutation rate	0.044	0.044	0.044	0.044	0.044	0.044
One-point rec. rate	0.3	0.3	0.3	0.3	0.3	0.3
Two-point rec. rate	0.3	0.3	0.3	0.3	0.3	0.3
Gene rec. rate	0.1	0.1	0.1	0.1	0.1	0.1
IS transposition rate	0.1	0.1	0.1	0.1	0.1	0.1
IS elements length	1,2,3	1,2,3	1,2,3	1,2,3	1,2,3	1,2,3
RIS transposition rate	0.1	0.1	0.1	0.1	0.1	0.1
RIS elements length	1,2,3	1,2,3	1,2,3	1,2,3	1,2,3	1,2,3
Gene transposition rate	0.1	0.1	0.1	0.1	0.1	0.1
Rand. const. mut. rate	0.01	–	0.01	–	0.01	–
Dc transposition rate	0.1	–	0.1	–	0.1	–
Dc IS elements length	1,2,3	–	1,2,3	–	1,2,3	–
Selection range	20%	20%	100%	100%	1000%	1000%
Precision	0%	0%	0%	0%	0%	0%
Avg best-of-run fitness	179.827	197.232	1914.8	1931.84	86215.27	89033.29
Avg best-of-run R-square	0.977612	0.999345	0.957255	0.995340	0.713365	0.811863
Success rate	16%	81%	–	–	–	–

3.2 First Approach: Direct Manipulation of Numerical Constants

To solve the sequence induction problem using random numerical constants, F = {+, -, *}, T = {a, ?}, the set of integer random constants R = {0, 1, 2, 3, 4, 5, 6, 7, 8, 9}, and "?" ranged over the integers 0, 1, 2, and 3. The parameters used per run are shown in the first column of Table 2. In this experiment, the first perfect solution was found in generation 45 of run 9 (the contribution of each sub-ET is indicated in square brackets):

$$y = [a^2] + [a] + [2a^4 + 4a^3] + [0] + [2a^2] + [1 + a] + [3a^4] \tag{3}$$

which corresponds to the target sequence (1).

As shown in the first column of Table 2, the probability of success for this problem is 16%, considerably lower than the 81% of the second approach (see Table 2, column 2). It is worth emphasizing that only the prior knowledge of the solution enabled us, in this case, to choose correctly the type and the range of the random constants.

To find a model to the "V" shaped function using random numerical constants F = {+, -, *, /, L, E, K, ~, S, C} ("L" represents the natural logarithm, "E" represents e^x, "K" represents the logarithm of base 10, "~" represents 10^x, "S" represents the sine function, and "C" the cosine) and T = {a, ?}. The set of rational random constants R = {0, 1, 2, 3, 4, 5, 6, 7, 8, 9}, and "?" ranged over the interval [-1, 1]. The parameters used per run are shown in the third column of Table 2. The best solution, found in run 50 after 4584 generations, is shown below (the contribution of each sub-ET is indicated in square brackets):

$$y = [\ln(0.99782a^2)] + [10^{\sin(1.27278a)}] + [10^{0.929a}] +$$

$$+ [0.77631 - 2.80112a^3] + [2.45714 + e^{0.981a} + e^a] \tag{4}$$

It has a fitness of 1989.566 and an R-square of 0.9997001 evaluated over the set of 20 fitness cases and an R-square of 0.9997185 evaluated against a testing set of 100 random points also chosen from the interval [-1, 1].

It is worth noticing that the algorithm does in fact integrate constants in the evolved solutions, but the constants are very different from the expected ones. Indeed, GEP (and I believe, all genetic algorithms with tree representations) can find the expected constants with a precision to the third or fourth decimal place when the target functions are simple polynomial functions with rational coefficients and/or when it is possible to guess pretty accurately the function set, otherwise a very creative solution would be found.

To predict sunspots using random numerical constants, the set of functions F = {4+, 4-, 4*, 4/} and T = {a, b, c, d, e, f, g, h, i, j, ?}. The set of rational random constants R = {0, 1, 2, 3, 4, 5, 6, 7, 8, 9}, and "?" ranged over the interval [-1, 1]. The parameters used per run are shown in the fifth column of Table 2. The best solution, found in run 92 after 4759 generations, is shown below:

$$y = \frac{2j^2}{h+i+j} + \frac{a+b+g}{0.995+0.847c+e} + \frac{1.903j+j^2}{i^2+j} \tag{5}$$

It has a fitness of 86603.2 and an R-square of 0.833714 evaluated over the set of 90 fitness cases.

3.3 Second Approach: Creation of Numerical Constants from Scratch

To solve the sequence induction problem without the facility to manipulate numerical constants, the function set was exactly the same as in the experiment with random constants. The terminal set consisted obviously of the independent variable alone.

As shown in the second column of Table 2, the probability of success using this approach is 81%, considerably higher than the 16% obtained using the facility to manipulate random constants. In this experiment, the first perfect solution was found in generation 44 of run 0 (the contribution of each sub-ET is shown in square brackets):

$$y = [2a] + [a^3 + a] + [a^4 + 3a^3 + 2a^2] + [4a^4] + [1] + [a^2 - a] + [0] \tag{6}$$

which is equivalent to the target sequence (1). In this case the algorithm created all the necessary constants from scratch by performing simple mathematical operations.

To find the "V" shaped function without using random constants, the function set is exactly the same as in the first approach. With this collection of functions, most of which extraneous, the algorithm is equipped with different tools for evolving highly accurate models without using numerical constants. The parameters used per run are shown in the fourth column of Table 2. In this experiment of 100 identical runs, the best solution was found in generation 4679 of run 10 (the contribution of each sub-ET is indicated in square brackets):

$$y = [\ln(2a^2) + 10^{\sin(a)}] + [2a + \sin(a) + a^2] +$$

$$+ [\cos(\cos(2a)) + e^{a^2}] + [e^{\sin(a)}] + [1 + e^a + e^{a^2}] \tag{7}$$

It has a fitness of 1990.023 and an R-square of 0.9999313 evaluated over the set of 20 fitness cases and an R-square of 0.9998606 evaluated against the same testing set used in the first approach, and thus is better than the model (4) evolved with the facility for the manipulation of random constants.

To predict sunspots without using random numerical constants, the function set is exactly the same as in the first approach. The parameters used per run are shown in the sixth column of Table 2. In this experiment of 100 identical runs, the best solution was found in generation 2273 of run 57:

$$y = j + \frac{d-i+3j}{b+e} + \frac{d+bj-ij}{a+2i} \tag{8}$$

It has a fitness of 89176.61 and an R-square of 0.882831 evaluated over the set of 90 fitness cases, and thus is better than the model (5) evolved with the facility for the manipulation of random constants.

It is instructive to compare the results obtained in both approaches. In all the experiments the explicit use of random constants resulted in a worse performance. In the sequence induction problem, success rates of 81% against 16% were obtained; in the "V" function problem average best-of-run fitnesses of 1931.84 versus 1914.80 and average best-of-run R-squares of 0.995340 versus 0.957255 were obtained; and in the sunspots prediction problem average best-of-run fitnesses of 89033.29 versus 86215.27 and average best-of-run R-squares of 0.811863 versus 0.713365 were obtained (see Table 2). Thus, in real-world applications where complex realities are modeled, of which nothing is known concerning neither the type nor the range of the numerical constants, and where most of the times it is impossible to guess the exact function set, it is more appropriate to let the system model the reality on its own without explicitly using random constants. Not only the results will be better but also the complexity of the system will be much smaller.

4 Conclusions

Gene expression programming is the most recent development on artificial evolutionary systems and one that brings about a considerable increase in performance due to the crossing of the phenotype threshold. In practical terms, the crossing of the phenotype threshold allows the unconstrained exploration of the search space because all modifications are made on the genome and because all modifications always result in valid phenotypes or programs. In addition, the genotype/phenotype representation of GEP not only simplifies but also invites the creation of more complexity. The elegant mechanism developed to deal with random constants is a good example of this.

In this work, the question of constant creation in symbolic regression was discussed comparing two different approaches to solve this problem: one with the explicit use of numerical constants, and another without them. The results presented here suggest that the latter is more efficient, not only in terms of the accuracy of the best evolved models and overall performance, but also because the search space is much smaller, reducing greatly the complexity of the system and, consequently, the precious CPU time.

Finally, the results presented in this work also suggest that, apparently, the term "constant" is just another word for mathematical expression and that evolutionary algorithms are particularly good at finding these expressions because the search is totally unbiased.

References

1. Banzhaf W (1994) Genotype-Phenotype-Mapping and Neutral Variation: A Case Study in Genetic Programming. In: Davidor Y, Schwefel H-P, and Männer R (eds) Parallel Problem Solving from Nature III, Lecture Notes in Computer Science, 866: 322-332. Springer-Verlag

2. Banzhaf W, Nordin P, Keller R E, Francone F D (1998) Genetic Programming: An Introduction: On the Automatic Evolution of Computer Programs and its Applications. Morgan Kaufmann
3. Cramer N L (1985) A Representation for the Adaptive Generation of Simple Sequential Programs. In: Grefenstette J J (ed) Proceedings of the First International Conference on Genetic Algorithms and Their Applications. Erlbaum
4. Ferreira C (2001) Gene Expression Programming: A New Adaptive Algorithm for Solving Problems. Complex Systems 13 (2): 87-129
5. Holland J H (1975) Adaptation in Natural and Artificial Systems: An Introductory Analysis with Applications to Biology, Control, and Artificial Intelligence. University of Michigan Press (second edition: MIT Press, 1992)
6. Koza J R (1992) Genetic Programming: On the Programming of Computers by Means of Natural Selection. MIT Press, Cambridge MA
7. Mitchell M (1996) An Introduction to Genetic Algorithms. MIT Press, Cambridge MA London
8. Weigend A S, Huberman B A, Rumelhart D E (1992) Predicting Sunspots and Exchange Rates with Connectionist Networks. In: Eubank S, Casdagli M (eds) Nonlinear Modeling and Forecasting, pages 395-432. Addison-Wesley, Redwood City CA

A Hybrid Genetic Algorithm for
Multi-Hypothesis Tracking

T.R.Vishnu Arun Kumar

Thiagarajar College of Engineering,
Madurai-625015, TamilNadu, India

Summary. Multi-hypothesis tracking is the process of tracking the paths traversed by multiple targets. The assignment of target detections in each scan to the existing tracks or forming new tracks is called the Data Association Problem. Since the tracking is continuous and there are multiple targets, the data association problem becomes a Combinatorial Problem . As it is impractical to find an optimal solution for combinatorial problems, Heuristic approaches are encouraged. These Approaches tend to produce sub-optimal solutions in a fixed amount of time. Recent researches show that Genetic Algorithms (GA) can be applied to combinatorial problems and the results are promising. In this paper, a near-optimal solution for the Multi-hypothesis tracking problem using a Hybrid Genetic Algorithm (HGA) has been discussed.

1 Introduction

The paths traversed by targets like missiles and aircrafts are constantly observed by multiple sensors and the *detections* are reported continuously. A set of detections from different scans that represents the same target, forms a *Track*. A specific combination of tracks forms *Hypothesis*. The detections in each scan are associated with those of previous scans appropriately to produce the exact tracks traversed by the targets. During a scan, the tracker receives a number of detections in which a single object might have been detected more than once. Under these circumstances, the number of different ways of associating detections into tracks may grow exponentially.

Many typical tracker applications require the calculation of the best hypothesis for each scan. As it is impractical to find the optimal hypothesis even for a small number of objects observed over time, sub-optimal search techniques are generally preferred [1]. *Genetic Algorithm* is a randomized search technique, which follows the process of natural evolution [8], [11]. Hybrid Genetic Algorithms perform better and converge faster when compared to traditional GA's as they use problem specific heuristics as the genetic operators for evolving the populations [9], [12]. Recent research works [2], [3] show that the application of Genetic Algorithms for Data Association Problem is more promising.The *Optimizer Algorithm* discussed in this paper has shown better results.

2 Multi-Hypothesis Tracker (MHT)

A single-hypothesis tracker calculates the cost of all possible hypotheses at each scan and selects the best hypothesis as the *Starting Hypothesis* for associating the detections of the next scan. This type of association will give only the local optimal hypothesis, as there is a chance of missing a better hypothesis, which may not coincide with the starting hypothesis.

A multi-hypothesis tracker calculates the cost of all possible hypotheses at each scan and declares the best hypothesis as the *tentative winner*. Besides the best hypothesis, a set of better hypotheses is also maintained. If left unchecked, the number of tracks and hypothesis will quickly expand beyond the resources of processing time and memory available. Hence the number of starting hypotheses is kept fixed and the most expensive hypotheses are *pruned out* periodically [1].

3 Genetic Algorithm

Genetic Algorithm (GA) has the following Salient Features:

- It is Adaptive in nature and so it can learn from experience *(Adaptive Genetic Algorithm)*.
- It exhibits *intrinsic parallelism*.
- It can be parallelized easily even on a loosely coupled Network without much *Communication Overhead*.
- It can also be used in the Optimization of *Non-differentiable functions*.
- It maintains a *Balance between Efficiency and Efficacy*.
- It determines *global optimal solution* in short interval of time.
- It is more *Efficient for Complex problems*.

The structure of the chromosome, genetic operators, initialization heuristics and the objective function used to evolve the population are described below. *Roulette wheel selection* is employed in the selection of fitter individuals from the population [10]. *Elitism* is also used to avoid loss of better hypotheses obtained during intermediate stage of evolution [7].

3.1 Chromosome Encoding

Each *Chromosome* represents a hypothesis and each *Gene* in the chromosome represents a detection. The chromosomes are encoded as strings representing the tracks obtained till the current scan in *chronological order*.

As tracking is continuous, the length of the chromosome will increase continuously for each scan. The length is managed by removing the associations of a fixed number of scans and storing in a separate *work space* [1]. The workspace can be referred, whenever the hypothesis is to be evaluated.

Consider the following scenario: Each scan reports 3 detections. Let there be 4 different targets detected in 3 scans. So totally there are 9 detections of 4 different targets at various intervals of time. Let the scan interval be 1 second. Let the velocity range be (10 - 30) m/s

For simplicity let the locations reported for each detection be (5,10), (23,15), (0,0), (10,30), (28,3), (5,19), (0,50), (30,20), (10,7) respectively. Then the correct associations of detections can be (1->4->7) (2->5->8) (3) (6->9). The above associations represent 4 tracks for the 4 different targets. The hypothesis (1->4->7) (2->5->8) (3) (6->9) is represented as chromosome *123124124* such that 1 representing target number 1 appears in gene positions 1, 4 & 7 in the chromosome.

The above strategy of representing chromosomes generates associations involving detections of successive scans, which are forward in time. This eliminates the need for the objective function to generate a *penalty* for associations between the detections belonging to the same scan as in [3]. This type of chromosomal encoding is named as *Chronocoding*. Furthermore the implementation of Chronocoding also eliminates the necessity of *Pruning Algorithm* and the *Initialization Heuristic* used in the previous version of the work.

3.2 Genetic Operators

The following are the Genetic Operators used in the Hybrid Genetic Algorithm.

Crossover Crossover is performed by selecting a Random Crossover site and The offspring is generated by arranging the genes after the crossover site in the order in which they appear in the other parent chromosome.

Mutation Mutation is performed by selecting a random gene and changing it to a random value between 1 and total number of targets.

shuffle Shuffle is performed by selecting two random genes and swapping them.

Inversion Inversion is performed by selecting two random genes and reversing all the genes present between them.

Genetic Operator	parent	Site	Child
Crossover	(123132214),(123423421)	4	(123123421),(123432214)
Mutation	(123132214)	6	(123134214)
Shuffle	(123132214)	5,6	(123123214)
Inversion	(123134214)	4,8	(123124314)

Table 1. Genetic Operators - An Illustration

3.3 Genetic Operators

The fitness of the chromosomes is evaluated by the objective function by using the Expected *Position Matching Criterion*. It generates a *higher* fitness value if the expected co-ordinate position of the target of a particular track matches with the actual co-ordinates of the target.

   Fitness Calculation: Let the velocity range be indicated by v_1 and v_2, actual velocity by ω and permissible change in velocity range by δv. Then the fitness of the chromosome can be calculated by the following Pseudocode. *if* $(v_1 <= \omega <= v_2)$
fitness = 100
else if $(v_1 - \delta v <= \omega <= v_1)$ *or* $(v_2 <= \omega <= v_2 + \delta v)$
fitness = 50
else
fitness = 10

   Thus the fitness of the chromosome (1->4->7) (2->5->8) (3) (6->9) is calculated as $200 + 200 + 0 + 100 = 500$

4 Tracker Algorithm

The Pseudocode of Multi-Hypothesis Tracker operation is described below:
   Tracker Algorithm Pseudocode

Do for each scan
(
1.Call HGA.
2.If the chromosome length is greater than the maximum length
       (L_{max}) allowed then call the *Pruning Algorithm*.
3.Display the Elite population (Best Hypotheses) of the current scan.
) *End Do*

5 Hybrid Genetic Algorithm (HGA)

A scan will report N detections to the tracker algorithm. The tracker will build the hypotheses by associating the detections of the current scan with the existing tracks or it will form new tracks using the HGA.

HGA Pseudocode
1.During the first scan, each detection represents a separate track.
2.*Do while* (There is no hypothesis representing exact
track in the Elite Population)
(
2.1.Apply *Disintegration Heuristic*.

2.2.*Do while(No. of Iterations < Maximum no. of Iterations(I_{max}))*
(

    2.2.1. Evaluate the fitness of the chromosomes.

    2.2.2. Select the parent chromosomes from the Population using *Roulette Wheel Selection.*

    2.2.3. Apply the genetic operators, *Optimizer Algorithm* and evolve the current population.

    2.2.4. Update *Elite Population* by replacing less fit chromosomes of the previous elite generation with the fitter chromosomes of the current generation.

) *End Do.*

) *End Do.*

3. The best hypothesis present in the *Elite population* represents the solution.

6 Heuristics

The HGA uses several *Problem Specific Heuristics* to achieve faster convergence. They are listed as follows

6.1 Disintegration Heuristic

The Disintegration heuristic is used only when there is no Hypothesis in the Elite population representing the exact tracks.

Disintegration Heuristic Pseudocode

1. Select two chromosomes from the existing population using *Roulette-Wheel Selection.*

2. Select the first gene of any one of the parents offspring.

3. Construct a table, which shows both the predecessor and successor of each detection in both the parent chromosomes.

4. Select the first gene from any one of the parents.

5. *Do for every detection in the hypothesis*

(

5.1. From the table select a detection, which is connected to the current gene of the offspring and has minimum number of associations.

5.2. Ties are broken randomly.

5.3. Mark the selected detection.

) *End Do*

6.2 Optimizer Algorithm

This is a *Greedy Approach* aiming at *Local Optimization.*

Optimizer Algorithm Pseudocode
1. Select certain number (N_{suc}) of chromosomes from the existing population using *Roulette- Wheel selection.*
2. Split each of the chromosomes into several fragments
3. Rearrange the detections in such a way that the fitness of the fragment is increased.
4. Integrate the fragments to form a complete hypothesis.

7 Simulation

The tracker algorithm is simulated for *NSWC Tracking Benchmark 1*, which has the following specification [4] , [5] , [6] .
   Radar Model

- Range Resolution = 50m
- Bandwidth ranges between 2.5 and 4.5 degrees.

   Targets

- Maximum acceleration = 7g (m/s2).

8 Tracker Environment

9 Results

The Tracker Algorithm is coded in VC++ and run in Pentium-III 450 MHz processor. A simulation of the entire process of Target Identification & Target tracking is also developed [13]], [14]. The time taken for determination of correct tracks at specific scans for *NSWC Tracking Benchmark 1* is tabulated below.

	Population Size= 20			Population Size= 20		
Scan No.	Worst Case	Avg. Case	Best Case	Worst Case	Avg. Case	Best Case
5	12.73	11.05	9.09	12.22	11.51	10.52
10	15.98	15.72	12.65	13.14	12.76	11.36
20	17.81	17.09	15.77	14.32	14.07	13.79
50	21.33	19.56	18.22	17.56	16.78	15.09

Table 2. Execution Time (in seconds)of Multi-hypothesis Tracker

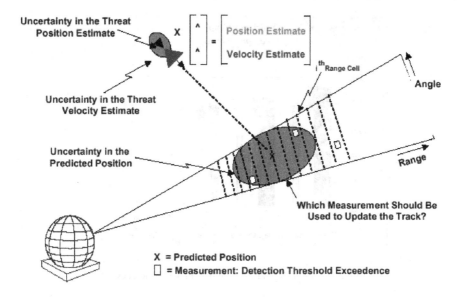

Fig. 1. Typical Tracker Environment

(Elite Population Size=20, Crossover Rate=0.7, Mutation Rate=0.25, Shuffle Rate=0.25, Inversion Rate=0.25, L_{max} =60, N_{suc}=5, N=3, I_{max}=100000)
where,
L_{max} - Maximum chromosome length allowed,
N_{suc} - Number of parents selected in Successor algorithm,
N - Number of detections in each scan.
I_{max} - Maximum number of iterations.

10 Conclusion

From the results as perceived from the figure, it is evident that the *Efficiency of the Tracker increases as the Population size increases*. This is obvious, as there is an increase in the number of search points in the solution domain. The Mutation Probability is kept as high as 0.25 so as to avoid *Premature Convergence*. The adverse effects of high mutation probability are avoided by maintaining *Elite Population*. Moreover, the implementation of the objective function depends on the type of the sensor, the targets and the geography of the environment. Hence the objective function for real world situations has to be implemented using problem specific knowledge.

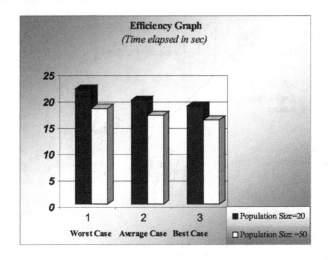

Fig. 2. Effect of Population size on Tracker Efficiency

11 Current Work

A Pseudo Parallel version of HGA has been planned along with the idea of Adaptive selection of genetic operators. Pseudo Parallelization is the process of parallel execution of the code on the same workstation by means of *Software Parallelism. Adaptive selection* of genetic operators can be achieved by modifying the operator probability depending on whether it increases or decreases the fitness of the chromosome at each generation. The Algorithm is being tested for other *Benchmark problems like NSWC Tracking Benchmark- II, ONR/NSWC Tracking Benchmark-III, IV, Pilot JCTN Benchmark, Benchmark 4.*

12 Acknowledgements

- The Research work presented in this paper is supported by, *TamilNadu State Council for Science and Technology (TNSCST)*.
- Special thanks to *DRDO (Defense Research & Development Organization) Panel* for their Valuable suggestions.

References

1. David B.Hillis, "Using a Genetic Algorithm for Multi - Hypothesis Tracking ", *Army Research Laboratory*, U.S. Government Work. Not protected by U.S. Copyright, IEEE, pp.112-117, 1997.
2. Harry Wechslerm and Jerzy W.Bala, "Learning To Detect Targets Using Scale-space and Genetic Search", *Centre for Artificial Intelligence*, Dept. of Computer Science, George Mason University, VA, USA, IEEE, pp 516-522, 1992.

3. H.C. Cobb and John J.Grefenstette, "Genetic algorithms for tracking changing environments", *ICGA - 5*, 1993.

4. Greg Watson, "Overview of ONR/NSWC Tracking Benchmark 4", *ONR-GTRI Target Tracking Workshop*, 1998.

5. Dale Blair, "Tracking Benchmarks: Concept, History, Results, and Impact", *ONR-GTRI Target Tracking Workshop*, 1998.

6. Dale Blair, "Technical Issues and Recent Advances in Target Tracking", *ONR-GTRI Target Tracking Workshop*, 1998.

7. P. Collard and J.P. Aurand, "An efficient Genetic Algorithm", in *ECAI*, Amsterdam, 1994.

8. Yuri Rabinovich and AVI Wigderson, "An analysis of a Simple Genetic Algorithm", Dept of Comp. Sci. Hebrew Univ., Israel, *IEEE*, pp 215 - 221, 1980.

9. David E. Goldberg, "Zen and the art of Genetic algorithms", *ICGA* 1989, pp.80 - 84.

10. Goldberg, D.E. and Deb. K., "A comparative analysis of selection schemes used in genetic algorithms", In G. Rawlins, ed., *Foundations of genetic algorithms*, Morgan Kaufmann, 1991.

11. Melanie Mitchell, "An introduction to genetic algorithms", *Prentice Hall of India Pvt. Ltd., New Delhi*, 1998.

12. Zbigniew Michalewicz, "Genetic algorithms + Data structures = Evolution Programs", *Springer - Verlag*, Newyork, 1998.

13. Adrian Low, "Introductory Computer Vision and Image processing", *McGraw Hill International Editions*, 1991.

14. Anil K. Jain, "Fundamentals of Digital Image Processing", *PHI Publications*, 1989.

Differential Evolution Algorithm With Adaptive Control Parameters

Junhong Liu and Jouni Lampinen

Laboratory of Information Processing, Lappeenranta University of Technology P.O. Box 20, FIN-53851 Lappeenranta, Finland
junhong.liu@lut.fi, jlampine@lut.fi

Summary. The Differential Evolution Algorithm is a floating-point encoded evolutionary algorithm for global optimization over continuous spaces. This algorithm so far uses empirically chosen fixed search parameters. This study is to introduce dynamic parameter control using fuzzy logic controllers whose inputs incorporate the relative function values and individuals of the successive generations to adapt the search parameters for the mutation operation and the crossover operation. Standard test functions are used to demonstrate. Based on experimental results, the Fuzzy Adaptive Differential Evolution Algorithm results in a faster convergence.

Key words: Differential Evolution; Evolutionary Algorithms; Fuzzy logic; Adaptation

1 Introduction

Setting the search parameter values in Evolutionary Algorithms (EA) can be distinguished into two types: parameter tuning and parameter control. Parameter tuning means the commonly practiced approach that amounts to finding good values for the parameters before the run of the algorithm and then running the algorithm using these values, which remain fixed during the run. Parameter control forms an alternative, as it amounts to starting a run with initial parameter values, which are changed during the run. Methods for changing the value of a parameter can be classified into one of three categories [1], *Deterministic Parameter Control, Adaptive Parameter Control and Self-Adaptive Parameter Control.* Fuzzy logic control methods have been used previously for dynamically computing appropriate settings, such as population size, crossover operator and mutation operator throughout the execution of Genetic Algorithm (GA) for avoiding the premature convergence problem by using the knowledge of the GA experts [2, 3].

The Differential Evolution algorithm (DE) as a branch of EA introduced by Price and Storn [4] has three control parameters. The mutation control parameter, F, is a real and constant factor that controls the amplification of the differential variation; the crossover control parameter, CR, controls which parameter contributes to which trial vector parameter in the crossover operation; and the population size, NP, is the number of the population members.

There are many papers about DE and its applications [4, 5, 6, 7, 8, 9, 10], where the ranges for mutation amplification, F, the crossover operator, CR, and the population size, NP, have been suitably pointed out. These parameter values determine whether the algorithm will find a near-optimum solution and whether it will find such a solution efficiently. Choosing the right parameter values is a time-consuming task. Several papers have discussed the influence of parameters on the performance of DE, for example, the stagnation discussion in [7] and taking into account the diversity in the population in [11].

However, there still exists a lack of knowledge on how to reasonably find the best values for the control parameters of DE for a given function. Since the interaction of control parameters with the DE performance is complex, a DE user should select parameter settings for DE for the problem at hand. Therefore, the trial-and-error method has to be used for finding good control parameters. In practice, the optimization run is performed multiple times with different settings. In some cases, the time for finding these parameters by trial-and-error is too long to be acceptable.

In a static DE, the control variables are kept fixed during the optimization process and have no response to parameter vectors and function values of current or former generations. In this paper a new Differential Evolution Algorithm that uses a fuzzy knowledge-based system [12, 13] to dynamically control DE parameters, such as F and CR, is proposed.

Firstly, the Differential Evolution Algorithm is provided and briefly, the possibility to implement fuzzy adapting for determining parameters of DE is analyzed, and then a fuzzy adapting idea is initially included in the determination of the parameters for DE. Linguistic fuzzy sets taking into consideration the uncertainty and non-linearity that appear in a function are developed and used to encode knowledge, expertise and experience on a computer, and to provide a novel and effective means to help expedite the convergence velocity of DE. Finally the Fuzzy Adaptive Differential Evolution Algorithm (FADE) is implemented.

2 Differential Evolution Algorithm

DE is a parallel direct search method that utilizes NP, D-dimensional, parameter vectors,

$$\mathbf{X}_{i,G}, \; i = 1, 2, \ldots, NP, \tag{1}$$

as a population for the generation G, where NP denotes the number of population members, and does not change during the optimization process so as to minimize the function,

$$f(\mathbf{X}) : \mathcal{R}^D \longrightarrow \mathcal{R}, \tag{2}$$

by choosing a suitable parameter vector,

$$\mathbf{X} = (x_1, x_2, \ldots, x_D) \,, \tag{3}$$

where $\mathbf{X}$ denotes a vector composed of D objective function parameters. The initial population is chosen randomly to cover the entire parameter space uniformly. As a rule, a uniform probability distribution will be assumed for all random decisions, if not otherwise stated. The crucial idea behind DE is a scheme for generating trial parameter vectors. Basically, DE generates new parameter vectors by adding the weighted difference between two population vectors to a third vector. If the resulting trial vector yields a lower or equal objective function value than a predetermined population member, the generated vector will replace the vector that it is compared with; otherwise, the old vector is retained.

According to previous reports [4, 5, 7, 8, 9, 10], F, CR and NP are kept fixed during the optimization process, and can be found by trial-and-error. The suggested initial choices are:

- $F \in [0.5, 1]$.
- $CR \in [0.8, 1]$.
- $NP = 10 \times D$ where D is the dimensionality of the discussed problem [5].

With a series of standard test functions the experiments about how the parameters influence the performance of DE were carried out [14]. The results from the experiments show that, with different functions (problems), $NP = 10 \times D$ is really helpful advice, but for F and CR, the situation can be complicated. At this point, a new method using fuzzy logic to adapt control parameters was introduced, which combine the features of DE and fuzzy logic and possessing the robustness and fuzziness.

3 Principle of Fuzzy Adapting

The idea of fuzzy adapting can be expressed using the control diagram in Fig. 1. The Fuzzy Logic Control system (FLC) is used as the basis of a fuzzy control mechanism. The following steps are used [12, 13].

3.1 System Parameters and Fuzzy Sets Membership Functions

The mean square root of differences concerning the successive generations over the whole population for function values (FC) and population members (PC) are inputs, the control parameters (F and/or CR) are outputs. The inputs were defined as:

Control Parameters
NP CR

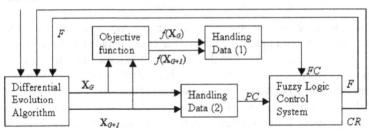

Fig. 1. Control diagram of DE with fuzzy adapting: PC - the parameter vector change in magnitude; FC - the function value change.

$$PC = \sqrt{\frac{1}{NP} \sum_{i=1}^{NP} \sum_{j=1}^{D} (x_{i,j}^{(n)} - x_{i,j}^{(n-1)})^2} , \tag{4}$$

$$FC = \sqrt{\frac{1}{NP} \sum_{i=1}^{NP} (f_i^{(n)} - f_i^{(n-1)})^2} , \tag{5}$$

$$d_{11} = 1 - (1 + PC) \times e^{-PC} , \tag{6}$$

$$d_{12} = 1 - (1 + FC) \times e^{-FC} , \tag{7}$$

$$d_{21} = 2 \times \left(1 - (1 + PC) \times e^{-PC}\right) , \tag{8}$$

$$d_{22} = 2 \times \left(1 - (1 + FC) \times e^{-FC}\right) , \tag{9}$$

where $f_i^{(n)}$ represents the ith component of the function value vector $\mathbf{f} \in \mathcal{R}^{NP}$ for the nth generation, $i = 1, 2, \ldots, NP$; $x_{i,j}^{(n)}$ represents the component in the ith row and j^{th} column of the parameter matrix $\mathbf{X}_{NP \times D}$ for the nth generation, $i = 1, 2, \ldots, NP$, $j = 1, 2, \ldots, D$; PC is called the parameter vector change in magnitude and depressed into the range of $[0, 1]$ as d_{11} and the range of $[0, 2]$ as d_{21}; FC is called the function value change and depressed into $[0, 1]$ as d_{12} and $[0, 2]$ as d_{22}.

PC and FC are treated as fuzzy variables with 6 elements for d_{i1} and d_{i2} of the ith FLC system; 3 elements are for F and CR individually. The values are then assigned as membership grades in 3 fuzzy subsets as follows:

- **S** is "small".
- **M** is "medium".
- **B** is "big".

Among a set of membership functions, Gaussian curve membership function, f_g, was chosen for each variable as shown in Fig. 2. The definition of each membership function is characterized by description given in Table 1.

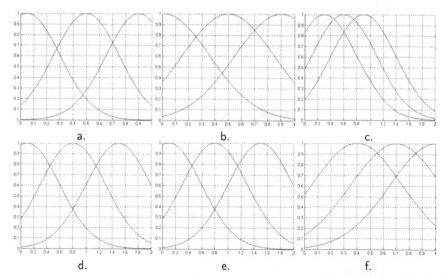

Fig. 2. Membership functions for inputs and outputs: (**a**) d_{11}; (**b**) d_{12}; (**c**) F; (**d**) d_{21}; (**e**) d_{22}; (**f**) CR

Table 1. Membership functions (d_{ij} = the jth input of the ith FLC system)

inputs,outputs	the first FLC	the second FLC
d_{i1}	$\mu_S(d_{11}) = f_g(d_{11}, 0.25, 0.05)$ $\mu_M(d_{11}) = f_g(d_{11}, 0.25, 0.5)$ $\mu_B(d_{11}) = f_g(d_{11}, 0.25, 0.9)$	$\mu_S(d_{21}) = f_g(d_{21}, 0.5, 0.1)$ $\mu_M(d_{21}) = f_g(d_{21}, 0.5, 0.8)$ $\mu_B(d_{21}) = f_g(d_{21}, 0.5, 1.5)$
d_{i2}	$\mu_S(d_{12}) = f_g(d_{12}, 0.35, 0.01)$ $\mu_M(d_{12}) = f_g(d_{12}, 0.35, 0.5)$ $\mu_B(d_{12}) = f_g(d_{12}, 0.35, 0.9)$	$\mu_S(d_{22}) = f_g(d_{22}, 0.5, 0.1)$ $\mu_M(d_{22}) = f_g(d_{22}, 0.5, 0.8)$ $\mu_B(d_{22}) = f_g(d_{22}, 0.5, 1.5)$
F	$\mu_S(F) = f_g(F, 0.5, 0.3)$ $\mu_M(F) = f_g(F, 0.5, 0.6)$ $\mu_B(F) = f_g(F, 0.5, 0.9)$	
CR		$\mu_S(CR) = f_g(CR, 0.35, 0.4)$ $\mu_M(CR) = f_g(CR, 0.35, 0.7)$ $\mu_B(CR) = f_g(CR, 0.35, 1.0)$

3.2 Fuzzy Rules

F and/or CR should be adapted based on PC and FC (Table 2). In brief the principles are:

- F (CR) should be large when PC or FC is large, indicating that the actual solution is far away from the expected.
- F (CR) should be small when PC and FC is small, showing that the actual solution is close to the expected.

Table 2. The fuzzy rules (S = small; M = medium; B = big)

rules	fuzzy sets		
	PC	FC	F, CR
1	S	S	S
2	S	M	M
3	S	B	B
4	M	S	M
5	M	M	M
6	M	B	B
7	B	S	B
8	B	M	B
9	B	B	B

3.3 Control Strategy

Mamdani's fuzzy inference method was used.

3.4 Defuzzification Strategy

The centroidal defuzzification technique (CDT) was used. It utilizes all the information in the fuzzy distribution to compute the centroids,

$$y^* = \frac{\sum_{j=1}^{n} \mu_z(w_j) \times w_j}{\sum_{j=1}^{n} \mu_z(w_j)}. \tag{10}$$

4 Implementation and Experimental Study

The setting of FLC presents a subjective view of an expert with respect to the objective characteristics of DE. The setting is viewable as the control surface of DE shown in Fig. 3. The change in behavior (convergence) of FADE (Test 4) in accordance with the fuzzy control action can be seen from the characteristics in Fig. 4 with the first test function in Sect. 5 as an example.

For testing the performance of an evolutionary algorithm, a set of standard test functions in Sect. 5 was used. In the tests, the settings in Table 3 were used. A test function denoted as Ackley's path function is used as an example for demonstration. Fig. 5 shows the average of 100 independent tests using traditional DE and FADE with different dimensionalities, like 2 and 50. In Fig. 5-b, the curve of DE lies above the curves of FADE, which means that for the given objective function value, the FADE algorithm needs less generations than DE does. Experiments based on the test functions in Sect. 5 also show that Fuzzy Adaptive Differential Evolution Algorithms

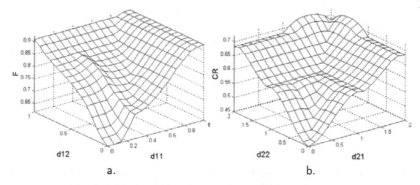

Fig. 3. Control surface of FADE: (**a**) Adapt F; (**b**) Adapt CR

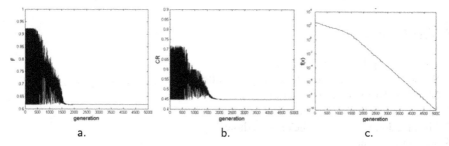

Fig. 4. Behavior of FADE for the sphere function: (**a**) F - generation (F starts from 0.9); (**b**) CR - generation (CR starts from 0.9); (**c**) $f(\mathbf{X})$ (function value) - gneration

Table 3. Settings for algorithms

method, parameters	settings for tests 1, 2, 3 and 4			
strategy	DE/rand /1/bin			
test problems	$\min f(\mathbf{X})$			
tests	1	2	3	4
number of individuals	$10 \times D$	$10 \times D$	$10 \times D$	$10 \times D$
crossover operator	0.9	0.9	FLC	FLC
mutation amplification	0.9	FLC	0.9	FLC

Test 1: the setting is according to the approximate values pointed out in the DE literature and kept constant per run. Tests 2, 3 and 4: the setting is according to the parameter values described in Fig. 1, the initial setting is identical to Test 1.

that adapt F and/or CR perform better than the static one when the dimensionality of the problem is high (for example, $D = 50$ in Table 4), but they do not show many advantages in the case of lower dimensionality (in Fig. 5-a) as shown by 2-dimension functions in Table 4.

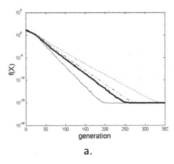
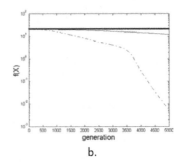

a.   b.

Fig. 5. Comparisons of DE with and without adaptation: (**a**) $D = 2$; (**b**) $D = 50$. Figure legend: ($\cdot\Delta$) DE; (–) DE adapting CR; (—) DE adapting F; (— –) DE adapting F and CR. Curves of DE and DE adapting CR are reprinted in (**b**)

5 Test functions [1]

1. First De Jong Function (sphere)
2. Second De Jong Function (Rosenbrock's valley)
3. Modified Third De Jong Function (step)
4. Modified fourth De Jong Function (quartic)
5. Fifth De Jong Function (Shekel's Foxholes)
6. Rastrigin's Function
7. Ackley's Path Function
8. Easom's function
9. Goldstein-Price's function
10. Six-hump camel back function

6 Conclusions

The Fuzzy Adaptive Differential Evolution Algorithm utilizing fuzzy logic control for parameter adaptation is proposed. Based on human knowledge and expertise, it is designated to expedite the convergence velocity of DE by the use of adaptive parameters.

The experimental results suggest that the proposed algorithm performs better than that using all the fixed parameters. The FADE algorithm related to the parameter, F, converged faster than the traditional DE specially when the dimensionality of the problem was high or the problem concerned was complicated. Current research demonstrates that FADE is a promising approach towards DE and fuzzy logic. This work presents a new point of view to DE parameter setting and can be understood as a new possible way for fuzzy parameter setting of DE. However, further work is considered essential in order to collect further evidence of the performance of FADE and its possible shortcomings.

---

[1] Complete functions are at http://www.lut.fi/~liu/func.ps.

Table 4. Results of experiments: comparison of DE and FADE

values	curves	values	curves
test function 1: $f(\mathbf{0}) = 0$ $D = 50; NP = 500$ $f_1 = 265.6947$ $f_2 = 50.1505$ $f_3 = 8.5802$ $f_4 = 1.2774e - 010$		test function 2: $f(\mathbf{1}) = 0$ $D = 50; NP = 500$ $f_1 = 1.1091e + 004$ $f_2 = 4.2865e + 003$ $f_3 = 628.4849$ $f_4 = 45.7620$	
test function 3: $f(\mathbf{X}) = 0$ $D = 50; NP = 500$ $f_1 = 263.5900$ $f_2 = 59.2300$ $f_3 = 16.7800$ $f_4 = 0$		test function 4: $f(\mathbf{0}) \leq 15$ $D = 30; NP = 300$ $f_1 = 31.8855$ $f_2 = 10.3965$ $f_3 = 9.7149$ $f_4 = 9.0983$	
test function 5: $f(-\mathbf{32}) \cong 0.998004$ $D = 2; NP = 20$ $f_1 = 1.0378$ $f_2 = 0.9980$ $f_3 = 0.9980$ $f_4 = 0.9980$		test function 6: $f(\mathbf{0}) = 0$ $D = 50; NP = 500$ $f_1 = 697.4709$ $f_2 = 476.5656$ $f_3 = 463.7412$ $f_4 = 296.6674$	
test function 7: $f(\mathbf{0}) = 0$ $D = 50; NP = 500$ $f_1 = 20.6084$ $f_2 = 20.0997$ $f_3 = 12.0502$ $f_4 = 0.0054$		test function 8: $f(\pi) = -1$ $D = 2; NP = 20$ $f_1 = -0.3604$ $f_2 = -1$ $f_3 = -0.2400$ $f_4 = -0.1800$	
test function 9: $f(0, -1) = 3$ $D = 2; NP = 20$ $f_1 = 3.0000$ $f_2 = 3.0000$ $f_3 = 3.0000$ $f_4 = 3.0000$		test function 10: $f(\mathbf{X}) = -1.0316$ $D = 2; NP = 20$ $f_1 = -1.0316$ $f_2 = -1.0316$ $f_3 = -1.0316$ $f_4 = -1.0316$	

Note: $f(\mathbf{X})$ = the optimum of the specified function. f_1: DE; f_2: DE adapting CR; f_3: DE adapting F; f_4: DE adapting F and CR. The horizontal axis is generation while the vertical one is function value. Figure legend is as that in Fig. 5.

References

1. Eiben A E, Hinterding R, and Michalewicz Z (1999) Parameter Control in Evolutionary Algorithms, IEEE Transactions on Evolutionary Computation, Vol. 3, No. 2, July
2. Lee M A, and Takagi H (1993) Dynamic Control of Genetic Algorithms using Fuzzy Logic Techniques. In: Proceedings of 5th International Conference on Genetic Algorithms, Urbana-Champaign, IL. pp. 76-83
3. Herrera F, Lozano M, Verdegay J L (1995) Tackling Fuzzy Genetic Algorithms. In: Winter G, Periaux J, Galan M, Cuesta P (eds) Genetic Algorithms in Engineering and Computer Science, pp. 167-189, John Wiley and Sons

4. Storn R and Price K (1995) Differential Evolution - a Simple and Efficient Adaptive Scheme for Global Optimization over Continuous Spaces. Technical report: TR-95-012, ICSI, March

5. Storn R (1996) On the Usage of Differential Evolution for Function Optimization. In: 1996 Biennial Conference of the North American Fuzzy Information Processing Society, Berkeley, CA, 19-22 Jun, pp. 519-523. IEEE, New York (USA).

6. Lampinen J (1999) A Bibliography of Differential Evolution Algorithm. Department of Information Technology, Lappeenranta University of Technology. Technical report, http://www.lut.fi/~jlampine/debiblio.htm

7. Lampinen J and Zelinka I (2000) On Stagnation of the Differential Evolution Algorithm. In: Proceedings of 6th International Conference on Soft Computing, June 7-9, Brno, Czech Republic, pp. 76-83. ISBN 80-214-1609-2

8. Storn R and Price K (1997) Differential Evolution - A simple evolution strategy for fast optimization. In: Dr. Dobb's Journal 22(4): 18-24 and 78, April

9. Storn R and Price K (1997) Differential Evolution - a Simple and Efficient Heuristic for Global Optimization over Continuous Spaces. In: Global Optimization, 11(4), pp. 341-359, December. Kluwer Academic Publishers

10. Price K V (1999) An Introduction to Differential Evolution. In: Corne D, Dorigo M and Glover F (eds) New Ideas in Optimization, pp. 79-108, McGraw-Hill, London (UK)

11. Lopez Cruz I L, Van Willigenburg L G, Van Straten G (2001) Parameter Control Strategy in Differential Evolution Algorithm for Optimal Control. In: Proceedings of the International Conference on Artificial Intelligence and Soft Computing, May 21-24, Cancun, Mexico, pp. 211-216. ACTA Press, Calgary (Canada). ISBN 0-88986-283-4, ISSN: 1482-7913

12. Bandemer H (1995) Fuzzy sets, fuzzy logic, fuzzy methods with applications. Wiley, Chichester (UK)

13. Zimmermann H J (1986) Fuzzy set theory - and its applications. Kluwer-Nijhoff, Boston (USA)

14. Liu J H and Lampinen J (2002) On Setting the control parameter of the Differential Evolution Algorithm. In: Proceedings of 8th International Conference on Soft Computing, June 5-7, Brno, Czech Republic, pp. 11-18. ISBN 80-214-2135-5

A Distributed-Population GA for Discovering Interesting Prediction Rules

Edgar Noda[1], Alex Alves Freitas[2], and Akebo Yamakami[1]

[1] School of Electrical and Computer Engineering (FEEC) - State University of Campinas (Unicamp), Brazil, edgar@dt.fee.unicamp.br, akebo@dt.fee.unicamp.br
[2] Computing Laboratory - University of Kent at Canterbury - Canterbury, CT2 7NF, UK, A.A.Freitas@ukc.ac.uk

Summary. In data mining, the quality of prediction rules basically involve three criteria: accuracy, comprehensible and interestingness. The majority of the rule induction, literature focuses on discovering accurate, comprehensible rules. In this paper we also take these two criteria into account, but we go beyond them in the sense that we aim at discovering rules that are interesting (surprising) for the user. The search is performed by distributed genetic algorithm (DGA) specifically designed to the discovery of interesting rules.

1 Introduction

A well-known data mining task is classification, which consists of predicting the class of an example (a record of a data set) out of a predefined set of classes, given the values of predictor attributes for that example [1] [6].This paper addresses a generalization of the classification task, called dependence modeling [4], where there are several goal attributes to be predicted. In this context, we addresses the discovery of prediction rules of the form:

IF (some conditions on the values of predictor attributes are true) *THEN* (predict a value for some goal attribute).

In our approach for dependence modeling the user specifies a small set of potential goal attributes, which she/he is interested in predicting. Although we allow more than one goal attribute, each prediction rule has a single goal attribute in its consequent (THEN part). However, different rules can have different goal attributes in their consequent.

In principle, the prediction rules discovered by a data mining algorithm should satisfy three properties, namely: predictive accuracy, comprehensibility and interestingness [5]. In this paper we propose a distributed-population genetic algorithm (GA) designed to discover a few rules that are both interesting and accurate. Both these criteria are included in the fitness function of the GA. In addition, designating, as the output of the GA, a small set of rules, which can be thought of as "knowledge nuggets" extracted from the data, facilitates the discovery of comprehensible knowledge. Discovered knowledge should also be comprehensible to the user [8]. Knowledge represented as high-level rules, as in the above-mentioned IF-THEN format, has the advantage of being intuitively compressible to the user.

Discovered knowledge should also be interesting to the user. By "interesting" we mean that discovered knowledge should be novel or surprising to the user. We emphasize that the notion of interestingness goes beyond the notions of predictive accuracy and comprehensibility. Discovered knowledge may be highly accurate and comprehensible, but it is uninteresting if it states the obvious or some pattern that was previously known by the user. A very simple, classical example shows the point. Suppose one has a medical database containing data about a hospital's patients. A data mining algorithm could discover the following rule from such a database: IF (patient is pregnant) THEN (patient is female). This rule has a very high predictive accuracy and it is very comprehensible. However, it is uninteresting, since it states an obvious, previously known pattern.

2 GA-Nuggets

In our previous work we have introduced a GA for dependence modeling, called GA-Nuggets [7]. This GA maintains a single, centralized population of individuals. In this paper we propose a major extension of that GA. It maintains a distributed population, consisting of several subpopulations, each of them evolving in an independent manner, with occasional migration between them. Subsection 2.1 briefly reviews the main aspects of GA-Nuggets [7], whereas the new distributed-population is described in subsection 2.2.

2.1 Single-Population GA-Nuggets

Each individual represents a candidate prediction rule of the form: IF *Ant* THEN *Cons*, where *Ant* is the rule antecedent and *Cons* is the rule consequent. *Ant* consists of a conjunction of conditions, where each condition is an attribute-value pair of the form $A_i = V_{ij}$, where A_i is the $i-th$ attribute and V_{ij} is the $j-th$ value of the domain of A_i. An individual is encoded as a fixed-length string containing z genes (see Fig. 1), where z is the number of attributes (considering both predictor and goal attributes). Only a subset of the attribute values encoded in the genome will be decoded into attribute values occurring in the rule antecedent. Therefore, although the genome length is fixed, its decoding mechanism effectively represents a variable-length rule.

Once the rule antecedent is formed, the algorithm chooses the best consequent for each rule in such a way that maximizes the fitness of an individual (candidate rule). In effect, this approach gives the algorithm some knowledge of the data-mining task being solved.

Fig. 1. Individual representation

The fitness function consists of two parts. The first one measures the degree of interestingness of the rule, while the second measures its predictive accuracy. The degree of interestingness of a rule, in turn, consists of two terms. One of them refers to the antecedent of the rule and the other to the consequent. The degree of interestingness of the rule antecedent (*AntInt*) is calculated by an information-theoretical measure [2] (see 1).

The computation of the rule consequent's degree of interestingness (*ConsInt*) is based on the idea that the prediction of a rare goal attribute value tends to be more interesting to the user than the prediction of a very common goal attribute value [3] (see 2). The computation of these two degrees of interestingness is described in detail in [7]]. The second part of the fitness function measures the predictive accuracy (*PredAcc*) of the rule (see 3).

$$\textbf{AntInt} = 1 - \left(\frac{\displaystyle\sum_{i=1}^{n} \textbf{InfoGain}(A_i)}{\log_2(|\textbf{dom}(G_k)|)} \right) \tag{1}$$

Where m_k is the number of possible values of the goal attribute G_k, ni is the number of possible values of the attribute A_i, $Pr(X)$ denotes the probability of X and $Pr(X|Y)$ denotes the conditional probability of X given Y.

$$\textbf{ConsInt} = (1 - \textbf{Pr}(G_{kl}))^{1/\beta} \tag{2}$$

Where $Pr(G_{kl})$ is the prior probability (relative frequency) of the goal attribute value G_{kl}, and β is a user-specified parameter, empirically set to 2 in our experiments. The exponent $1/\beta$ in the equation [5] can be regarded as a way of reducing the influence of the rule consequent interestingness in the value of the fitness function.

$$\textbf{Fitness} = \frac{w_1 \cdot \frac{\textbf{AntInt} + \textbf{ConsInt}}{2} + w_2 \cdot \textbf{PredAcc}}{w_1 + w_2} \tag{3}$$

Where w_1 and w_2 are user-defined weights. In our experiment they are set to 1 and 2, respectively.

GA-Nuggets uses a tournament selection method with tournament size 2. The algorithm uses uniform crossover extended with a "repair" procedure. After the standard crossover is done, the algorithm checks if any invalid individual was created. If so, a repair procedure is performed to produce valid-genotype individuals. The mutation operator randomly transforms the value of an attribute into another value belonging to the domain of that attribute.

There are two operators, called condition-insertion and condition-removal, who control the size of the rules being evolved by randomly inserting/removing a condition into/from a rule antecedent. The probability of applying each of these operators depends on the current number of conditions in the rule antecedent. The larger the number of conditions in the current rule antecedent, the smaller the probability of applying the condition-insertion operator.

2.2 Distributed-Population GA-Nuggets

In this new version of GA-Nuggets, the entire population is divided into p subpopulations, where p is the number of goal attributes. In each subpopulation all individuals are associated with the same goal attribute. The individual representation of the distributed-population version of GA-Nuggets is similar to the individual representation of the single-population version of GA-Nuggets, described in subsection 2.1. The only difference is that the goal attribute is fixed for all individuals of the same subpopulation. Each subpopulation evolves independently from the others (except for some occasional migrations).

One advantage of this distributed population approach, with a fixed goal attribute for each subpopulation, is to reduce the number of crossovers performed between individuals predicting different goal attributes. Since crossover is restricted to individuals of the same subpopulation, crossover swaps genetic material of two parents, which represent candidate rules predicting the same goal attribute.

Distributed-population GA-Nuggets has a migration procedure where, from time to time, an individual of a subpopulation is copied into another subpopulation, as follows. The subpopulations evolve in a synchronous manner, so that in each subpopulation the $i - th$ generation is started only after the $(i - 1) - th$ generation has been completed in all subpopulations, for $i = 2, \dots, g$, where g is the number of generations (which is the same for all subpopulations).

Migration takes place every m generations. Each population sends individuals to all the other subpopulations. More precisely, in each subpopulation S_i, $i = 1, \dots, p$, the migration procedure chooses $(p - 1)$ individuals to be migrated. Each of those $p - 1$ migrating individuals will be sent to a distinct subpopulation S_j, $j = 1, \dots, p$, $j \neq i$. The choice of the individuals to be migrated is driven by the fitness function, taking into account the fact that different subpopulations are associated with different goal attributes. In each subpopulation S_i the migration procedure knows, for each individual, not only the actual value of its fitness in that subpopulation (called its home fitness), but also what would be the value of the fitness of that individual if it were placed in another subpopulation S_j, $j = 1, \dots, p, j \neq i$, predicting a value of the $j - th$ goal attribute. The latter is called the foreign fitness of the individual in subpopulation S_j. Subpopulation S_i sends to subpopulation S_j a copy of the individual with maximum foreign fitness in S_j.

On the other hand, each subpopulation $S_i, i = 1, \dots, p$, receives $p - 1$ individuals, each coming from a different subpopulation $S_j, j = 1, \dots, p, j \neq i$. Among these incoming $p - 1$ individuals, only one is accepted by subpopulation S_i. The accepted individual is the one with the largest fitness value. This is equivalent to a tournament selection among the incoming individuals.

The fitness function of distributed-population GA-Nuggets is the same as the fitness function of single-population GA-Nuggets, as defined by formula 3. In distributed-population GA-Nuggets the application of the selection method and genetic operators is independently performed in each of the subpopulations. Each subpopulation uses the same selection method and genetic operators (described in subsection 2.1), which are applied only to the local individuals in that subpopulation.

3 Computational Results and Discussion

The data sets used to evaluate the previously described algorithms were obtained from the UCI repository of machine learning databases ($http: //www.ics.uci.edu/AI /Machine - Learning.html$). The data sets used are Zoo, Car Evaluation, Auto Imports and Nursery. They are normally used for evaluating algorithms performing the classification task.

The zoo database contains 101 instances and 18 attributes. In the pre processing phase the attribute containing the name of the animal was removed, since this attribute has no generalization power. In our experiments the set of potential goal attributes used was predator, domestic and type. The car evaluation dataset contains 1728 instances and 6 attributes. The attributes buying and car acceptability were used as potential goal attributes. The auto-imports 85M dataset contains 205 instances and 26 attributes. The attribute normalized-losses and 12 instances were removed because of missing values. The attributes symbolling, body-style and price, were chosen as goals. The nursery school data set contains 12960 instances and 9 attributes. In our experiments, the attributes used as potential goal attributes were finance, social and health.

We emphasize that in our approach for dependence modeling we do not aim at classifying the whole test set. Rather, the goal is to discover a few interesting rules to be shown to a user. We can think of the discovered rules as the most valuable "knowledge nuggets" extracted from the data. These knowledge nuggets are valuable even if they do not cover the whole test set. In other words, the value of the discovered rules depends on their predictive accuracy on the part of the test set covered by those rules, but not on the test set as a whole. For each data set we have run a 10-fold cross-validation procedure [6] to evaluate the quality of the rules discovered by two algorithms, namely: single-population GA-Nuggets (section 2.1), and distributed-population GA-Nuggets (section 2.2). The computational experiments measured both the predictive accuracy (accuracy rate in the test set) and the degree of interestingness of the rules discovered by the two algorithms.

3.1 Predicative accuracy

We now compare the results, for all datasets, of the two algorithms concerning the predictive accuracy issue. The first column is the name of the goal attribute, follow by the possible values. Then the table is subdivided for each algorithm with the coverage and the respective predictive accuracy.

Tables 1, 2, 3, and 4 shows the results in the zoo, car evaluation, auto-imports, and nursery datasets. Assuming the single-population algorithm as a baseline the sign $(+/-)$ indicates when the distributed-population algorithm significantly outperforms / was outperformed by the baseline (talking into account the standard deviations).

Distributed-population GA-Nuggets obtained somewhat better results than single-population GA-Nuggets. In only one case the single-population GA-Nuggets found rules with significantly higher accuracy. Distributed GA-Nuggets significantly outperformed the rule induction algorithm in five cases.

Table 1. Predictive Accuracy (%) in the Zoo data set

Goal Attrib	Attrib Value	GA Cov	GA Pred Acc	DGA Cov	DGA Pred Acc
Predator	False	4.4	50.5 ± 8.9	3.2	48.0 ± 8.2
	True	2.8	75.0 ± 11.2	2.4	84.0 ± 11.1
Domestic	False	5.2	97.1 ± 5.2	6.2	90.5 ± 4.4
	True	0.8	0.0 ± 0.0	0.8	0.0 ± 0.0
Type	1	6.4	100.0 ± 0.0	6.4	100.0 ± 0.0
	2	3.6	100.0 ± 0.0	3.6	100.0 ± 0.0
	3	0.2	0.0 ± 0.0	1.1	$95.0 \pm 13.0 \ (+)$
	4	2.2	100.0 ± 0.0	2.2	100.0 ± 0.0
	5	0.5	100.0 ± 0.0	0.8	100.0 ± 0.0
	6	1.1	90.0 ± 10.0	1.1	90.0 ± 10.0
	7	2.0	83.3 ± 10.2	2.0	85.0 ± 11.0

Table 2. Predictive Accuracy (%) in the Car Evaluation data set

Goal Attrib	Attrib Value	GA Cov	GA Pred Acc	DGA Cov	DGA Pred Acc
Buying	V-High	1.2	60.0 ± 16.3	1.0	50.0 ± 16.7
	High	2.5	4.5 ± 4.5	2.2	7.5 ± 3.8
	Med	2.5	2.5 ± 2.5	2.3	5.0 ± 3.3
	Low	2.3	100.0 ± 0.0	2.0	100.0 ± 0.0
Accept	Unacc	10.4	100.0 ± 0.0	10.4	100.0 ± 0.0
	Acc	0.1	0.0 ± 0.0	0.0	0.0 ± 0.0
	Good	0.0	0.0 ± 0.0	0.1	0.0 ± 0.0
	V-Good	0.0	0.0 ± 0.0	0.1	0.0 ± 0.0

3.2 Degree of Interestingness

Similar to the predictive accuracy tables, the degree of interestingness results for all datasets are presented with the attribute goal, possible value, the consequent degree

Table 3. Predictive Accuracy (%) in the Auto Imports data set

Goal Attrib	Attrib Value	GA Cov	GA Pred Acc	DGA Cov	DGA Pred Acc
Simb.	-3	0.0	0.0 ± 0.0	0.0	0.0 ± 0.0
	-2	0.0	0.0 ± 0.0	0.8	0.0 ± 0.0
	-1	1.2	55.0 ± 13.8	1.6	63.3 ± 14.4
	0	2.2	96.0 ± 2.7	2.0	98.0 ± 2.0
	1	1.7	70.0 ± 15.3	2.3	70.0 ± 10.2
	2	1.2	63.3 ± 14.4	1.3	90.0 ± 10.0 (+)
	3	1.2	70.0 ± 15.3	1.9	$70.0 \pm 12,6$
Body	Hardtop	0.6	0.0 ± 0.0	0.4	0.0 ± 0.0
	Wagon	0.6	0.0 ± 0.0	1.6	13.3 ± 5.4 (+)
	Sedan	0.6	60.0 ± 16.3	2.1	82.5 ± 9.9 (+)
	Hatch	2.6	76.7 ± 6.7	2.8	71.7 ± 5.4
	Convert.	0.6	40.0 ± 16.3	1.0	25.0 ± 8.3
Price	Low	11.4	100.0 ± 0.0	13.4	100.0 ± 0.0
	Average	3.2	90.0 ± 4.1	3.7	81.7 ± 9.7
	High	1.4	72.5 ± 12.6	1.3	90.0 ± 10.0

Table 4. Predictive Accuracy (%) in the Nursery data set

Goal Attrib	Attrib Value	GA Cov	GA Pred Acc	DGA Cov	DGA Pred Acc
Finance	Conv.	2.2	80.0 ± 13.3	3.4	100.0 ± 0.0 (+)
	Inconv.	3.4	100.0 ± 0.0	3.9	100.0 ± 0.0
Social	Non-prob	3.2	1.1 ± 1.1	2.2	0.0 ± 0.0
	Slightly prob	27.7	6.4 ± 4.3	2.0	0.0 ± 0.0 (−)
	Problem.	4.4	100.0 ± 0.0	10.2	100.0 ± 0.0
Health	Recomm.	0.0	0.0 ± 0.0	0.0	0.0 ± 0.0
	Priority	291.6	0.0 ± 0.0	0.2	0.0 ± 0.0
	Not Recomm.	54.5	12.8 ± 9.8	15.8	41.8 ± 14.4 (+)
	Spec priority	4.6	100.0 ± 0.0	10.0	100.0 ± 0.0
	Very recomm.	10.8	100.0 ± 0.0	8.6	100.0 ± 0.0

of interestingness and the two antecedent interestingness values. Table 5, 6, 7, and 8 shows the results in the zoo, car evaluation, auto imports, and nursery datasets.

Distributed-population GA-Nuggets obtained results considerably better than single-population GA-Nuggets. With respect to rule interestingness, the former algorithm significantly outperformed the latter in 22 out of 44 cases - considering all the discovered rules in all the four data sets - whereas the reverse was true in just five out of 44 cases. In the other cases the difference between the two algorithms was not statistically significant.

We have also observed that distributed-population GA-Nuggets has performed a more cost-effective search than single-population GA-Nuggets, in the sense that in general the former has obtained good solutions in earlier generations, by comparison with the latter. (Both versions of the GA had the same total population size, so that the comparison was fair.) Hence, overall, considering the results in the four data sets, distributed-population GA-Nuggets represents an improvement over single-population GA-Nuggets.

Table 5. Interestingness (%) in the Zoo data set

Goal Attrib	Attrib Value	Cons Int.	GA AntInt	DGA AntInt
Predator	False	74.4	97.5 ± 0.4	95.9 ± 1.0 (−)
	True	66.8	94.9 ± 0.5	96.4 ± 0.4 (+)
Domestic	False	35.7	96.3 ± 0.5	96.9 ± 0.6
	True	93.3	96.9 ± 0.7	97.9 ± 0.4
Type	1	77.1	94.7 ± 0.2	94.6 ± 0.1
	2	89.0	93.9 ± 0.3	93.9 ± 0.3
	3	97.5	93.2 ± 0.6	92.3 ± 0.2 (−)
	4	94.3	93.4 ± 0.2	94.7 ± 0.3 (+)
	5	97.9	94.3 ± 0.4	94.0 ± 0.3
	6	95.9	93.4 ± 0.3	92.4 ± 0.4 (−)
	7	94.9	95.3 ± 0.1	95.1 ± 0.2

Table 6. Interestingness (%) in the Car Evaluation data set

Goal Attrib	Attrib Value	Cons Int.	GA AntInt	DGA AntInt
Buying	V-High	86.6	99.4 ± 0.0	99.4 ± 0.0
	High	86.6	99.4 ± 0.0	99.4 ± 0.0
	Med	86.6	99.3 ± 0.0	99.4 ± 0.0 (+)
	Low	86.6	98.8 ± 0.0	99.0 ± 0.0 (+)
Accept	Unacc	54.7	96.5 ± 0.0	96.4 ± 0.0 (−)
	Acc	88.3	93.2 ± 0.0	93.3 ± 0.0 (+)
	Good	97.9	94.3 ± 0.0	94.3 ± 0.0
	V-Good	98.1	94.3 ± 0.0	94.3 ± 0.0

4 Conclusion and future works

In this paper we have presented two algorithms for discovering "knowledge nuggets" - rules that have both a good predictive accuracy and a good degree of interestingness. The algorithms were developed for discovering prediction rules in the dependence modeling task of data mining.The algorithms presented in this paper are actually two different versions of a genetic algorithm. One of these versions uses a single population of individuals, whereas the other version uses a distributed population of individuals.The other characteristics of the GA were kept the same, as much as

Table 7. Interestingness (%) in the Auto Imports data set

Goal Attrib	Attrib Value	Cons Int.	GA AntInt	DGA AntInt
Simb.	-3	100.0	99.3 ± 0.1	100.0 ± 0.0 (+)
	-2	99.2	98.3 ± 0.1	99.0 ± 0.3 (+)
	-1	94.1	97.7 ± 0.1	97.8 ± 0.1 (+)
	0	82.1	97.7 ± 0.2	97.5 ± 0.1
	1	85.8	97.8 ± 0.2	97.9 ± 0.1
	2	91.6	97.4 ± 0.2	98.1 ± 0.1 (+)
	3	93.8	98.1 ± 0.1	98.7 ± 0.1 (+)
Body	Hardtop	97.9	97.5 ± 0.3	98.3 ± 0.4 (+)
	Wagon	93.6	97.6 ± 0.2	98.1 ± 0.3
	Sedan	72.3	96.5 ± 0.5	97.8 ± 0.5 (+)
	Hatch	82.1	97.1 ± 0.3	97.5 ± 0.1
	Convert.	98.4	98.1 ± 0.2	98.6 ± 0.1 (+)
Price	Low	64.8	94.2 ± 0.5	96.8 ± 0.1 (+)
	Average	80.8	92.9 ± 0.9	95.1 ± 0.3 (+)
	High	96.3	90.8 ± 0.4	96.1 ± 0.2 (+)

Table 8. Interestingness (%) in the Nursery data set

Goal Attrib	Attrib Value	Cons Int.	GA AntInt	DGA AntInt
Finance	Conv.	71.1	99.8 ± 0.0	99.9 ± 0.0 (+)
	Inconv.	70.3	99.8 ± 0.0	99.9 ± 0.0 (+)
Social	Non-prob	81.7	99.7 ± 0.0	99.9 ± 0.0 (+)
	Slightly prob	81.6	99.8 ± 0.0	99.9 ± 0.0 (+)
	Problem.	81.6	99.7 ± 0.0	99.8 ± 0.0 (+)
Health	Recomm.	81.7	94.9 ± 0.0	94.9 ± 0.0
	Priority	99.9	99.7 ± 0.0	99.9 ± 0.0 (+)
	Not Recomm.	98.7	96.3 ± 0.7	94.6 ± 0.4 (−)
	Spec priority	81.9	93.5 ± 0.3	93.4 ± 0.3
	Very recomm.	82.9	94.1 ± 0.3	94.3 ± 0.3

possible, in the two versions, in order to allow us to compare the two versions in a manner as fair as possible.

This comparison was performed across four public-domain, real-world data sets. The computational experiments measured both the predictive accuracy and the degree of interestingness of the rules discovered by the three algorithms.

Overall the computational results indicate a somewhat advantage to the distributed approach, with respect to predictive accuracy. With respect to the degree of interestingness of the discovered rules, the distributed-population version of the GA obtained results considerably better than the single population algorithm.

One direction for future research consists of developing a new version of the distributed-population GA where each subpopulation is associated with a goal attribute value, rather than with a goal attribute as in the current distributed version. It will be interesting to compare the performance of this future version with the performance of the current distributed version, in order to empirically determine the cost-effectiveness of these approaches, considering their pros and cons discussed in subsection 3.

References

1. Fayyad U M, Piatetsky-Shapiro G and Smyth P (1996) From data mining to knowledge discovery: an overview. In: U.M. Fayyad, et al.(eds) Advances in Knowledge Discovery and Data Mining. AAAI/MIT Press.
2. Freitas A A (1998) On objective measures of rule surprisingness. Lecture Notes in Artificial Intelligence. Springer. 1510:1–9.
3. Freitas A A (1999) A genetic algorithm for generalized rule induction. In: R. Roy et al. Advances in Soft Computing - Engineering Design and Manufacturing. Springer, 340–353.
4. Freitas A A (2000) Understanding the crucial differences between classification and discovery of association rules - a position paper. ACM SIGKDD Explorations, 2(1):65–69.
5. Freitas A A (2002) Data Mining and Knowledge Discovery with Evolutionary Algorithms. Springer-Verlag Berlin.
6. Hand D J (1997) Construction and Assessment of Classification Rules. John Wiley & Sons.
7. Noda E, Freitas A A, and Lopes H S (1999) Discovering interesting prediction rules with a genetic algorithm. Proc. of the Congress on Evolutionary Computation (CEC-99), IEEE Press 1322–1329.
8. Spiegelhalter D J, Michie D, and Taylor C C (1994) Machine Learning, Neural and Statistical Classification. Ellis Horwood, New York.

Confidence interval based crossover using a L_1 norm localization estimator for real-coded genetic algorithms

César Hervás-Martínez, Nicolás García-Pedrajas, and Domingo Ortiz-Boyer

Department of Computing and Numerical Analysis
University of Córdoba, Spain
({chervas, npedrajas, ma1orbod}@uco.es)

Summary. In this work we propose a new multiparent crossover operator for real-coded genetic algorithms based on the extraction of the statistical features (localization and dispersion of genes) of the best individuals of the population. For the construction of the proposed crossover we determine a confidence interval using an estimator of the localization parameter of the gene's distribution of the best individuals of the population using an L_1 norm. This crossover will be called *Confidence Interval based Crossover using L_1 norm* (CIXL1). We construct bilateral confidence intervals for each gene from the statistical distribution of each localization parameter; then we define a crossover operator using as parents the localization parameters and the lower and upper limits of the confidence interval. A theoretical study shows the statistical features of the offspring obtained using this crossover provided that we are in an evolutionary stage with high selective pressure.

The comparison of the proposed operator with BLX-α, UNDX-1, Extended Fuzzy and Seed crossovers for the optimization of different test functions shows the efficiency of the operator both in the optimal value achieved and in the convergence rate.

Key words: Real-coded genetic algorithms; Multi-parent crossover; Confidence intervals; L_1 norm.

1 Introduction

Real-coded genetic algorithms (RCGAs) are used as an approach for solving numeric optimization problems in continuous domains ([1], [2], [3]). These algorithms are specially interesting for the optimization of multimodal and/or epistatic functions with many variables. One of the most important aspects of these algorithms is the design of the crossover operator. This operator generates new individuals from two members of the population (arity 2 crossovers, [2], [4]) or more than two members (arity $n > 2$ or multiparent crossovers, [5], [6]).

Our work is focussed in the design of a new arity n crossover which extracts the most relevant statistical features of localization and dispersion of the fittest individuals of the population. As a working hypothesis we consider that the distribution of the genes of the fittest individual are continuous during all the evolutionary process in order to estimate the localization and dispersion parameters ([7], [8]).

Under this assumption, and due to the fact that we do not know the specific distribution of the genes of the best individuals, we will use a localization estimator based on L_1 norm. This estimator is the sample median of the genes of the best n individuals on each generation, it follows a distribution that does not depend on the distribution of the genes and that is a binomial with parameters n and $1/2$.

From this localization estimator and its associated distribution we construct bilateral confidence intervals for each gene, and form three parents for the crossover operation. The first one is made up by the lower bounds of the confidence intervals, the second one is made by the upper bounds of the confidence intervals, and the third one is made up by the localization estimator itself, that is, the sample median of the genes.

This article is organized as follows: Section 2 presents the crossover operator based on L_1 norm; Section 3 shows the results of the comparison of this operator with other multiparent crossover operators in the optimization of different functions; finally, Section 4 states the conclusions of our work.

2 Multiparent crossover using L_1 norm

We will consider the i-th gene without loss of generality. Let β be the set of N individuals that form the population and let $\beta^* \subset \beta$ be the subset of the best n individuals, and q the number of genes of each chromosome. Let us assume that the genes, β_i, of the chromosomes of the individuals in β^* are independent random variables with a continuous distribution $H(\beta_i)$, and a localization parameter of the form μ_{β_i}, then we have a model $\beta_i = \mu_{\beta_i} + e_i$, for $i = 1, \ldots, q$, being e_i a random variable.

If we assume that the genes of the n fittest individuals form a random sample $(\beta_{i_1}, \beta_{i_2}, \ldots, \beta_{i_n})$ of the distribution of the fittest individuals of the population β_i^b, then it can be written:

$$\beta_{ij}^b = \mu_{b_i} + e_{ij}, \quad j = 1, 2, \ldots, n. \tag{1}$$

Using this model we analyze an estimator of the localization parameter μ_{b_i} for the i-th gene based in the minimization of the dispersion function induced by the L_1 norm, $D_1(\mu_{b_i}) = \sum_{j=1}^{n} |\beta_{ij} - \mu_{b_i}|$. The estimator is the sample median $\hat{\mu}_{b_i} = M_{b_i}$, of the distribution of the β_i^b ([9]). The sample median estimator is a better localization estimator than the sample mean when the form of the H distribution is not known. The confidence interval for a sample of size n, with a confident coefficient of $1 - \alpha$, is constructed using the Neyman method, due to the fact that the binomial distribution associated with the median does not depend on the H distribution. With this method we have:

$$P(\beta_{i(k_1)} > M_{b_i}) = \sum_{j=0}^{k_1-1} \binom{n}{j} \left(\frac{1}{2}\right)^n = \alpha/2, \qquad (2)$$

and

$$P(M_{b_i} > \beta_{i(k_2)}) = \sum_{j=k_2}^{n} \binom{n}{j} \left(\frac{1}{2}\right)^n = \alpha/2. \qquad (3)$$

From these equations we have a confidence interval of the form, being $k_1 = k+1$ and $k_2 = n - k$:

$$I_{1\alpha}(\mu_{b_i}) = \left[\beta_{i(k+1)}, \beta_{i(n-k)}\right], \qquad (4)$$

where $\beta_{i(k+1)}$ and $\beta_{i(n-k)}$ are the gene values at position $k+1$ and $n-k$ once the sample of genes has been sorted. The value of k is determined from the underlying binomial distribution. As we have stated, the proposed interval does not depend on the distribution of the genes. This feature is very important as the distribution of the genes of the best individuals will probably change along the evolution.

As the binomial distribution is discrete, it is possible that we cannot obtain discrete values of $\beta_{i(k+1)}$ and $\beta_{i(n-k)}$ that verify $P(\beta_{i(k+1)} \leq \mu_{b_i} \leq \beta_{i(n-k)}) = 1 - \alpha$. This effect is specially important if the number of best individuals considered is small.

In such cases a nonlinear interpolation method is used([9]) for obtaining the values of the lower bound, using $\beta_{i(k)}$ and $\beta_{i(k+1)}$, and upper bound, using $\beta_{i(n-k)}$ and $\beta_{i(n-k+1)}$, of the confidence interval.

The interpolation method is the following. Let $(1 - \alpha) = \gamma$ be the desired confidence interval, and the two possible intervals taken from the binomial tables be $(\beta_{i(k)}, \beta_{i(n-k+1)})$, with a confidence coefficient of γ_k, and $(\beta_{i(k+1)}, \beta_{i(n-k)})$ with a confidence coefficient of γ_{k+1}, where $\gamma_{k+1} \leq \gamma \leq \gamma_k$. Then, the interpolated bounds of the interval are:

$$\hat{\mu}_{b_{iL}} = (1 - \lambda)\beta_{i(k)} + \lambda\beta_{i(k+1)}, \qquad (5)$$

and

$$\hat{\mu}_{b_{iU}} = (1 - \lambda)\beta_{i(n-k+1)} + \lambda\beta_{i(n-k)}, \qquad (6)$$

where $\lambda = \frac{(n-k)\gamma_i}{k+(n-2k)\gamma_i}$, and $\gamma_i = \frac{\gamma_k-\gamma}{\gamma_k-\gamma_{k+1}}$.

2.1 Crossover operation

Once we have constructed the described confidence interval, we build three virtual individuals that will act as parents in the crossover. The first one is formed by all the lower bounds of the confidence interval of each gene of the chromosome, it is called CIL; the second one is formed by all the upper bounds of the confidence interval of each gene of the chromosome, it is called CIU; and the third one is formed by all the medians of the values of each gene of the chromosome, it is called CIM.

These individuals divide the domain $[a_i, b_i]$ of the i-th gene into three intervals: $I_i^L \equiv [a_i, CIL_i)$, $I_i^M \equiv [CIL_i, CIU_i]$, and $I_i^U \equiv (CIU_i, b_i]$.

The *Confidence Interval based Crossover operator using L_1 norm* (CIXL1) obtains an offspring β^s, from an individual of the population, $\beta^k = (\beta_1^k, \ldots, \beta_p^k)$, and the three individuals defined above. We consider the fitness of the four individuals, function $f(\cdot)$, and the position of the genes of β^k within the three intervals defined above. The three possible cases are:

Case 1: $\beta_i^f \in I_i^L$. If $f(\beta^f) > f(CIL)$ then $\beta_i^s = r(\beta_i^f - CIL_i) + \beta_i^f$ else $\beta_i^s = r(CIL_i - \beta_i^f) + CIL_i$.

Case 2: $\beta_i^f \in I_i^M$. If $f(\beta^f) > f(CIM)$ then $\beta_i^s = r(\beta_i^f - CIM_i) + \beta_i^f$ else $\beta_i^s = r(CIM_i - \beta_i^f) + CIM_i$.

Case 3: $\beta_i^f \in I_i^R$. If $f(\beta^f) > f(CIU)$ then $\beta_i^s = r(\beta_i^f - CIU_i) + \beta_i^f$ else $\beta_i^s = r(CIU_i - \beta_i^f) + CIU_i$.

where r is a random number belonging to [0, 1].

One of the most important aspects of the effect of the proposed crossover on the population is the change that it produces in the localization and dispersion parameters of the probability density function of the population. We are going to analyze the changes produced by the crossover depending on the subinterval of the gene.

First, let us assume that the gene β_i^k involved in the crossover is within the confidence interval $\beta_i^k \in I_i^M$. We consider that in the j-th generation β_i^k is a random variable with $E(\beta_i^k) = \mu_j$ and $V(\beta_i^k) = \sigma_j^2$, and the i-th gene of the fittest individuals follow a distribution $\beta_i^b \in H$ with $E(\beta_i^b) = \mu_{b_i}$ and $V(\beta_i^b) = \sigma_{b_i}^2$. We also consider that the two distributions are independent. The median and the variance of the sample median of the fittest individuals are: $E(M_{b_i}) = \mu_{b_i}$, and $V(M_{b_i}) = \frac{1}{4nf^2(0)}$, where f is the density function of the random variable e_i. As the underlying distribution function H is unknown we must estimate the standard deviation of the sample median. We propose an asymptotic estimator of the standard deviation based on the length, L, of the confidence interval([9]), that is:

$$\frac{1}{2n^{1/2}f(0)} = \frac{L}{2}z_{\alpha/2}, \tag{7}$$

where $L = \hat{\mu}_{b_{iU}} - \hat{\mu}_{b_{iL}}$, if we take into account case 2. We have two possibilities: $f(\beta^k) \geq f(CIM)$ with a probability p, and $f(\beta^k) < f(CIM)$ with a probability $1 - p$. In the first case $\beta_i^s = (1+r)\beta_i^k - rCIM_i$, so the mean and variance of the distribution of β_i^s are:

$$E(\beta_i^s) = (1+r)E(\beta_i^k) - rE(CIM_i) = (1+r)\mu_i - r\mu_{b_i}, \tag{8}$$

and

$$V(\beta_i^s) = (1+r)^2\sigma_i^2 + r^2\frac{L^2}{4z_{\alpha/2}^2}. \tag{9}$$

In the first stages of the evolution $\sigma^2 >> \sigma_b^2$ and $\mu \neq \mu_b$ as the set of the fittest individuals is a subset of the whole population. Along the evolution, the selection pressure will produce that $\mu \rightarrow \mu_b$ and $\sigma^2 \rightarrow \sigma_b^2$, yielding:

$$E(\beta_i^s) = (1+r)\mu - r\mu_b = \mu_b, \tag{10}$$

and

$$V(\beta_i^s) = (1+r)^2\sigma_b^2 + r^2\frac{L^2}{4z_{\alpha(2)}^2}. \tag{11}$$

In the second case, if we work the same way we obtain similar results:

$$E(\beta_i^s) = (1+r)\mu_b - r\mu_b = \mu_b, \tag{12}$$

and

$$V(\beta_i^s) = (1+r)^2\frac{L^2}{4z_{\alpha/2}^2} + r^2\sigma_b^2. \tag{13}$$

The distribution of the generated offspring will be a mixture of distributions. This mixture will follow a distribution of parameters:

$$E(\beta_i^s) = p\mu_{b_i} + (1-p)\mu_{b_i} = \mu_{b_i}, \tag{14}$$

and

$$V(\beta_i^s) = p^2(1+r)^2\sigma_b^2 + p^2r^2\frac{L^2}{4z_{\alpha/2}^2} + (1-p)^2(1+r)^2\frac{L^2}{4z_{\alpha/2}^2} + (1-p)^2r^2\sigma_b^2. \tag{15}$$

So, we can conclude that, if $\beta_i^k \in I_i^M$ and along the evolution the values of p and r are appropriate, then the distribution of the offspring will have the same mean than the mean of the fittest individuals, and a variance less than the variance of the fittest individuals. The variance of the offspring will also depend on the distribution of the best individuals of the population, and the performance of the crossover will strongly depend on such distribution.

3 Experimental setup

Taking into account the theorem of *no free lunch*, we have used a set of well characterized functions instead of a large a number of functions in order to test the goodness of the proposed crossover. We have used the four functions extracted from the test set proposed in [10] that are shown in the following table:

Function	Expression	Range
Hypersphere	$f_1(\mathbf{x}) = \sum_{i=1}^{q} x_i^2$	$x_i \in [-5.12, 5.12]$
Rastrigin	$f_2(\mathbf{x}) = \sum_{i=1}^{q} (x_i^2 - 10\cos(2\pi x_i) + 10)$	$x_i \in [-5.12, 5.12]$
Schwefel	$f_3(\mathbf{x}) = \sum_{i=1}^{q} \left(\sum_{j=1}^{i} x_j\right)^2$	$x_i \in [-65.536, 65.536]$
Ackley	$f_4(\mathbf{x}) = 20 + e - 20e^{\left(-0.2\sqrt{1/q\sum_{i=1}^{q} x_i^2}\right)}$ $-e^{\left(1/q\sum_{i=1}^{q}\cos(2\pi x_i)\right)}$	$x_i \in [-30, 30]$

For all the functions the minimum is in $\mathbf{x}_m = (0, 0, \ldots, 0)$ and $f(\mathbf{x}_m) = 0$. The dimensionality is a factor that has effects on the complexity of the functions ([11]). In order to establish the same degree of difficulty, we have chosen a dimensionality of $q = 30$ for all the functions.

Hypersphere is a simple, continuous, strongly convex, unimodal and separable function. Rastrigin is a continuous, scalable, multi-modal and separable function, its contour is made up by a large number of local minima whose value increase with the distance to the global minimum. Schwefel is a continuous, unimodal and nonseparable function. Its main difficulty is that its gradient is not oriented along its axis due to the epistaxis of its variables. So, a search using the gradient is very slow. Ackley is a continuous, multimodal, and nonseparable function, where the exponential term covers its surface with a lot of local minima. This function has a moderated complexity because, although a search algorithm using the steepest descent method will probably be trapped in a local optimum, a search strategy analyzing a wider area could find better solutions. Ackley function provides a test case where it is needed for the search strategy to establish an adequate balance between exploration and exploitation.

We have compared CIXL1 with two of the most interesting current crossover of arity two: the BLX-α ([2]) crossover with $\alpha = 0.5$ and Extended Fuzzy ([12]) with $S2$ strategy; and with two multiparent crossovers: UNDX ([13]) with three parents and Seed ([14]) with a number of parents that is defined by the parameter m. The values of m used are: $m = 16$ for Hypersphere, $m = 8$ for Rastrigin, $m = 2$ for Schwefel, and $m = 6$ for Ackley. These values are the best obtained by the authors of this crossover ([14]).

The selection of the mutation operator is closely related to the crossover operator and the problem to solve. Using the previous results in [15], [12], [16], [13] and [14], we have chosen the following mutation operators: for the Hypersphere and Schwefel functions the non-uniform mutation ([17]), for the Rastrigin and Ackley functions the non-uniform mutation for arity 2 crossovers, and the discrete and continuous modal mutation respectively for multiparent crossovers ([18]).

The parameters used for CIXL1 are ([15]) $n = 5$ and $1 - \alpha = 0.70$, for all the functions, except for Rastrigin where we used $1 - \alpha = 0.95$.

The genetic algorithm has a fixed size population of 100 individuals randomly initialized, a crossover probability of 0.6, a probability of gene mutation of 0.05, and a tournament selection method with 2 opponents and elistism. Each experiment will be repeated 10 times with different random seeds and the population will evolve for 5000 generations.

3.1 Results

As we can see on Table 1 the best results for Rastrigin, Schwefel and Ackley are obtained using the CIXL1 operator. For Hypersphere the Seed crossover obtained the best results, nevertheless, the performance of CIXL1 is also very good.

Table 2 shows the results of a Tamhane statistical test for the comparison of the results of all the crossovers applied. For three out of four problems the results of CIXL1 are significantly better than the results obtained with the other operators with a significance level 0f 0.01. The results for Hypersphere cannot be compared as the variance of the experiments is 0. Table 3 shows the result of a Student t test between CIXL1 and the second best crossover for every problem.

Table 1. Results of the five crossover operators on the test functions

Crossover	Mean	SD	Best	Worse	Mean	SD	Best	Worse
			Hipersphere				Rastrigin	
CIXL1	1.22e-16	0.00e+00	1.22e-16	1.22e-16	7.12e-11	1.82e-10	0.00e+00	5.74e-10
BLX-α	2.97e-15	1.51e-15	1.33e-15	5.33e-15	3.43e+01	7.98e+00	2.09e+01	4.77e+01
Ext. Fuzzy	8.26e-11	4.92e-11	2.31e-11	9.10e-11	3.38e+01	6.68e+00	1.79e+01	4.18e+01
UNDX	9.55e-11	2.36e-10	0.00e+00	7.66e-10	2.74e+01	5.74e+00	1.74e+01	3.98e+01
Seed	0.00e+00	0.00e+00	0.00e+00	0.00e+00	2.38e-03	4.32e-04	1.70e-03	3.03e-03
Crossover			Schwefel				Ackley	
CIXL1	1.34e-05	4.02e-22	1.34e-05	1.34e-05	2.78e-08	2.76e-08	9.03e-09	9.41e-08
BLX-α	5.51e-01	2.11e-01	1.64e-01	9.65e-01	1.84e-07	8.12e-08	7.9e-08	3.5e-07
Ext. Fuzzy	2.21e+01	9.15e+00	1.15e+01	3.96e+01	3.14e-05	1.20e-05	1.26e-05	5.14e-05
UNDX	3.77e+00	2.60e+00	1.24e+00	9.74e+00	4.75e-01	6.94e-02	3.8e-01	5.9e-01
Seed	1.89e+01	1.17e+01	4.81e+00	4.63e+01	3.16e-02	4.49e-03	2.60e-02	4.18e-02

Table 2. The table shows the difference, $I - J$, between the mean with CIXL1, I, and the other crossover J, and the significance of this difference using a Tamhane test

Crossover	Sphere		Rastrigin		Schwefel		Ackley	
	$(I-J)$	Sign.	$(I-J)$	Sign.	$(I-J)$	Sign.	$(I-J)$	Sign.
BLX-α	-2.85e-15	-	-3.43e+01	0.000	-5.51e-01	0.000	-1.56e-07	0.001
Ext. Fuzzy	-8.26e-11	-	-3.38e+01	0.000	-2.21e+01	0.000	-3.14e-05	0.000
UNDX	-9.55e-11	-	-2.74e+01	0.000	-3.77e+00	0.013	-4.75e-01	0.000
Seed	1.22e-16	-	-2.38e-03	0.000	-1.89e+01	0.002	-3.16e-02	0.000

Figure 1 shows the convergence rate of each function using the different crossovers. For the Hypersphere function the convergence rate of CIXL1 is slightly worse that the convergence rate of Seed. For Rastrigin and Ackley functions, CIXL1 and Seed show similar convergence rates, nevertheless the solution obtained by CIXL1 is closer to the global optimum. For Schwefel function, CIXL1 converges faster than all the other crossovers. For this function Seed shows a premature convergence to a local optimum very far from the global solution.

Table 3. student *t* test between the results of CIXL1 and the second best crossover for each test function

Function	Crossover	Sig. Levene test	Mean difference	Standard Error	Sig. t test
Hypersphere	Seed	-	1.22e-16	-	-
Rastrigin	Seed	0.000	-2.38e-03	1.36e-04	0.000
Schwefel	BLX-α	0.003	-5.51e-01	6.67e-02	0.000
Ackley	BLX-α	0.014	-1.56e-07	2.71e-08	0.000

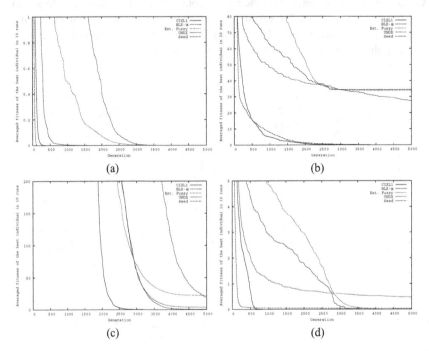

(a)   (b)

(c)   (d)

Fig. 1. (a) Averaged fitness of the best individual in 10 runs for Hypersphere(b), Rastrigin (c), Schwefel (d), and Ackley

4 Conclusions and future work

In this work we have proposed a new crossover operator that uses the information obtained from the localization and dispersion parameters of the subset of the best individuals of the population, in order to establish a balance between the exploration and exploitation of the search space. This operator generates offsprings in the region where the best individuals of the population are localized, sampling this way a region where is probable to obtain fitter individuals. This region is established by the confidence interval of each gene.

We have compared the performance of CIXL1 with that of commonly used crossover operators, namely, BLX-α, Extended Fuzzy, UNDX, and Seed for the optimization of unimodal, multimodal, and epistatic functions. The comparison shows the robustness of the proposed crossover, that also shows a fast convergence to values close to the global optimum, without being trapped in local minima.

It is obvious that the number of best individuals and the confidence coefficient have an important influence over the performance of the proposed crossover. Now, we are working in a sensibility analysis that can guide us in the most suitable values of these parameters depending on the kind of problem to solve. A more deep knowledge about the distribution of the genes along the evolution would be also very interesting for improving the results of this new crossover operator.

Acknowledgement. The authors would like to thank E. Sanz-Tapia for her helping in the final version of this paper. This work has been financed in part by the projects TIC2001-2577 and TIC2002-04036-C05-02 of the Spanish CICYT.

References

1. A. Wright. Genetic algorithms for real parameter optimization. In G. J. E. Rawlin, editor, *Foundations of Genetic Algorithms 1*, pages 205–218, San Mateo, 1991. Morgan Kaufmann.
2. L. J. Eshelman and J. D. Schaffer. Real-coded genetic algorithms and interval-shemata. In L. Darrell Whitley, editor, *Foundation of Genetic Algorithms 2*, pages 187C3.3.7:1–C3.3.7:8.–202, San Mateo, 1993. Morgan Kaufmann.
3. F. Herrera, M. Lozano, and J. L. Verdegay. Tackling real-coded genetic algorithms: Operators and tools for behavioural analysis. *Artificial Inteligence Review*, 12:265–319, 1998. Kluwer Academic Publisherr. Printed in Netherlands.
4. F. Herrera, E. Herrera-Viedma, E. Lozano, and J. L. Verdegay. Fuzzy tools to improve genetic algorithms. In *Second European Congress on Intelligent Techniques and Soft Computing*, pages 1532–1539, 1994.
5. H. Kita, I. Ono, and S. Kobayashi. Multi-parental extension of the unimodal normal distribution crossover for real-coded genetic algorithms. In Peter J. Angeline, Zbyszek Michalewicz, Marc Schoenauer, Xin Yao, and Ali Zalzala, editors, *Proceedings of the Congress of Evolutionary Computation*, volume 2, pages 1581–1588, Mayflower Hotel, Washington D.C., USA, 6-9 July 1999. IEEE Press.
6. A. E. Eiben. Experimental results on the effects of multi-parent recombination: an overview. In L. D. Chambers, editor, *Practical Handbook of Genetic Algorithms Complex Coding Systems*, volume 3. CRC Press, 1999.
7. D. Ortiz, C. Hervás, and J. Muñóz. Genetic algorithm with crossover based on confidence interval as an alternative to tradicional nonlinear regression methods. In Michel Verleysen, editor, *The 9th European Symposium on Artificial Neural Networks*, pages 193–198, Bruges, Belgium, April 2001a. D-Facto.
8. D. Ortiz, C. Hervás, and J. Muñóz. Genetic algorithm with crossover based on confidence interval as an alternative to least squares estimation for nonlinear models. In *The 4th Metaheuristics International Conference*, Porto, Portugal, July 2001b. Kluwer Academic Publishers.
9. T. P. Hettmansperger and J. W. McKean. *Robust Nonparametric Statistical Methods*. Arnold John/Wiley and Sons, London/New York, 1998.
10. A. E. Eiben and Th. Bäck. An empirical investigation of multi-parent recombination operators in evolution strategies. *Evolutionary Computation*, 5(3):347–365, 1997.

11. J. H. Friedman. An overview of predictive learning and function approximation. In V. Cherkassky, J. H. Friedman, and H. Wechsler, editors, *From Statistics to Neural Networks, Theory and Pattern Recognition Applications*, volume 136 of *NATO ASI Series F*, pages 1–61. Springer-Verlag, 1994.

12. F. Herrera and M. Lozano. Gradual distributed real-coded genetic algorithms. *IEEE Transactions on Evolutionary Computation*, 4(1):43–63, April 2000.

13. I. Ono and S. Kobayashi. A real-coded genetic algorithm for function optimization using unimodal normal distribution crossover. In *7th International Conference on Genetic Algorithms*, pages 246–253, Michigan, USA, July 1997. Michigan State University, Morgan Kaufman.

14. S. Tsutsui and A. Ghosh. A study of the effect of multi-parent recombination in real coded genetic algorithms. In *Proc. of the ICEC*, pages 828–833, 1998.

15. D. Ortiz. *Operadores de cruce basados en intervalos de confianza en algoritmos genéticos con codificación real*. PhD thesis, E.T.S.I. Informática, 2001.

16. L. J. Eshelman. The CHC adaptive search algorithm: How to safe search when engaging in non-traditional genetic recombination. In *Foundations of Genetic Algorithms*, pages 256–283. Morgan Kaufman Publisher, San Mateo, 1991.

17. Z. Michalewicz. *Genetic Algorithms + Data Structures = Evolution Programs*. Springer-Verlag, New York, 1992.

18. H. M. Voigt and T. Anheyer. Modal mutations in evolutionary algorithms. In *The First IEEE Conference on Evolutionary Computation*, pages 88–92, 1994.

Cooperative Particle Swarm Optimization for Robust Control System Design

Renato A. Krohling[1], Leandro dos S. Coelho[2], and Yuhui Shi[3]

[1] Universidade Federal do Espírito Santo, Departamento de Engenharia Elétrica, Campus de Goiabeiras, C.P. 01-9001, CEP 29060-970, Vitória, ES, Brazil renato@ele.ufes.br
[2] Pontifícia Universidade Católica do Paraná, Laboratório de Automação e Sistemas, Rua Imaculada Conceição, 1155, CEP 80215-030, Curitiba, PR, Brazil lscoelho@rla01.pucpr.br
[3] EDS Embedded Systems Group, 1401 E. Hoffer Street, Kokomo, IN 46902 USA Yuhui.Shi@EDS.com *

Summary. In this paper, a novel design method for robust controllers based on Cooperative Particle Swarm Optimization (PSO) is proposed. The design is formulated as a constrained optimization problem, i.e., the minimization of a nominal H_2 performance index subject to a H_∞ robust stability constraint. The method focuses on two (PSOs): One for minimizing the performance index, and the other for maximizing the robust stability constraint. Simulation results are given to illustrate the effectiveness and validity of the approach.

1 Introduction

The robust control system design presented in this paper is formulated as a mixed H_2/H_∞ optimal control problem which consists of the design of a controller which internally stabilizes the plant and satisfies the robust stability constraint on the ∞ norm. The mixed H_2/H_∞ problem has received a great deal of attention from the theoretical viewpoint of design methods [1,2,3]. In the last few years some research interest has been devoted to develop practical design method for this relevant problem [4,5,6,7].

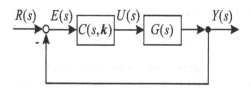

Fig. 1. Feedback control system.

* Renato A. Krohling was sponsored by the Brazilian Research Council (CNPq) under Grant 301009/99-6

In [4], the mixed H_2/H_∞ optimal PID control design problem was formulated as a minimization of the integral of the squared-error (ISE) performance index subject to a H_∞ robust stability constraint. For the solution of that problem a hybrid method consisting of a genetic algorithm(GA) with binary encoding for minimization of the ISE performance index together with a numerical algorithm for evaluating the H_∞ robust stability constraint was proposed. Another solution method for the problem was proposed in [5] which is based on a genetic/interval approach, consisting of a GA for minimization of the ISE performance index and interval methods for evaluating the H_∞ robust stability constraint.

In this paper, based on previous work [6,7] one considers the design of robust controllers for uncertain plant formulated as constrained optimization problem by using Particle Swarm Optimization (PSO). Firstly, the synthesis of mixed H_2/H_∞ fixed-structure controller is formulated as the minimization of a nominal H_2 performance index subject to a H_∞ robust stability constraint. Next, for the solution of the highly non-linear constrained minimization problem for which a closed solution can not be obtained, a new method based on two PSOs is proposed: One for minimizing the ISE performance index, and the other for maximizing the robust stability constraint. Simulation results are presented to show the effectiveness and validity of the novel approach.

The rest of this paper is organized as follows: section 2 describes the problem; section 3 is devoted to an explanation of particle swarm optimization; in section 4 the design method using two PSOs is presented; in section 5 a design example is given and section 6 presents some conclusions.

2 Problem Description

In the context of single-input single-output linear time-invariant systems [8], consider the feedback control system shown in Figure 1. The fixed-structure controller is described by a rational transfer function $C(s, \mathbf{k})$, where $\mathbf{k} = [k_1, k_2, \ldots, k_m]$ stands for the vector of the controller parameters. It is assumed that the plant to be controlled is described by the nominal transfer function $G_0(s)$ and undergoes a perturbation described by $\triangle G(s)$, which is assumed to be stable and limited by:

$$|\triangle G(jw)| < |\triangle W_m(jw)| \quad \forall \, w \in [0, \infty),$$

where the function $W_m(s)$ is stable and known. In this paper is considered that the plant is described by the multiplicative uncertainty model given by:

$$G(s) = G_0(s)(1 + \triangle W_m(s)).$$

It is required that $\triangle G(s)$ does not cancel unstable poles of $G_0(s)$ in forming $G(s)$. The Condition for robust stability is stated, as follows [8]: If the nominal control system $(\triangle G(s) = 0)$ is stable with the controller $C(s, \mathbf{k})$, then the controller $C(s, \mathbf{k})$ guarantees robust stability of the control system, if and only if the following condition is satisfied:

$$\left\| \frac{C(s,\mathbf{k})G_0(s)\triangle W_m(s)}{1+C(s,\mathbf{k})G_0(s)} \right\|_\infty \tag{1}$$

Applying the definition of the H_∞ norm on the condition for robust stability results:

$$\left\| \frac{C(s,\mathbf{k})G_0(s)W_m(s)}{1+C(s,\mathbf{k})G_0(s)} \right\|_\infty =$$

$$= \max_{w\in[0,\infty)} \left(\frac{(C(jw,\mathbf{k})G_0(jw)W_m(jw)).(C(-jw,\mathbf{k})G_0(-jw)W_m(-jw))}{(1+C(jw,\mathbf{k})G_0(jw)).(1+C(-jw,\mathbf{k})G_0(-jw))} \right)^{0.5}$$

$$= \max_{w\in[0,\infty)} (\alpha(w,\mathbf{k}))^{0.5}$$

where

$$\alpha(w,\mathbf{k}) = \frac{\alpha_z(w,\mathbf{k})}{\alpha_n(w,\mathbf{k})} = \frac{(C(jw,\mathbf{k})G_0(jw)W_m(jw)).(C(-jw,\mathbf{k})G_0(-jw)W_m(-jw))}{(1+C(jw,\mathbf{k})G_0(jw)).(1+C(-jw,\mathbf{k})G_0(-jw))}$$

Thus the condition for robust stability in the frequency domain is represented by:

$$\max_{w\in[0,\infty)} (\alpha(w,\mathbf{k}))^{0.5} < 1 \tag{2}$$

The function $\alpha(w,\mathbf{k})$ can also be expressed in the following form:

$$\alpha(w,\mathbf{k}) = \frac{\alpha_z(w,\mathbf{k})}{\alpha_n(w,\mathbf{k})} = \frac{\sum_{j=0}^{p}\alpha_{zj}(\mathbf{k})w^{2j}}{\sum_{i=0}^{q}\alpha_{ni}(\mathbf{k})w^{2i}} \tag{3}$$

Both polynomials $\alpha_z(w,\mathbf{k})$ and $\alpha_n(w,\mathbf{k})$ have only even powers of w, and the coefficients $\alpha_{zj}(\mathbf{k})$ and $\alpha_{ni}(\mathbf{k})$ are functions of $\mathbf{k}$.

The mixed H_2/H_∞ optimal control problem consists of minimizing the H_2 nominal performance index subject to the H_∞ robust stability constraint. Let the nominal H_2 performance be the integral of squared error (ISE) given by:

$$J = \int_0^\infty e^2(t)dt \tag{4}$$

The error transfer function $E(s)$ for a control system as the one shown in Figure 1 is given by:

$$E(s) = \frac{R(s)}{1+C(s,\mathbf{k})G_0(s)} \tag{5}$$

where $R(s)$ is a unit step signal. Using the Parseval theorem [9], an alternative representation in the frequency domain for J is given by:

$$J = \frac{1}{2\pi j} \int_{-j\infty}^{+j\infty} E(s)E(-s)dt \tag{6}$$

Let $E(s) = \frac{C(s)}{D(s)}$ with $C(s) = \sum_{i=0}^{n-1} c_i s^i$, $D(s) = \sum_{i=0}^{n} d_i s^i$, where n is the degree of the polynomial $D(s)$. The degree of the polynomial $C(s, \mathbf{k})$ must be smaller or equal to $n-1$ for the integral to be finite.

Equation (6) can be written as:

$$J_n = -\frac{1}{2\pi j} \int_{-j\infty}^{+j\infty} \frac{(\sum_{i=0}^{n-1} c_i s^i).(\sum_{i=0}^{n-1} c_i (-s)^i)}{(\sum_{i=0}^{n} d_i s^i).(\sum_{i=0}^{n} d_i (-s)^i)} ds \tag{7}$$

Equation (7) can be solved using the residue theorem. Expressions for J_n as functions of the coefficients of the error transfer function $E(s)$, are found in [9].

The evaluation of the performance index ISE depends on the stability of the closed loop system. Consequently, the value of J_n is always positive as it should be for stable systems. In [10] is indicated that stability and evaluation of J_n can be performed simultaneously. If the stability test fails, then the computation of J_n can be terminated. Since $E(s)$ contains the parameters of the controller, the value of J_n can be minimized adjusting the vector of parameters $\mathbf{k}$. Then the mixed H_2/H_∞ optimal control problem can be stated as follows:

$$\min_{\mathbf{k}} J_n(\mathbf{k}) \quad \text{subject to} \quad \max_{w}(\alpha(w, \mathbf{k}))^{0.5} < 1 \tag{8}$$

Hence, the problem of the synthesis of the controller is now one of how to solve the constrained minimization problem above. In the following, we investigate a very promising technique of evolutionary computation, i.e., particle swarm optimization to solve the optimization problem given by the expression (8).

3 Particle Swarm Optimization

Particle swarm optimization (PSO) is an evolutionary algorithm [11,12]. PSO is initialized with a population of random solutions. Unlike most of the evolutionary algorithms, each potential solution (individual) in PSO is also assigned a randomized velocity, and the potential solutions, called *particles*, are then "flown" through the problem space.

Each particle keeps track of its coordinates in the problem space, which are associated with the best solution (fitness) it has achieved so far. (The fitness value is also stored.) This value is called *pbest*. Another *'best'* value that is tracked by the global version of the particle swarm optimizer is the overall best value, and its location, obtained so far by any particle in the population. This location is called *gbest*.

The particle swarm optimization concept consists of, at each time step, changing the velocity (accelerating) of each particle toward its *pbest* and *gbest* locations (global version of PSO). Acceleration is weighted by a random term, with separate

random numbers being generated for acceleration toward *pbest* and *gbest* locations. The procedure for implementing the global version of PSO is listed as follows:
List 1:

1. Initialize a population (array) of particles with random positions and velocities in the *n*-dimensional problem space.
2. For each particle, evaluate its fitness value.
3. Compare each particle's fitness evaluation with the particle's *pbest*. If current value is better than *pbest*, then set *pbest* value equal to the current value, and the *pbest* location equal to the current location in *n*-dimensional space.
4. Compare fitness evaluation with the population's overall previous best. If current value is better than *gbest*, then reset *gbest* to the current particle's array index and value.
5. Change the velocity and position of the particle according to equations (9) and (10) respectively [13,14]:

$$\mathbf{v_i} = \xi.\mathbf{v_i} + c_1.\text{rand}().(\mathbf{p_i} - \mathbf{v_i}) + c_2.\text{Rand}().(\mathbf{p_g} - \mathbf{v_i}) \qquad (9)$$

$$\mathbf{x_i} = \mathbf{x_i} + \mathbf{v_i} \qquad (10)$$

6. Loop to step 2) until a criterion is met, usually a sufficiently good fitness or a maximum number of iterations (generations).

where $\mathbf{x_i} = [x_{i1}, x_{i2}, \ldots, x_{in}]^T$ stands for the position of the *i*-th particle, $\mathbf{v_i} = [v_{i1}, v_{i2}, \ldots, v_{in}]^T$ stands for the velocity of the *i*-th particle and $\mathbf{p_i} = [p_{i1}, p_{i2}, \ldots, p_{in}]^T$ represents the best previous position (the position giving the best fitness value) of the *i*-th particle. The index *g* represents the index of the best particle among all the particles in the group. Variable ξ is the inertia weight, c_1 and c_2 are positive constants; rand() and Rand() are two random functions in the range [0,1].

Particles's velocities on each dimension are clamped to a maximum velocity *Vmax*. If the sum of accelerations would cause the velocity on that dimension to exceed *Vmax*, which is a parameter specified by the user, then the velocity on that dimension is limited to *Vmax*. *Vmax* is an important parameter that determines the resolution with which the regions around the current solutions are searched. If *Vmax* is too high, the PSO facilitates global search, and particles might fly past good solutions. If *Vmax* is too small, on the other hand, the PSO facilitates local search, and particles may not explore sufficiently beyond locally good regions. In fact, they could become trapped in local optima, unable to move far enough to reach a better position in the problem space [15]. The first part in equation (9) is the momentum part of the particle. The inertia weight ξ represents the degree of the momentum of the particles. The second part is the "cognition" part, which represents the independent thinking of the particle itself. The third part is the "social" part, which represents the collaboration among the particles. The constants c_1 and c_2 represent the weighting of the "cognition" and "social" parts that pull each particle toward *pbest* and *gbest* positions. Thus, adjustment of these constants changes the amount of "tension" in the system. Low values allow particles to roam far from already found better regions before being tugged back, while high values result in abrupt movement toward, or past, already found better regions.

Early experience with particle swarm optimization (trial and error, mostly) led us to set the acceleration constants c_1 and c_2 equal to 2.0 for almost all applications. $Vmax$ is often set at about 10-20% of the dynamic range of the variable on each dimension. The population size selected is problem-dependent. Population size of 20-50 are probably most common. It is learned that small population sizes are acceptable for PSO to be optimal in terms of minimizing the total number of evaluations (population size times the number of generations) needed to obtain a sufficient good solution.

4 The Proposed Method

For background material on cooperative coevolutionary algorithms, see for instance [16,17]. Based on previous works [6,7], we employ two PSOs. PSO$_1$ to minimize the nominal H_2 performance index $J_n(\mathbf{k})$ and PSO$_2$ to maximize the robust stability constraint $(\alpha(w,\mathbf{k}))^{0.5}$ as depicted in Figure 2.

Initially, PSO$_1$ is started with the controller parameters within the search domain as specified by the designer. These parameters are then sent to PSO$_2$ which is initialized with the variable frequency w. PSO$_2$ maximizes the robust stability during a fixed number of generations for each particle of PSO$_1$. Next, if the maximum value of the robust stability constraint is larger than one, a penalty will be added to the corresponding individual of PSO$_1$. Particles of PSO$_1$ which satisfy the robust stability constraint will not be penalized.

Let the performance index be $J_n(\mathbf{k})$ then the value of the objective function for each particle of PSO$_1$ $\mathbf{k_i}$ with $(i = 1, \ldots, \mu_1)$ is determined by:

$$F_1(\mathbf{k_i}) = (J_n(\mathbf{k_i}) + P(\mathbf{k_i})) \tag{11}$$

where μ_1 denotes the population size of PSO$_1$. The penalty function is discussed in the following.

Let the robust stability constraint be $\max(\alpha(w,\mathbf{k}))^{0.5}$. The value of the objective function for each particle of PSO$_2$ w_j with $(j = 1, \ldots, \mu_2)$ is determined by:

$$F_2(w_j) = \alpha(w, \mathbf{k_i}) \tag{12}$$

where μ_2 denotes the population size of PSO$_2$.

The penalty for the individual $\mathbf{k_i}$ is calculated by means of the penalty function given by:

$$P(\mathbf{k_i}) = \begin{cases} M_2, & \text{if} \quad \mathbf{k_i} \text{ is unstable} \\ M_1 \cdot \max(\alpha(w,\mathbf{k_i})), & \text{if} \quad (\alpha(w,\mathbf{k_i}))^{0.5} > 1 \\ 0, & \text{if} \quad (\alpha(w,\mathbf{k_i}))^{0.5} < 1 \end{cases} \tag{13}$$

If the individual $\mathbf{k_i}$ does not satisfy the stability test applied to the characteristic equation of the system, then $\mathbf{k_i}$ is an unstable individual and it is penalized with a

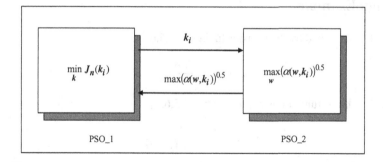

Fig. 2. Optimization with two cooperative PSOs.

very large positive constant M_2. In case that the particle $\mathbf{k_i}$ satisfies the stability test, but not the robust stability constraint, it is an infeasible individual and is penalized with $M_1 . \max \alpha(w, \mathbf{k_i})$, where M_1 is a positive constant. Otherwise, the particle $\mathbf{k_i}$ is feasible and is not penalized.

The method can be summarized as follows: Given the plant with transfer function $G_0(s)$, a fixed-structure controller with transfer function $C(s, \mathbf{k})$ and a weighting function $W_m(s)$, determines the error transfer function $E(s)$ and the robust stability constraint $\alpha(w, \mathbf{k})$. It needs also to specify the lower and upper bounds of the controller parameters.

It is suitable to describe the method in the form of an algorithm given in the following list:

List 2:

1. Initialize the two populations PSO_1 $\mathbf{k_i}$ with $(i = 1, \ldots, \mu_1)$ and PSO_2 w_j with $(j = 1, \ldots, \mu_2)$ within the corresponding ranges.
2. For each particle $\mathbf{k_i}$ of the PSO_1, run the second PSO_2 for a number maximum of iterations $iter_{2max}$ in order to calculate the maximum value of $\alpha(w, \mathbf{k_i})$ given by g_{2best}. If no particle of PSO_1 satisfy the constraint $(\alpha(w, \mathbf{k_i}))^{0.5} < 1$ then a feasible solution is assumed to be non-existent and the algorithm stops. In this case, a new controller structure has to be assumed.
3. Run PSO_2 for each particle w a number maximum of iterations $iter_{2max}$ in order to calculate the maximum value of $F_2(w_j)$ according to the Eq. (12) given by g_{2best}.
4. Run PSO_1 for each particle $\mathbf{k_i}$ in order to calculate the value of $F_1(\mathbf{k_i})$ according to the equations (11) and (13).
5. If the maximum number of iterations of PSO_1 $iter_{1max}$ is not reached, then set $iter_1 = iter_1 + 1$ go to step 3, otherwise stop, and the solution of the optimization problem $\mathbf{k}^*$ is given by g_{1best}.

5 Design Example

The plant is described by the transfer function [4,5,6,7]:

$$G(s) = \frac{1.8}{s^2(s+2)}$$

The weighting function is expressed by [4,5,6,7]:

$$W_m(s) = \frac{0.1}{s^2 + 0.1s + 10}$$

The fixed-structure controller is given by following transfer function [5,6,7]

$$C(s,\mathbf{k}) = \frac{s^2 + 2k_4k_5s + k_5^2}{(s+k_2).(s+k_3)}$$

For the control system as shown in Figure 1, the error transfer function $E(s)$, the performance index J_5, and the robust stability constraint are given in [6,7]. The parameters of the controller are searched within the following bounds: $k_1 = [1,1000], k_2 = [1,100], k_3 = [1,100], k_4 = [0,1], k_5 = [0.1,100]$. The following PSO parameters are used: Population size of PSO_1 $\mu_1 = 50$, population size of PSO_2 $\mu_2 = 30$, penalty constants $M_1 = 100, M_2 = 100000$.

The method using PSOs described in the previous section, has been applied to the design of the robust controller. The best controller parameters for this example obtained by the minimization of the performance index $J_5(\mathbf{k})$ subject to the robust stability constraint $\max(\alpha(w,\mathbf{k}))^{0.5}$ using the proposed method is given by $\mathbf{k}^* = [1000; 14.418; 14.569; 1; 0.541]$. The minimal value of the performance index after 100 generations is $J_5(\mathbf{k}^*) = 0.1506$.

The results obtained by the proposed method is equivalent to those reported in [5,6,7]. In order to test the controller obtained by the proposed method, a unit step signal is applied to the system as shown in Figure 1. The closed loop response for both cases: a) nominal plant $G_0(s)$ and b) the uncertain plant $G(s)$ is depicted in Figure 3. It can be noticed that the controller obtained by the proposed method can control the perturbed plant $G(s)$ successfully.

6 Conclusions

This paper presents a method for synthesis of robust controllers with fixed-structure. The problem is formulated as an optimization problem with a constraint of type H_∞ norm. A method based on two PSOs is presented: One PSO is used to minimize the nominal H_2 performance index and the other to maximize the H_∞ robust stability constraint.

The validity of the novel method is demonstrated by a design example and the simulation result is compared with those reported in the literature. The results obtained so far demonstrate clearly the potential of intelligent search methods based

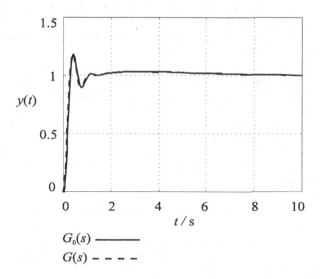

Fig. 3. The closed-loop step response.

on swarm intelligence for synthesis of robust controller. Particle swarm optimization can also be applied to design other kinds of controllers formulated as constrained optimization problems, i.e., including other performance criteria and constraints. Future work in this direction will be dedicated.

References

1. Bernstein, B.S. and Haddad, W.M. (1994) LQG Control with a Performance Bound: A Riccati Equation Approach. IEEE Trans. Automatic Control 34(3), pp. 293-305
2. Doyle, J.C., Zhou, K., Glover, K., and Boddenheimer, B. (1994) Mixed and Performance Objectives II: Optimal Control. IEEE Trans. Automatic Control 39(8), pp. 1575-1587
3. Snaizer, M. (1994) An Exact Solution to General SISO Mixed Problems via Convex Optimization. IEEE Trans. Automatic Control 39(12), pp. 2511-2517
4. Chen, B.-S., Cheng, Y.-M., and Lee, C.-H. (1995) A Genetic Approach to Mixed Optimal PID Control. IEEE Control Systems Magazine, pp. 51-60
5. Lo Bianco, C.G., and Piazzi, A. (1997) Mixed Fixed-Structure Control via Semi-Infinite Optimization. In Proc. of the 7th IFAC International Symposium on CACSD, Gent, Belgium, pp. 329-334
6. Krohling, R.A. (1998) Genetic Algorithms for Synthesis of Mixed Fixed-Structure Controller. In Proc. of the 13th IEEE International Symposium on Intelligent Control, ISIC'98, Gaithersburg, USA, pp. 30-35
7. Krohling, R.A., Coelho, L.S., and Coelho, A.A.R. (1999) Evolution Strategies for Synthesis of Mixed Fixed-Structure Controllers. In Proceedings of the IFAC World Congress, Beijing, China, pp. 471-476

8. Doyle, J.C. Francis, B.A., and Tannenbaum, A. R. (1992) Feedback Control Theory. Macmillan Publishing, New York, USA

9. Newton, G.C., Gould, L.A., and Kaiser, J.F. (1957) Analytic Design of Linear Feedback Controls. John Wiley & Sons, New York

10. Jury, E.I. (1974) Inners and the Stability of Dynamic Systems. John Wiley, New York

11. Eberhart, R.C., and Kennedy, J. (1995) A New Optimizer Using Particle Swarm Theory. In Proc. of the 6th. Int. Symposium on Micro Machine and Human Science, Nagoya, Japan. IEEE Service Center, Piscataway, NJ, pp. 39-43

12. Kennedy, J., and Eberhart, R.C. (1995) Particle swarm optimization. In Proc. of the IEEE Int. Conf. on Neural Networks IV, IEEE Service Center, Piscataway, NJ, pp.1942-1948

13. Shi, Y., and Eberhart, R.C. (1998) Parameter Selection in Particle Swarm Optimization. In Proc. of the Conf. on Evolutionary Programming VII: EP98, Springer-Verlag, New York, pp. 591-600

14. Shi, Y. and Eberhart, R.C. (1998) A Modified Particle Swarm Optimizer. In Proc. of the IEEE International Conference on Evolutionary Computation, Piscataway, IEEE Service Center, NJ, pp. 69-73

15. Fan, H.-Y., and Shi, Y. (2001) Study of Vmax of the Particle Swarm Optimization Algorithm. In Proc. of the Workshop on Particle Swarm Optimization. Purdue School of Engineering and Technology, IUPUI, Indianapolis,IN

16. Potter, M. A. and De Jong, K. (1994) A Cooperative Coevolutionary Approach to Function Optimization. In Proc. of the 3rd Conference on Parallel Problem Solving from Nature. Springer-Verlag, New York, pp. 257

17. Potter, M. A. (1997) The Design and Analysis of a Computational Model of Cooperative Coevolution. Doctoral dissertation, George Mason University, 1997

1-D Parabolic Search Mutation

C. Robertson & R.B. Fisher

School of Informatics
University of Edinburgh
Edinburgh, EH1 2QL, UK
craigr@dai.ed.ac.uk

Summary. This document describes a new mutation operator for evolutionary algorithms based on a 1-dimensional optimisation strategy. This provides a directed, rather than random, mutation which can increase the speed of convergence when approaching a minimum. We detail typical comparative results of unimodal and polymodal optimisations with and without the operator.

1 Introduction

1.1 Canonical EA Definition

The general scheme for an EA optimisation consists of a number of operators used in succession :

- *Initialization* Sets of parameters (*genes*) encoded as strings (*chromosomes*) are initialized, generally randomly, within their domain constraints.
- *Evaluation* The strings are then assessed using the evaluation function to give their fitness values. Often these fitness values are stored alongside the chromosomes themselves in some data structure.
- *Selection* Selection schemes determine which of the strings should be propagated into the next generation. These schemes often generate a probability of propagation for each string.
- *Reproduction* A new set of candidate strings are generated from the old ones. Some method is used to take two strings from the old population (parents) to produce two new strings (offspring).
- *Mutation* Elements of strings are changed randomly under some probabilistic selection scheme.
- *Replacement* Offspring replace their parents in the new population if they have a better fitness value.

A canonical EA [1] is a subset of reproductive population algorithms. These are algorithms that keep a collection of candidate solutions that are iteratively improved over successive generations. The following characteristics are those that represent a large subset of those found in the literature:

- All candidate solution vectors are the same length.
- The population of solutions is always the same size as the algorithm progresses.
- Reproduction between populations always uses two parent chromosomes, chosen by some probabilistic selection strategy.
- Reproduction is performed by crossover operations.
- Mutation is possible, according to some predefined population-wide mutation probability.

1.2 Canonical EA Operators

The main operator of EAs is the crossover, which carries useful information forwards through population generations. There are many probabilistic ways of selecting chromosomes to crossover. Most are based on some ranking of chromosomes by their fitness value. One example of this is to generate a probability of selection by cumulative summing of the ascending sorted normalised fitness values and using the values as a probability of selection.

Another method is to pick two sets of two chromosomes from the population at random and choose the best of each pair, known as tournament selection.

Crossover itself is generally a matter of overwriting two chromosomes (parents) with the two others (offspring). The offspring are formed by overwriting a random number genes from one parent onto the chromosome of the other. This is done symmetrically to form the two offspring, as shown in figure 1.

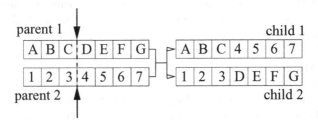

Fig. 1. Canonical 1-point EA Crossover

1.3 *Building Block Hypothesis* and the Problem of Premature Convergence

The reason put forward by Holland (and reported in [7]) for convergence in evolutionary algorithms is that they identify good building-blocks and eventually combine

---

[1] Following and extending the definition given for canonical GA, given in [5].

these into bigger building blocks. Premature convergence happens when all chromosomes throughout a population become the same. This means that all crossover operations will yield offspring identical to their parents. If only crossover is used as an operator, it is clear that premature convergence would result very quickly as the best building blocks reached the same chromosome. In order to optimise it is necessary to generate new information. To get to an optimum from this situation though, even using mutation, is a very tedious process. It is also necessary to take timely action to ensure population diversity is maintained as well.

The function of mutation is to add new material into populations and thereby avoid just this premature convergence issue. It is possible to perform a random walk over the whole solution space using only mutation. Some even have argued that cross-over is not strictly necessary in EAs [12] since repeated mutations allow a random walk through the solution space.

1.4 Extensions: Types of Mutation

In this paper, we do not concern ourselves with the multiplicity of crossover operations although we acknowledge that some of them are likely to achieve faster convergence in specific problem domains. We present below a list of mutation operators, which is extensive but not exhaustive.

- *Random* This is the mutation operator as used in a canonical EA. In a chromosome selected by some probabilistic scheme, a random gene is set to a new value randomly assigned inside that gene's domain limits. In essence, this scheme adds new information into the gene pool, although at random. It has a destructive effect on a generational EA because it moves a solution a random amount in a single dimension.
- *N mutations* Essentially the same as one, performed N times. Adds N times as much information but is N times as destructive to an individual chromosome. Moves the solution chromosome in a random N dimensional direction. Similar to an annealing process.
- *Creep* This operator, and those that follow are extensions to the random mutation operator. In creep mutation, we move the selected chromosome a known amount in a single dimension, either forward or backwards, subject to its domain constraints. It is more stable than random mutation although, depending on creep size, likely to force a gene to its domain limits.
- *Directed Hill Climbing Mutation* This mutation requires some memory of previous mutations. If the last mutation of this chromosome produced a better fitness function, them perform the same mutation until it becomes worse, then back off.
- *Domain Limit Mutation* This operator randomly pushes the gene to one or other of its domain limits. This is useful if the solution lies close to the limits of its domain.
- *Non-uniform mutation* This mutation operator is essentially creep mutation with a variable step size. It allows a space to be searched uniformly at first then more locally as generations proceed.

One important consideration when designing mutation operators is *atomic versus complex mutation*. A useful rule when designing mutations is that it is better to have mutations that do one action rather than mutations that do a combination of actions. The reason for this is that the complex mutations can usually be built from a combination of the atomic ones.

2 Algorithm

2.1 New Mutation Operator: Parabolic Search Strategy

Parabolic mutation is a new form of mutation which is an approach to providing the best single step information for a single gene mutation.

1. Define a 1D search envelope of size τ.
2. A single gene is chosen at random from a given chromosome, with value x_1 and corresponding chromosome fitness y_1.
3. Create two new chromosomes by copying the original and replacing the selected gene with values of $x_0 = (x_1 - \tau)$ and $x_2 = (x_1 + \tau)$.
4. Evaluate the new chromosomes and get their fitnesses, $y_0 = f(x_0)$ and $y_2 = f(x_2)$
5. Make a decision based on table I.

Table 1. Decision table for parabolic mutation

Fitness state	1-D Landscape	Minimizer	Maximizer
$y_0 < y_1 < y_2$	increasing	choose x_0	choose x_2
$y_0 < y_1 > y_2$	hilltop	choose $\min(x_0, x_2)$	fit parabola choose x_{best}=max
$y_0 > y_1 > y_2$	decreasing	choose x_2	choose x_0
$y_0 > y_1 < y_2$	valley	fit parabola choose x_{best}=min	choose $\max(x_0, x_2)$

with parabolic fitting [10] as seen in figure 2:
Let

$$m_0 = \frac{(y_1 - y_0)}{(x_1 - x_0)}$$

$$m_1 = \frac{(y_2 - y_1)}{(x_2 - x_1)}$$

$$k = \frac{(m_1 - m_0)}{(x_2 - x_0)}$$

The maximum or minimum is then:

$$x_{best} = \frac{\left(x_0 + x_1 - \frac{m_0}{k}\right)}{2}$$

This will find either maximum or minimum values depending on the values of (x_j, y_j), as shown in figure 2.

6. Replace the gene by the new value and return the chromosome to the population.

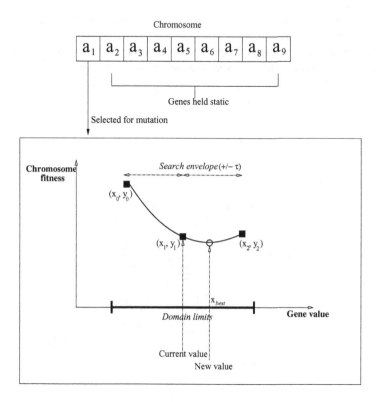

Fig. 2. Parabolic Fit Mutation

Note that if the space around the original value of the gene, x_1, is strictly increasing or strictly decreasing we have essentially a directed creep mutation. When our values straddle a local minimum, however, we can have very fast convergence to that minimum, depending on the size of τ.

2.2 Test Context

We use a canonical EA, as shown in figure 3. Each of the modules is self-contained and is explained below.

- *Initialization* Each gene in each chromosome is randomly initialized within its domain constraints.

- *Evaluation* Each chromosome is evaluated with a specified fitness function.
- *New Population Generation* The new population is formed in four steps:
  - *Elitist.* This is a common method of ensuring that the best chromosome from the last population is preserved intact. It is simply copied as the first member of the new population.
  - *Crossover.* We use tournament selection to produce the next tranche of population.
  - *Mutation.* We switch between parabolic search mutation and random mutation every iteration.
  - *Re-initialization.* Some of the new population is entirely re-randomized each iteration. This is normally a very low percentage of the chromosomes in the population.

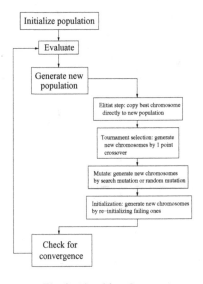

Fig. 3. Algorithm Context

3 Example Applications

3.1 Setup

In order to test speed of convergence, the new operator was employed to optimise some well known functions (outlined in Yao [8]). Example times for convergence in iterations and seconds are given, together with some discussion where relevant. The evolutionary strategy he used $(\mu, \lambda) = (30, 200)$-ES is not directly mappable to our scheme, instead, we use a population of 200 (except where stated) with replacement as outlined in section I.

Note that for unimodal functions, the issues of *speed* of convergence as well as quality of solution are important for our applications. For polymodal solutions, *any* equally important solution is useful. We do not claim that we are able to find either all solutions or the global optimum. For a discussion of how we can find all optima, see section 5.

In all cases, we give the average results over ten runs for clear convergence to an optimum, in iterations (i.e no further improvement). Note that we do not give timings as all experiments took less than 5 seconds on a 1GHz Athlon PC system running the Linux/GNU operating system.

We also give graphs showing averaged fitness function values of the best chromosome in the Appendix. Note that the fitness function numbering is as in Yao [8].

3.2 Single Minimum Function Optimization

The following functions were all optimised trivially (in one population iteration) using the new operator in the context specified:

- Sphere model:

$$f_1(x) = \sum_{i=1}^{30} x_i^2 \tag{1}$$

with $-100 \le x_i \le 100$, $\min(f_1) = f_1(0,...,0) = 0$

- Schwefel's Problem No.1:

$$f_2(x) = \sum_{i=1}^{30} |x_i| + \prod_{i}^{30} |x_i| \tag{2}$$

with $-10 \le x_i \le 10$, $\min(f_2) = f_2(0,...,0) = 0$

- Schwefel's Problem No.2:

$$f_3(x) = \sum_{i=1}^{30} \left(\sum_{j=1}^{i} (x_j) \right)^2 \tag{3}$$

with $-100 \le x_i \le 100$, $\min(f_3) = f_3(0,...,0) = 0$

3.3 Function Optimization with Many Local Minima

The following functions, which have an increasing number of optima of varying importance, were optimised.

Sine Product Function

$$f(\mathbf{x}) = sin(x_0) \times sin(x_1) \qquad (4)$$

$$\text{where } x_i \in [-2\pi, 2\pi]$$

This function has two maxima and two minima of equal importance in the given range.

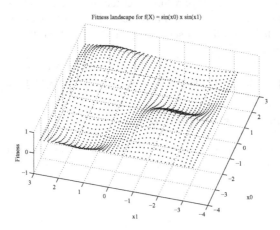

Fig. 4. Sin × sin Function

Product Function

$$f(\mathbf{x}) = x_0 \times x_i \times ... x_{10} \qquad (5)$$

$$\text{where } x_i \in [-5, 5]$$

This function has 2 maxima and two minima of equal importance in the given range.

Himmelblau Function

Himmelblau is a set of quartic form functions, generally the following is used:

$$f(\mathbf{x}) = (x_0^2 + x_1 - 11)^2 + (x_0 + x_1^2 - 7)^2 \qquad (6)$$

$$\text{with } x_i \in [-5, 5]$$

For which there is a set of known local optima of approximately the same importance:

$f(3, 2) = 0$
$f(-3.78, -3.28) = 0.0054$
$f(-2.81, 3.13) = 0.0085$
$f(3.58, -1.85) = 0.0011$

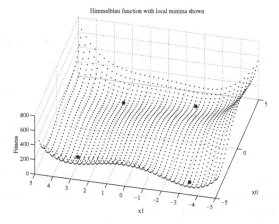

Fig. 5. Himmelblau Function

Six Hump Camel Function, f_{16}

This is a relatively simple function with six omptima inside the given domain.

$$f_{16}(x) = 4x_1^2 - 2.1x_1^4 + \frac{1}{3}x_1^6 + x_1x_2 - 4x_2^2 + 4x_2^4 \qquad (7)$$

with $-5.0 \le x_i \le 5.0$, $\min(f_{16}) = f_{16}(\pm 0.08983, \mp 0.7126) = -1.0316285$

Generalized Rosenbrock's Function, f_5

$$f_5(x) = \sum_{i=1}^{29} [100(x_{i+1} - x_i^2)^2 + (x_i - 1)^2 \qquad (8)$$

with $-30 \le x_i \le 30$, $\min(f_5) = f_5(1, 1, .., 1) = 0$

Quartic Function with Noise, f_7

$$f_7(x) = \sum_{i=1}^{30} ix^4 + random[0, 1) \qquad (9)$$

with $-1.28 \le x_i \le 1.28$, $\min(f_7) = f_7(0, ..., 0) = 0$

Generalized Rastrigin's Function, f_9

$$f_9(x) = \sum_{i=1}^{30} [x_i^2 - 10cos(2\pi x_i) + 10)] \qquad (10)$$

with $-5.12 \le x_i \le 5.12$, $\min(f_9) = f_9(0, ..., 0) = 0$

In this case we used a population of 50 chromosomes.

4 Results

Graphs of convergence are shown in the Appendix (figures 6 to 12). It can be seen that in general the mutation search method works as well or better, that is faster, than regular mutation. In certain circumstances, for example the first three fitness functions, convergence is achieved in a single iteration, indicating that for this kind of (essentially gradient descent) problem search mutation is extremely fast. In other unimodal problems it is generally much faster than regular mutation. It should be noted that because we used regular mutation every few iterations as well, also we benefited from a form of annealing.

- *Sphere model, Schwefel's Problem No.1, Schwefel's Problem No.2.* These were all optimised in a single population iteration using the the new mutation operator. There is no graph shown in the appendix.
- *Sine Product Function.* One optimum for this function was found after a single iteration in both the case of maximization and minimization. A graph of convergence is shown in figure 6 in the appendix. This graph is of interest because the relative time for convergence using the regular operator is much slower and it does not reach the optimum, only approaching it in 5000 iterations.
- *Product Function.* Given the same set of starting chromosomes, the new operator performs as expected and convergences relatively quickly. Shown in figure 7.
- *Himmelblau Function.* Another instance of single iteration convergence with the new operator. It should be mentioned that finding the global minimum in this experiment is luck, rather than a product of the methodology. Shown in figure 8.
- *Six Hump Camel Function, f_{16}.*
  This result shows that this fitness function is particularly complex and varies very rapidly inside its domain limits. This kind of function is not well suited to the operator but it performed as well as the regular one. Both converged very quickly. Shown in figure 9.
- *Generalized Rosenbrock's Function, f_5.*
  Both operators performed similarly but the space is very complex with rapid gradient changes. Various sizes of search window would probably have brought even greater benefits in this example. Shown in figure 10.
- *Quartic Function with Noise, f_7.*
  Fast convergence was achieved with both operators. Notice that the noise causes even the same chromosome to have different evaluations with each iteration. Shown in figure 11.
- *Generalized Rastrigin's Function, f_9.*
  The error converged relatively quickly and to a lower value than the canonical EA. Shown in figure 12.

5 Conclusions and Further Work

We have found that a search mutation is as good or better than regular random mutation in every case we have tested. It has been particularly noticeable that in some contexts it can be used to solve optimisations in a single iteration. This is because the quality of the building blocks is good. Rather than being random they give the best possible information at the time.

In other cases, where the function is not particularly amenable to parabolic fitting, the operator acts as creep mutation until approaching a minimum, where the parabola fitting ensures fast convergence onto the local optimum.

Extension to Multi-Objective Optimisations

If it is necessary to find multiple optima or search for a global optimum there are two methods available to do this, non-random initialization and enforced diversity. If a good estimate of the solution is available, it makes no sense to randomly initialise a population, seeding around this solution is a much better practice. In large and complicated but regular spaces (i.e. locally continuous), one or two passes can be made to heuristically search for good areas in which to perform the optimisation. An example would be to place chromosomes on a regular grid, assess their fitness and then assign a number of search chromosomes to each area based on relative fitness.

One method of ensuring a good spread of chromosomes throughout the space is to test the diversity of the population and reject similar chromosomes. This method is not widely adopted since good measures of diversity are difficult to formulate. This is a problem on which further work is now being done.

Acknowledgements

The work presented in this paper was funded by UK EPSRC grant GR/M97138.

References

1. Robertson C., Fisher R. B., Werghi N., Ashbrook A. P. "An Evolutionary Approach to Fitting Constrained Degenerate Second Order Surfaces". in Evolutionary Image Analysis, Signal Processing and Telecommunications, Proc. First European workshop on evolutionary computation in image analysis and signal processing (EvoIASP99). Goteborg, Sweden, pp 1-16, Springer LNCS 1596, May (1999).
2. Robertson C., Fisher R. B., Werghi N., Ashbrook A. P., "An Improved Algorithm to Extract Surfaces from Complete Range Descriptions", Proc. World Manuf. Conf, WMC'99 (ISMT'99) - Sept. , pp 592-598, ICSC Academic Press, Durham (1999)
3. Robertson C., Fisher R. B., Werghi N., Ashbrook A. P., "Object reconstruction by incorporating geometric constraints in reverse engineering", Computer-Aided Design, Vol 31(6), pp 363-399, (1999).
4. Eggert D, Fitzgibbon A., Fisher R. B., "Simultaneous Registration of Multiple Range Views For Use In Reverse Engineering of CAD Models", Computer Vision and Image Understanding, Vol 69, No 3, pp 253-272, March (1998).
5. Rawlins G. J. E. (ed), *Foundations of Genetic Algorithms*, Morgan Kaufmann Inc., San Mateo, CA, (1991), pp3.

328   C. Robertson & R.B. Fisher

6. Michalewicz Z., *Genetic Algorithms + Data Structures = Evolution Programs*, Third Edition, Springer, (1996).
7. Fogel D. B., "An Introduction to Simulated Evoutionary Optimization", in IEEE Transactions on Neural Networks, Vol.5, No.1, January, pp3-14., (1994)
8. Chen K. , Parmee I. C., Gane C. R., "Dual Mutation Strategies for Mixed-integer Optimisation in Power Station Design.", Proceedings of IEEE International Conference on Evolutionary Computation, Indiana University, April , pp. 385-390., (1997)
9. Roy R., Parmee I. C., "An Overview of Evolutionary Computing for Multimodal Function Optimisation", Proceedings of WSC2. June (1997).
10. Press W.H., Teukolsky S.A., Vetterling W.T. , and Flannery B.P.. *Numerical Recipes in Fortran*. Cambridge University Press, (1992).
11. Yao X. and Liu Y., "Fast evolution strategies," Control and Cybernetics. 26(3):467-496, (1997).
12. Benson K., "Evolving Automatic Target Detection Algorithms theat Logically Combine Decision Spaces", Proc. British Machine Vision Conference, M. Mirmehdi and B. Thomas (eds), BMVA Press, pp 685., (2000).

Appendix - Graphs of Convergence

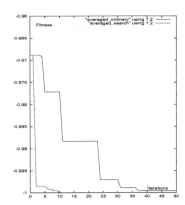

Figure 6: Convergence for Sine ×
sine Function

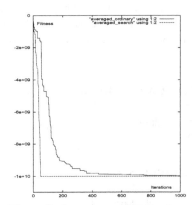

Figure 7: Convergence for Product
Function with 10 Genes

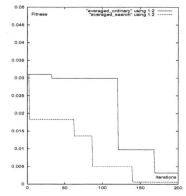

Figure 8: Himmelblau Function
Convergence

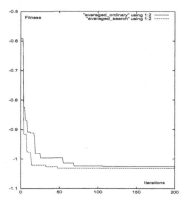

Figure 9: Six Hump Camel
Function Convergence

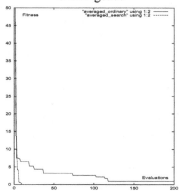

Figure 10: Generalized
Rosenbrock's Function
Convergence

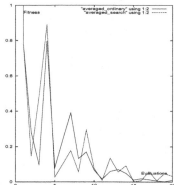

Figure 11: Noisy Quartic Function
Convergence

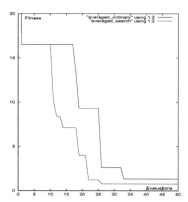

Figure 12: Generalized Rastrigin's Function Convergence

Global optimization of climate control problems using evolutionary and stochastic algorithms

Carmen G. Moles[1], Adam S. Lieber[2], Julio R. Banga[1], and Klaus Keller[3]

[1] Process Engineering Group, Instituto de Investigaciones Marinas (CSIC), 36208 Vigo, Spain cmoles@iim.csic.es, julio@iim.csic.es
[2] Mission Ventures, San Diego, CA 92130, U.S.A aslieber@alumni.princeton.edu
[3] Department of Geosciences, The Pennsylvania State University, 16802-2714 University Park, PA, U.S.A. kkeller@geosc.psu.edu

Summary. Global optimization can be used as the main component for reliable decision support systems. In this contribution, we explore numerical solution techniques for nonconvex and nondifferentiable economic optimal growth models. As an illustrative example, we consider the optimal control problem of choosing the optimal greenhouse gas emissions abatement to avoid or delay abrupt and irreversible climate damages. We analyze a number of selected global optimization methods, including adaptive stochastic methods, evolutionary computation methods and deterministic/hybrid techniques.

Differential evolution (DE) and one type of evolution strategy (SRES) arrived to the best results in terms of objective function, with SRES showing the best convergence speed. Other simple adaptive stochastic techniques were faster than those methods in obtaining a local optimum close to the global solution, but mis-converged ultimately.

Key words: global optimization, optimal control, climate thresholds, climate change detection

1 Introduction

The optimization of dynamic systems is an important class of problems arising in most scientific and engineering areas. In fact, optimal control problems are the subject of many research efforts in fields such as economics, physics, and virtually all engineering branches, with problems ranging from optimal trajectory determination for spaceships to the computation of optimal operating policies for chemical plants. The objective of optimal control is to find a set of control variables (which are functions of time) in order to maximize the performance of a given dynamic system, as measured by some functional, and all this subject to a set of path constraints. The dynamics of most systems are usually described in terms of differential equations or, as in this paper, equations in differences.

Global warming is undoubtedly a hot and controversial topic, receiving major attention from scientists, policy makers and the general public. The climate of our planet is predicted to warm because human activities are changing the chemical composition of the atmosphere due to emissions of greenhouse gases, mainly carbon dioxide, methane, and nitrous oxide. Presently, there is a considerable interest in analyzing the costs of reducing carbon emissions and designing policies for sustainable energy strategies. Here, we consider the optimal control problem of choosing the optimal greenhouse gas emissions abatement to avoid or delay abrupt and irreversible climate damages. The climate-economy system that is presented in this study is not smooth and shows significant hysteresis responses [7], [19], [24], which introduce local solutions into the economic model. Due to the local maxima in the objective function, the traditional local optimization algorithms fail to obtain the global optimum. As an alternative, we show how several global optimization methods can provide satisfactory solutions with reasonable computational efforts.

The paper is organized as follows: section 2 discusses the optimization of dynamic systems and its formulation. Section 3 classifies the different global optimization methods in addition to a brief description of those utilized in this contribution. Section 4 formulates the optimal climate control problem and sections 5 and 6 present some results and conclusions.

2 Optimization of Dynamic Systems

In general, optimal control problems can be ultimately formulated (e.g. via control parameterization) as non-linear programming (NLP) problems subject to the dynamics of the system (acting as equality constraints) and possibly to other inequality constraints. The numerical solution of these problems is usually a very challenging task due to the highly non-linear and frequently discontinuous nature of the dynamics of most processes. In fact, these characteristics make the NLP multimodal, i.e. there is not a unique global solution. Instead, a number of local solutions are possible, so the problem becomes much more complicated since the absolute best (global) solution must be found in a reliable and efficient way. In fact, standard gradient-based techniques for NLPs (e.g. Sequential Quadratic Programming (SQP)) are of local nature, i.e. they will converge to one of the local solutions, and in fact without giving any information about its local nature.

In order to surmount these difficulties, the so-called Global Optimization (GO) methods must be used in order to ensure proper convergence to the truly best possible solution. Unfortunately, solving GO problems is a non trivial exercise, and the current state of the art is far from being fully satisfactory. However, several types of GO methods can find suitable near-optimal solutions in many instances, as it will be discussed in the next section.

3 Global Optimization Methods

Basically, global optimization methods can be roughly classified as deterministic ([12], [15] and [23]) and stochastic strategies ([1], [13] and [29]). It should be noted that, although deterministic methods can guarantee global optimality for certain GO problems, no algorithm can solve general GO problems with certainty in finite time [13]. In fact, although several classes of deterministic methods (e.g. branch and bound) have sound theoretical convergence properties, the associated computational effort increases very rapidly (often exponentially) with the problem size.

In contrast, many stochastic methods can locate the vicinity of global solutions with relative efficiency, but the cost to pay is that global optimality can not be guaranteed. However, in practice the user can be satisfied if these methods provide him/her with a very good (often, the best available) solution in modest computation times. Furthermore, stochastic methods are usually quite simple to implement and use, and they do not require transformation of the original problem, which can be treated as a black box. This characteristic is especially interesting since very often the researcher must link the optimizer with a third-party software package where the process dynamic model has been implemented. Finally, many stochastic methods lend themselves to parallelization very easily, which means that medium-to-large scale problems can be handled in reasonable wallclock time.

In this study, we have considered a set of selected GO methods which can handle black-box models. The selection has been made based on their published performance and on our own experiences considering their results for a set of GO benchmark problems. Although none of these methods can guarantee optimality, at least the researcher can solve a given problem with different methods and take a decision based on the set of solutions found. Usually, several of the methods will converge to essentially the same (best) solution. It should be noted that although this result can not be regarded as a confirmation of global optimality (it might be the same local optimum), it does give the user some extra confidence. Further, it is usually possible to have estimates of lower bounds for the cost function and its different terms, so the goodness of the 'global' solution can be evaluated (sometimes a 'good enough' solution is sufficient).

The GO methods which we have considered are:

- **ICRS**: an stochastic GO method presented by Banga and Casares [3], improving the Controlled Random Search (CRS) method of Goulcher and Casares [10]. Basically, ICRS is a sequential (one trial vector at a time), adaptive random search method which can handle inequality constraints via penalty functions.
- **LJ**: another simple stochastic algorithm, described by Luus and Jaakola [21]. LJ considers a population of trial vectors at each iteration. Constraints are also handled via penalty functions.
- **DE**: the Differential Evolution method, as presented by Storn and Price [27]. DE is a heuristic, population-based approach to GO. The original code of the DE algorithm [27] did not check if the new generated vectors were within their bound constraints, so we have slightly modified the code for that purpose.
- **SRES**: the Evolution Strategy using Stochastic Ranking [25] is a (μ,λ)-ES evolutionary optimization algorithm which uses stochastic ranking as the constraint handling technique. The stochastic ranking is based on the bubble-sort algorithm

and is supported by the idea of dominance. It adjusts the balance between the objective and penalty functions automatically during the evolutionary search.

- **GLOBAL**: this is a hybrid GO method by Csendes [9], which essentially is a modification of the algorithm by Boender et al. [6]. This method uses a random search followed by a local search routine. Initially, it carries out a clustering phase were the Single Linkage method is used. Next, two different local search procedures can be selected for a second step. The first (LOCAL) is an algorithm of Quasi-Newton type that uses the DFP (Davison-Fletcher-Powell) update formula. The second, more appropriate for problems with discontinuous objective functions or derivatives, is a robust random search method (UNIRANDI) by Jarvi [17].

- **GCLSOLVE**: a deterministic GO method, implemented in Matlab as part of the optimization environment TOMLAB [14]. It is a version of the DIRECT algorithm [18] that handles nonlinear and integer constraints. GCLSOLVE runs for a predefined number of function evaluations and considers the best function value found as the global optimum.

- **MCS**: the Multilevel Coordinate Search algorithm by Huyer and Neumaier [16], inspired by the DIRECT method [18], is an intermediate between purely heuristic methods and those allowing an assessment of the quality of the minimum obtained. It has an initial global phase after which a local procedure, based on a SQP algorithm, is launched. These local enhancements lead to quick convergence once the global step has found a point in the basin of attraction of a global minimizer.

With this selection we have also tried to achieve a balanced representation of the most promising approaches to global optimization, namely adaptive stochastic methods (ICRS and LJ), evolutionary computation methods (DE and SRES), and deterministic/hybrid approaches (GCLSOLVE, MCS and GLOBAL). It should be noted that several of these methods are closely related despite their different origins: e.g. ICRS is very similar to simple evolution strategies, which are the heart of SRES. Thus, our purpose is also to provide the interested reader with a sound comparison of these alternatives considering a challenging and interesting problem.

For the sake of fair comparison, we have considered Matlab implementations of all these methods. The main reason to use Matlab is that it is a convenient environment to postprocess and visualize all the information arising from the optimization runs of the different solvers, allowing careful comparisons with little programming effort. Further, new methods (or modifications to existing ones) can be easily prototyped and evaluated. However, as a drawback, it is well known that Matlab programs usually are one order of magnitude (or more) slower than equivalent compiled Fortran or C codes. In order to minimize this effect, we have implemented the more costly part of the problem (i.e. system dynamic simulation plus objective function) in compiled f77 modules, which are callable from the solvers via simple gateways. Since most stochastic methods use 90% (or more) of the computation time in system simulations (especially if their complexity level is medium to large), this procedure ensures good efficiency while retaining the main advantages of the Matlab environment.

Finally, in order to illustrate the comparative performance of multi-start local methods for this type of problems, a multi-start code (named **ms-FMINCON**) was also implemented in Matlab making use of the FMINCON code, which is part of the MATLAB Optimization Toolbox [11]. FMINCON is indicated for unconstrained functions. Its default algorithm is a quasi-Newton method that uses the formula of Broyden, Fletcher, Goldfarb and Shanno (BFGS) for updating the approximation of the Hessian matrix. Its default line search algorithm is a safeguarded mixed quadratic and cubic polynomial interpolation and extrapolation method.

4 Optimal Climate Control Problem

The global optimization problem presented in this paper is based on the Dynamic Integrated model of Climate and the Economic (so-called DICE model [22]). It integrates the economics, carbon cycles, climate science and impacts allowing the weighing of the costs and benefits of taking steps to slow greenhouse warming.

We have also considered the changes introduced to the initial DICE model (see Nordhaus, 1994 [22]) by Keller et al.[19]. Specifically, we considerer damages caused by an uncertain environmental threshold imposed by an ocean circulation change (known as a North Atlantic Thermohaline circulation (THC) collapse). Ocean modelling studies suggest that the THC may collapse when the equivalent CO_2 concentration $P_{CO_{2,e}}$ (the concentration of CO_2 and all other greenhouse gases expressed as the concentration of CO_2 that leads to the same radiative forcing) rises above a critical value $(P_{CO_{2,e,crit}})$ (e.g. [26]). This phenomenon is represented in the model by imposing a threshold specific climate damage (θ_3) for all times after the THC has collapsed. The $P_{CO_{2,e}}$ is approximated by an exponential fit to previously calculated stabilization levels in the DICE model.

The DICE model is a long-term dynamic model of optimal economic growth linking economic activity and climate change. Individual utility (U) arises in the model from an increasing transform of per capita consumption per year (c), in this case the Bernoullian utility function $U(c) = ln(c)$. The social welfare function in the model is simply the sum of these individual utilities over the exogenously given population (L)

$$U(t) = L(t) \ln c(t) \tag{1}$$

Because of the temporal nature of the model (and the optimal control problem), utility must be weighed between generations. To compare present and future utility a "pure rate of social time preference" (ρ) is assigned. The objective function that the model seeks to maximize, then, is the discounted sum of individual utility from generalized consumption over a given time horizon, expressed as U*:

$$U^* = \sum_{t=t_o}^{t^*} U(t) \, (1+\rho)^{-t} \tag{2}$$

Consumption is limited by the output in a given time period and by the desire not to consume all output but rather invest output into growing consumption in future periods. Total consumption (C) is the difference between gross world product (Q) and gross investment (I):

$$C(t) = Q(t) - I(t) \tag{3}$$

While consumption adds to an individual's utility, investment adds to the capital stock (K), which otherwise depreciates at a proportional "rate of decay." The growth of this capital stock via gross investment allows output to increase in the next period by a Cobb-Douglas production function of capital and labor (equivalent to population in this model):

$$Q(t) = \Omega(t)\,A(t)\,K(t)^{\gamma}\,L(t)^{1-\gamma} \tag{4}$$

Multifactor productivity (A) and the elasticity of output with respect to capital (γ) are both exogenously defined parameters, while the output scaling factor (Ω) relates to abatement costs and climate damages for that time period in a manner discussed below.

The DICE model links economic production with physical systems by assuming that carbon emissions (E) into the atmosphere are proportional to economic activity, or world product. Emissions depend on the carbon intensity of this output (σ) and the level of carbon emissions abatement (μ), where this control rate is the only policy instrument for mitigating climate change:

$$E(t) = [1 - \mu(t)]\,\sigma(t)\,Q(t) \tag{5}$$

From these emissions each period a fraction (β) is added to the atmospheric carbon stock (M), while the rest is absorbed by natural sinks in the biosphere and ocean. Additionally, a fraction (δ_m) of the atmospheric carbon stock above pre-industrial carbon dioxide levels (590 Gt. C) decays at each time step into the deep ocean:

$$M(t) = 590 + \beta\,E(t-1) + (1 - \delta_M)[M(t-1) - 590] \tag{6}$$

Atmospheric carbon dioxide contribute to a change (F) in radiative forcing from preindustrial levels:

$$F(t) = 4.1\,\frac{\ln(M(t)/590)}{\ln(2)} + O(t) \tag{7}$$

where $O(t)$ represents the radiative forcing due to other greenhouse gases such as methane and Chloro Fluoro Carbons (CFCs), specified in forcing equivalent terms.

The additional radiative forcing from the buildup of these greenhouse gases causes an increase in global mean surface temperature from pre-industrial levels (T). The relationship is modeled by:

$$\begin{aligned} T(t) = T(t-1) + (1/R_1)[F(t) - \lambda T(t-1) \\ - (R_2/\tau_{12})(T(t-1) - T^*(t-1))], \end{aligned} \tag{8}$$

and

$$T^*(t) = T^*(t-1) + (1/R_2)[(R_2/\tau_{12})$$
$$(T(t-1) - T^*(t-1))]. \tag{9}$$

In these equations T^* is the upper ocean temperature, R_1 and R_2 represent the respective thermal capacities of the oceanic mixed layer and the deep ocean, respectively, (τ_{12}) is the heat transfer rate from the upper ocean to the deep ocean, and (λ) is related to the "climate sensitivity". "Climate sensitivity" describes the equilibrium mean surface temperature response to a doubling of atmospheric carbon dioxide.

The DICE model closes the connection between climate and the economy by specifying a damage function (D) that reveals the economic damages in terms of lost output from realized surface temperature change,

$$D(t) = \theta_1 \, T(t)^{\theta_2} \tag{10}$$

and a cost function (TC) that estimates the cost of complying with a given carbon abatement policy in terms of lost output:

$$TC(t) = b_1 \, \mu(t)^{b_2} \tag{11}$$

In these equations θ_1, θ_2, b_1, and b_2 are empirically derived model parameters. Given that the calculated abatement costs and climate damages are specified as fractions of gross world product, total output is now re-scaled with the scaling (Ω):

$$\Omega(t) = [1 - TC(t)]/(1 + D(t)) \tag{12}$$

This factor is approximated in the DICE model by the equation

$$\Omega(t) = 1 - TC(t) - D(t) \tag{13}$$

which approximates the explicit scaling factor closely for relatively small climate damages.

Model parameter values are used from the original DICE model, with one exception. We adopt a climate sensitivity of 3.6 degrees Celsius per doubling of CO_2 instead of the previously used 2.9 degrees as our standard value. Based on the analysis of climate data and the expert opinion of the Intergovernmental Panel on Climate Change (IPCC), Tol and de Vos [28] estimate the values of the median and the standard deviation of the climate sensitivity as 3.6 and $1.1\,^o C$.

The objective is the maximization of discounted sum of individual utility U^*, as defined in equation 2. The decision variables are the investment and CO_2 abatement over time. Although results are reported for the period from year 1995 until 2155, a longer time horizon of 470 years is used to avoid end effects, resulting in 94 decision variables (47 for abatement and 47 for investment).

5 Results and Discussion

Each of the GO methods considered has several adjustable search parameters which can greatly influence their performance, both in terms of efficiency and robustness.

We have followed the published recommendations for each method (see references [2], [4], [5], [8], [9], [21] and [27]), together with the feedback obtained after a few preliminary runs. All the computation times reported here were obtained using a low cost platform, PC Pentium III/866 MHz.

For the problem considered here, Keller et al. [19] found a best value of $C^* = 26398.83$ using DE (after several restarts from the best solution and very long iterations). In this work, the best result (see table 1) was also obtained with the DE method, which converged to essentially the same value, $C^* = 26398.71$. Further, the SRES method converged to $C^* = 26398.64$ but about 10 times faster than DE.

The MCS method also converged to a quite good result ($C^* = 26397.00$) in less than 250 s, i.e. again one order of magnitude faster than DE. The ICRS method was able to arrive to a reasonably good value in just 10 min of computation. Table 1 presents the best objective value obtained by each method, in addition to the corresponding CPU times and the required number of function evaluations. No results are presented regarding the GLOBAL algorithm, which did not converge successfully, probably due to the relatively large dimension of the problem.

Table 1. Optimization results (best result for each method)

	DE	SRES	MCS	ICRS	GCLSOLVE	LJ
N eval	$3.5 \cdot 10^6$	$3.5 \cdot 10^5$	71934	386860	65000	20701
CPU time,s	6652.76	640.43	246.29	600.54	62272.64	57.32
C^*	26398.71	26398.64	26397.00	26383.71	26377.06	26375.83

The comparison of single figures of final objective function values and the associated computation times can be totally misleading. In order to provide with a more fair comparison of the different methods, a plot of the convergence curves (objective function values represented as relative error versus computation time) is presented in Fig. 1, where the best three curves per method from a set of 30 are plotted. It can be seen that the ICRS presented the most rapid convergence initially, but was ultimately surpassed by DE and SRES. The latter arrived to the close vicinity of the best known solution much faster then DE, so it could be regarded as the method of choice for this type of problems.

When solving a global optimization problem, it is usually interesting to examined its multimodal nature. In order to illustrate its non-convexity, the problem was also solved using the multi-start (ms-FMINU) approach, considering 100 random initial vectors, which were generated satisfying the bounds on the decision variables. This strategy converged to a large number of local solutions, as depicted in the histogram shown in Fig. 2. It is very significant that despite the huge computational effort associated with the 100 runs, the best value found ($C^* = 23854.71$) was still far from the solutions of the GO methods, obtained with much smaller computation times. These results illustrates well the inability of the multi-start approach to handle highly multimodal problems like this one.

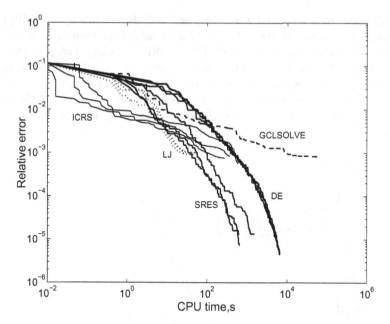

Fig. 1. Convergence Curves

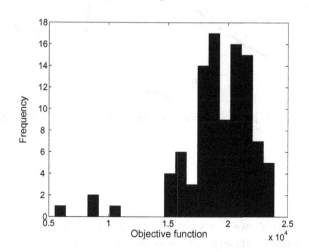

Fig. 2. Histogram of solutions found by the multi-start strategy

Finally, the solutions found were compared with that of Keller et al. [19] by inspecting the decision variables values at the different optima. Plots of the decision variables for the best solutions (i.e. for the DE and SRES algorithms) are presented in Fig. 3 and Fig. 4. It should be noted that there are somewhat significative differences in the optimal abatement and investment policies obtained, although the objective function values are very similar. This indicates a very low sensitivity of the cost function with respect to that control variables, a rather frequent result in dynamic optimization problems.

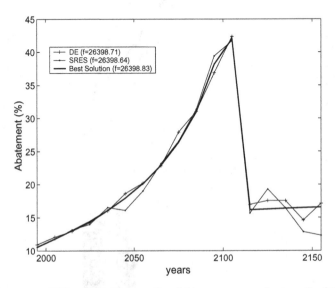

Fig. 3. Abatement of CO_2 policy (best results of this paper versus best result of Keller et al. (2000) [20])

6 Conclusions

In this study, we have shown how a challenging optimal control problem regarding the design of optimal greenhouse gas emissions can be efficiently solved by using several global optimization (GO) methods. Despite the main drawback of stochastic GO methods (i.e., inability to guarantee global optimality), several of these methods were capable of reaching very good solutions in moderate computation times.

Evolutionary strategies (represented by the SRES method) presented the fastest convergence to the vicinity of the best known solution, closely followed by several other hybrid and stochastic methods. Differential Evolution (DE) arrived to the best solution, although at a rather large computational cost. Simple adaptive stochastic methods were not able to arrive to very refined results, but they presented an interesting first period of fast convergence which suggest new hybrid approaches.

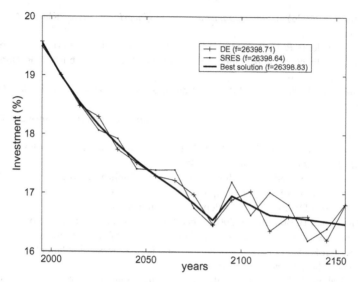

Fig. 4. Investment policy (best results of this paper versus best result of Keller et al. (2000) [20])

References

1. Ali M, Storey C, Törn A (1997) Application of stochastic global optimization algorithms to practical problems. *J. Optim. Theory Appl.*, 95(3):545–563
2. Balsa-Canto E, Alonso A A, Banga J R (1998) Dynamic optimization of bioprocesses: deterministic and stochastic strategies. *Presented at ACoFop IV (Automatic Control of Food and Biological Processes) Göteborg-Sweden.*, 21-23 September
3. Banga J R, Casares J (1987) ICRS: Application to a wastewater treatment plant model. In I. S. S. 100, editor, *Process Optimisation*, IChemE Symposium Series No. 100, pages 183–192. Pergamon Press, Oxford, UK
4. Banga J R, Irizarry R, Seider W D (1998) Stochastic optimization for optimal and model-predictive control. *Comput. Chem. Eng.*, 22(4-5):603–612
5. Banga J R, Seider W D (1996) Global optimization of chemical processes using stochastic algorithms. In: State of the Art in Global Optimization, C. A. Floudas and P. M. Pardalos (eds.), Kluwer Academic Pub., Dordrecht, The Netherlands, pages 563–583
6. Boender C, Kan A R, Timmer G, Stougie L (1982) A stochastic method for global optimization. *Math. Programming*, 22:125
7. Broecker W (2000) Abrupt climate change: Causal constraints provided by the paleoclimate record. *Earth-Science Reviews*, 51(1-4):137–154
8. Carrasco E, Banga J R (1997) Dynamic optimization of batch reactors using adaptive stochastic algorithms. *Ind. Eng. Chem. Res.*, 36(6):2252–2261
9. Csendes T (1998) Nonlinear parameter estimation by global optimization - efficiency and reliability. *Acta Cybernetica*, 8(4):361–370
10. Goulcher R, Casares J (1978) The solution of steady-state chemical engineering optimisation problems using a random-search algorithm. *Comput. Chem. Eng.*, 2:33–36
11. Grace A (1994) *Optimization Toolbox User's Guide.* The Math Works Inc

12. Grossmann I (1996) *Global Optimization in Engineering Design.* Kluwer Academic Publishers
13. Guus C, Boender E, Romeijn H (1995) *Stochastic Methods,* chapter In Horst, R. Pardalos, P.M. eds. Handbook of Global Optimization. Kluwer Academic Publishers, R. Horst and P.M. Pardalos edition
14. Holmström K (1999) The TOMLAB optimization environment in matlab. *Adv. Model. Optim.,* 1:47
15. Horst R, Tuy H (1990) *Global Optimization: Deterministic Approaches.* Springer-Verlag, Berlin, Germany
16. Huyer W, Neumaier A (1999) A global optimization by multilevel coordinate search. *Journal of Global Optimization,* 14:331–355
17. Jarvi T (1973) A random search optimiser with an application to a maxmin problem. page 3. Publications of the Inst. of Appl. Math. Univ. of Turk
18. Jones D (2001) DIRECT. In Encyclopedia of Optimization
19. Keller K, Bolker B, Bradford D (2002) Uncertain climate thresholds in economic optimal growth models. *Journal of Environmental Economics and Management,* in review
20. Keller K, Tan K, Morel F, D.F B (2000) Preserving the ocean circulation: Implications for climate policy. *Climatic Change,* 47:17–43
21. Luus R, Jaakola T (1973) Optimisation of non-linear functions subject to equality constraints. *IEC Process. des. Dev.,* 12:380–383
22. Nordhaus W (1994) *Managing the Global Commons: The Economics of Climate Change.* The MIT Press, Cambridge, Massachusetts
23. Pinter J (1996) *Global Optimization in Action. Continuous and Lipschitz Optimization: Algorithms, Implementations and Applications.* Kluwer Academics Publishers, Dordrecht
24. Rahmstorf S (1997) Risk of the sea-change in the atlantic. *Nature,* 388:825–826
25. Runarsson T, Yao X (2000) Stochastic ranking for constrained evolutionary optimization. *IEEE Transactions on Evolutionary Computation,* 4:284–294.
26. Stocker T, Schmittner A (1997) Influence of CO2 emission rates of the stability of the thermohaline circulation. *Nature,* 388:862–865
27. Storn R, Price K (1997) Differential evolution - a simple and efficient heuristic for global optimization over continuous spaces. *Journal of Global Optimization,* 11:341–359
28. Tol R, de Vos A (1998) A bayesian statistical analysis of the enhanced greenhouse effect. *Climatic Change,* 38:87–112
29. Törn A, Ali M, Viitanen S (1999) Stochastic global optimization: Problem classes and solution techniques. *Journal of Global Optimization,* 14:437

Global optimization of climate control problems using evolutionary and stochastic algorithms

Carmen G. Moles[1], Adam S. Lieber[2], Julio R. Banga[1], and Klaus Keller[3]

[1] Process Engineering Group, Instituto de Investigaciones Marinas (CSIC), 36208 Vigo, Spain cmoles@iim.csic.es, julio@iim.csic.es
[2] Mission Ventures, San Diego, CA 92130, U.S.A aslieber@alumni.princeton.edu
[3] Department of Geosciences, The Pennsylvania State University, 16802-2714 University Park, PA, U.S.A. kkeller@geosc.psu.edu

Summary. Global optimization can be used as the main component for reliable decision support systems. In this contribution, we explore numerical solution techniques for nonconvex and nondifferentiable economic optimal growth models. As an illustrative example, we consider the optimal control problem of choosing the optimal greenhouse gas emissions abatement to avoid or delay abrupt and irreversible climate damages. We analyze a number of selected global optimization methods, including adaptive stochastic methods, evolutionary computation methods and deterministic/hybrid techniques.

Differential evolution (DE) and one type of evolution strategy (SRES) arrived to the best results in terms of objective function, with SRES showing the best convergence speed. Other simple adaptive stochastic techniques were faster than those methods in obtaining a local optimum close to the global solution, but mis-converged ultimately.

Key words: global optimization, optimal control, climate thresholds, climate change detection

1 Introduction

The optimization of dynamic systems is an important class of problems arising in most scientific and engineering areas. In fact, optimal control problems are the subject of many research efforts in fields such as economics, physics, and virtually all engineering branches, with problems ranging from optimal trajectory determination for spaceships to the computation of optimal operating policies for chemical plants. The objective of optimal control is to find a set of control variables (which are functions of time) in order to maximize the performance of a given dynamic system, as measured by some functional, and all this subject to a set of path constraints. The dynamics of most systems are usually described in terms of differential equations or, as in this paper, equations in differences.

Global warming is undoubtedly a hot and controversial topic, receiving major attention from scientists, policy makers and the general public. The climate of our

2 Problem Definition

The Minimum Rectilinear Steiner Tree (MRST) problem arises in global routing and wiring estimation where we seek low-cost topologies to connect the pins of signal nets [5]. Research on the MRST problem has been guided by several fundamental results. Hanan [6] has shown that there always exists an MRST with Steiner points chosen from the intersections of all the horizontal and vertical lines passing through all the points in P and this result was generalized by Snyder [7] to all higher dimensions. However a result by Garey and Johnson [8] establishes that despite this restriction on the solution space, the MRST problem remains NP-complete, this has given rise to numerous heuristics as surveyed by Hwang, Richards and Winter [9].

The MRST problem can be defined as below: Given n points in the Euclidean plane, we seek a tree that connects them all. If the edges in this tree are to be selected from all possible edges, that is from the complete graph on the points, we have the familiar problem of finding a spanning tree in an undirected graph. If the edges of the tree must be horizontal and vertical, the additional points where the edges meet are called Steiner points, and the resulting tree is a rectilinear Steiner tree. A shortest such tree on a set of given points is a Minimal Rectilinear Steiner Tree(MRST)[10].

3 Global Router Algorithm

Before the global routing process begins a routing graph is extracted from the given placement. Routing is done based on this graph. Computing a global route for a net corresponds to finding a corresponding path in the routing graph. Each edge represents a routing channel and the vertex is the intersection of the two channels. First the vertices that represent the terminal of the net are added to the routing graph. Then the shortest route for the net is found which is nothing but the Steiner Problem in Graphs. Then the global routing problem is formulated as finding the Steiner Minimal Tree on the routing graph.

The algorithm is divided into two phases:
Phase 1:

The Steiner tree algorithm we developed is based on the shortest path heuristic. We developed a simple genetic algorithm for the construction of a minimum spanning tree, which is used in the generation of the Steiner minimum tree.

The algorithm for SPANNING_TREE() is given below:

```
SPANNING_TREE( )
Begin
        Initialize parameters :  generation count , crossover
            and mutation probabilities
        Initialize parent population randomly
        Apply repair heuristics to the parent population
```

```
/* Repair heuristics are to test cycle and self loops*/
While ( termination condition not reached)
   Begin
      Select parents based on the total length of
         the Spanning tree
      Apply crossover and mutation
      /* Single Point Crossover and Exchange Mutation*/
      Evolve new population
      Replace previous population by new population
   End
End
```

The Cycle and Self loop tester are implemented to remove cycles and self loops from the population and hence make the population a set of feasible solutions.

Phase 2:

The Steiner points are restricted to the Hanan grid [6], formed by the intersections of vertical and horizontal lines passing through the initial demand points. There are various heuristics available to construct a MRST, and most of them use MST as a starting point.

The I-Steiner algorithm [11] constructs the MRST by evaluating all possible Steiner points for their impact on MST cost. The algorithm operates on a series of passes, in each pass the single Steiner point which provides the greatest improvement in spanning tree cost is selected and added to the set of demand points. Points are added until no further improvement can be obtained. In experiments with random point sets, this algorithm obtains nearly 11

The heuristics we have used here for the construction of MRST is the "BOI" or "edge-removal technique" of Borah, Owen and Irwin [12]. In this approach, an edge and a vertex pair that are close to each other in an MST are determined. A cycle is introduced, if the vertex is connected to some location on the edge; this cycle can be eliminated by removing a second edge along the cycle. The BOI approach determines a set of these pairings along with edges that are to be removed. The edges are removed and new connections are inserted until no improvements can be obtained. The approach has low complexity with performance comparable to that of I-Steiner.

The algorithm is given below:

```
STEINER_TREE( )
Begin
    Build the routing graph G
    For (each Net)
    Begin
        Initialize weights for edges.
        Find the minimum cost spanning tree T.
        For each (vertex, edge) pair of the spanning tree
```

```
     Begin
          Find the optimum Steiner point to connect this edge to the vertex
          (at a suitable point)
          Find the longest edge on the generated cycle
          Compute the cost of the modified tree, and store the pair in a list,
                        if the cost is less than the MST
          End
     While the list is not empty
     Begin
          Remove the pair from the list which results in lowest cost
          Re compute the longest edge on the cycle and the cost of the tree
          If the edges to be replaced are in the tree and the cost is less
          modify the tree
     End
  End
End
```

4 Hybrid CHC Genetic Algorithm

A Genetic Algorithm is an intelligent optimization technique used to solve many complex optimization problems. A simple GA evolves a solution from an initial population of random solutions using crossover and mutation. An SGA often fails to produce an optimal solution within an acceptable time because of the random initialization. Hence hybridization techniques are used to initialize the population to reduce the search space. A hybrid GA combines techniques particular to a problem with the simple GA. The simplest hybrid GA applies problem-specific information to operate on fairly good organisms and then operates as conventional GA. In this paper we present a Hybrid CHC Genetic Algorithm for the global routing of Macro cell layouts. A CHC genetic algorithm uses intelligent reproduction techniques and techniques to come out of local optima.

CHC, which is a non-traditional genetic algorithm is used in this paper, it differs from a classic GA as follows

- For a population of N, it guarantees that the best individuals found so far shall always survive by putting the children and parents together leading to 2N individuals and selecting the best N individuals for further processing.
- To avoid premature convergence, two similar individuals separated by a small Hamming distance are not allowed to mate.
- Mutation isn't needed during normal processing.
- An external mutation operator re-initializes the population when the population has converged or search has stagnated.

The pseudo-code for the Hybrid CHC genetic algorithm is shown below:

HCHC( )

```
Begin
    Set the generation count c = 1
    Set threshold value
    Initialize and evaluate parent population P(t)
    While (Termination Condition is not reached) do
            c = c + 1
            Apply selection
            Perform crossover on selected individuals to get Offspring C(t)
            Evaluate Offspring C(t)
            Select new parent population from previous parent
               population P(t-1) and the new child population C(t)
               using Elitism
        If P(t) is equal to P(t-1)
            then decrease the threshold value
        If Convergence occurs
            Reinitialize population by external mutation.
    End
End
```

To guarantee that the best individuals found so far will always survive, elitist selection makes the newly created children compete with the members of the parent
population for survival by merging and ranking them according to their fitness values. If the GA's population is N, 2N candidates will compete for the N positions in
the next generation. A mechanism is used to slow premature convergence, It checks
the Hamming distance between two potential parents before their crossover. If half
of the Hamming distance does not exceed a convergence threshold, the two individuals are not allowed to mate. Whenever no children are accepted into the next
generation either because no potential mates were mated or because none of the children were better than the worst member of the parent population, the convergence
threshold is decreased. Premature convergence happens when the threshold drops
to zero. An outside mutation is used to re-initialize the population by keeping the
best individual found so far and initializing the others randomly. The convergence
threshold is also reset. The CHC genetic algorithm generally does well with small
population sizes. This is because CHC is able to maintain population diversity in a
small population size without sacrificing implicit parallelism by combining a highly
disruptive crossover operator with a conservative selection procedure that preserves
any progress ever made.

5 Solution Encoding

Most GAs use binary strings for encoding due to simple implementation and simple
crossover and mutation operators. However integer encoding is also used, especially
in complex problems. Our Encoding consists of an array of integers that represent
the edges of the graph in the case of a Minimum Spanning Tree and the edges of a
Steiner graph in the case of a Steiner Minimal Tree.

6 CHC Genetic Operators

Reproduction Operator
The crossover operator is used to generate offspring from the parents chosen by
the selection schemes. We use Uniform crossover. It is implemented along with the
repair algorithm. The repair algorithm act on the infeasible chromosomes and convert
them to feasible ones.
   Mutation
We use Random mutation, which randomly changes a gene in the chromosome In
CHC genetic algorithm when premature convergence occur we reinitialize the whole
population by means of random mutation.

7 Implementation and Results

The HCHC algorithm is implemented and three standard test problems B1,B3 and B9
from the problem sets of J.E.Beasley [13] are tested. The results are comparable and
are tabulated in this section. A simple GA with one point crossover and exchange mutation is also implemented and the results are compared with the HCHC algorithm.
In HCHC, Uniform crossover and Random mutation are used for reproduction. The
probabilities of crossover and mutation are set at 0.5 and 0.03 respectively for both
SGA and HCHC. The convergence threshold is set to 7.5. The best results out of 10
independent runs for the test problems are given below:
   Test Problem : B1 Nodes : 50 Edges : 63 Optimum : 82 Population Size for SGA
: 110 Population Siz ⌐ ˥˥ᴄ˥˥ᴄ ˓˓

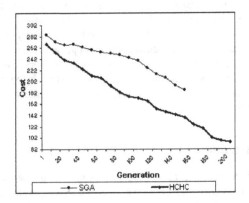

Fig. 1. Comparison of cost reduction in SGA and HCHC algorithms for B1

Test Problem : B3
Nodes:50 Edges:63 Optimum : 138
Population Size for SGA : 110

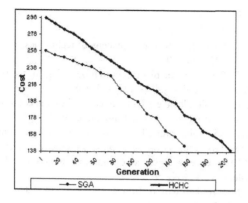

Fig. 2. Comparison of cost reduction in SGA and HCHC algorithms for B3

Population Size for HCHC: 90
Test Problem : B9
Nodes:75 Edges:94 Optimum : 220
Population Size for SGA : 110
Population Size ⌐ ⎍⎍⎍⎍ ⎍⎍

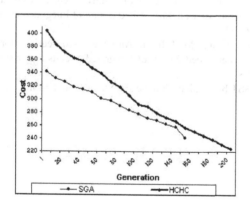

Fig. 3. Comparison of cost reduction in SGA and HCHC algorithms for B9

8 Conclusion

In this paper we have presented a HCHC genetic algorithm for macro-cell global routing. The HCHC algorithm shows much better performance over SGA, and generates comparable results for the Beasley's problem set. We further propose to improve the implementation for higher number of nodes and compare the results with the other existing algorithms.

References

1. K. Shahookar and P. Mazumder, VLSI cell placement techniques, in ACM Computing surveys, vol 23, pp. 143-220, Jun 1991
2. J. P. Cohoon and W. D. Paris, Genetic placement in proceedings of the IEEE International conference on Computer aided design, pp. 422-425, 1986
3. R. M. Kling, Placement by simulated evolution master's thesis in coordinated science lab, college of Engg. , Univ of Illinois at Urbana Campaign, 1987
4. H. Esbensen, A macro cell global router based on two genetic algorithms, in Proc of European Design Automation Conf, Euro-DAC, pp. 428-433, Grenoble, France Sep 1994.
5. B. T. Preas and M.J.Lorenzetti, Physical design automation of vlsi systems, Benjamin/Cummings, Menlo Park, CA, 1988
6. M. Hanan, On Steiner's problem with rectilinear distance, SIAM Journal, Applied mathematics, 14(1966) pp. 255-265
7. T. I. Snyder, On the exact location od stiner points in general dimension, SIAM J, Comp, 21(1992) pp. 163-180
8. M. Garey and D. S. Johnson, The Rectilinear stiner problem is NP-Complete, SIAM J, Applied mathematics, 32(1977) pp. 826-834
9. F. K. Hwang, D. S. Richards and P. Winter, The Steiner tree problem, North Holland, 1992
10. Bryant A. Julstrom, Seeding the population: Improved performance in a genetic algorithm for the rectilinear Steiner problem, ACM 089791 - 647 - 6 /94/0003, 1994
11. A. B. Kahng and G. Robins, A new class of iterative Steiner tree heuristics with good performance , IEEE Trans on Computer Aided design of Integrated circuits and systems, 11(7), 1992, pp. 893-902
12. M. Borah, R. M. Owens and M. J. Irwin, An edge based heuristic for Steiner routing, IEEE trans on Computer Aided design of Integrated circuits and systems, 13(12), Dec 1994, pp. 1563-1568
13. J. E. Beasley, An SST-based algorithm for the Steiner problem in graphs, Netwoks, 19, 1989, pp. 1-16.

Genetic Models for the Rational Exploitation of Resources

Adina Florea[1] and Cosmin Carabelea[2]

[1] University "Politehnica" of Bucharest, Romania, E-mail: adina@cs.pub.ro
[2] ENS des Mines de Saint-Etienne, France, E-mail: carabelea@emse.fr

Summary. The paper presents a multi-agent system aimed to solve the problem of rational exploitation of natural renewable resources by self-interested agents, a problem known also as the Tragedy of Commons. The system is geared towards a particular instance of this problem, namely the FishBanks game. The agents use genetic algorithms to build plans for fishing in several fishing banks and to buy or sell ships, in order to maximize their profit. We investigate the use of several genetic models based on cooperative co-adapted species to model the multi-agent world from the point of view of a particular agent and to predict society and environment evolution. The system is evaluated under different assumptions regarding the agent profiles and environment features, and the corresponding experimental results are discussed. An auction model for buying and selling ships is also proposed.

1 Introduction

The paper presents a multi-agent system with self-interested agents that try to solve the problem of common exploitation of renewable resources, problem know also as the Tragedy of Commons [4]. The agents in our system have to develop plans of resource exploitation to maximize their profit but have no knowledge of other agents' plans. The multi-agent system we propose is designed to solve a particular instance of the problem, namely the FishBanks game, first developed by Meadows [6]. The game consists of a number of fishing companies, each having its own ships and sharing the same set of fishing areas, called banks, which are renewable resources. A company must decide to which fishing banks it will send its ships during several fishing seasons in order to maximize its profit. Profit may be also increased by buying new ships to augment fishing capacity or by selling unprofitable ones. The desire to maximize profit motivates the companies to increase their level of fishing, the collective effect of which may lead to the depletion of common resources and the eventual profit loss.

The aim of the work presented here is to show how a distributed planning problem may be solved by means of a genetic approach to planning and to compare and evaluate several genetic representations. The agents use genetic algorithms to develop plans for resource exploitation and ship trading by modeling both their own

behavior and the presumed behavior of the other agents in the system. Several genetic models for evolving plans are considered under different assumptions: resource
exploitation is free or is banned by a central authority if the level drops under the critical limit; companies have an ecologically oriented profile or a profit oriented profile; the company profiles are known or unknown, the companies maintain a constant
number of ships or are allowed to buy and sell ships. In this last case, we propose an
auction for trading ships in which a company selects its bidding price based on presumed models of other companies. For each of the assumptions, the performance of
the solution is evaluated using criteria such as health of the environment, individual
accumulated profit of the companies, and average profit of all companies, measuring
thus the society welfare.

The remainder of the paper is structured as follows. Section 2 presents the problem specific details and the agent model in the multi-agent system we propose; Section 3 deals with the genetic planning representations, algorithm details and obtained
results; Section 4 presents the modifications required to model ship buying and selling and the auction model, Section 5 contains a brief presentation of related approaches. Finally, Section 6 presents conclusions and future work.

2 Multi-agent system model

We first state problem-specific details and then describe the multi-agent system to
model the problem. There are M companies, each company having a number of ships
NS_i $1 \leq i \leq M$. Some companies are ecologically oriented and are building fishing
plans which preserve the minimum ecological level; some other companies are profit
oriented and are interested only in making profit. The companies are able to catch
fish in several fishing banks of type either inshore or deep-sea, or they may keep
their ships at the port. Each ship has an associated fishing capacity (Fc) and an
associated cost for fishing in deep-sea (c_1), fishing inshore (c_2) or being kept at the
port (c_3), with $c_1 > c_2 \gg c_3$. There are P fishing banks with an initial amount of fish,
an associated regeneration coefficient, and a minimum amount of fish which allows
regeneration to take place. The companies can send ships to fishing banks only at the
beginning of each of K fishing seasons. Fish regeneration takes place at the end of
every season.

A company is represented by a *Company Agent* (CA) which has a representation
of itself consisting of the number of ships it owns, the associated costs of sending
ships to fish or keep them in port, and the assumed goals of the company, namely
maximum profit or profit and ecological balance. A Company Agent has also a representation of the environment containing the features of the fish banks, the number
and identity of the other companies in the system, the number of ships each company
owes but may or may not know their profile.

The Environment Agent (EA) keeps track of the environment evolution and has
knowledge about which companies are active in the system, the fish bank characteristics, and the amount of fish caught in different banks at the end of every season.
The EA may assume only this passive role or may act as a guardian of the fishing

resources by imposing a banning on fishing in depleted areas. The environment agent is acting also as the facilitator of the system: every new company entering the system must register to the EA, declare the number of ships it owes, and may querying the facilitator about other CAs.

The Auction Agent (AA) is responsible for managing fish selling at market price at the end of every season, company bank accounts, selling and buying ships between companies in organized auctions or from the shipyard. All fish caught by the companies are assumed to be sold at the market price. The profit of a company is computed as the difference between the prices obtained by selling the caught fish at market price and the cost of sending ships to fish or keep them in port, at the end of every season. Selling ships may also increase a company's profit.

Agent communication takes place between Company Agents and the Environment Agent or between companies and the Auction Agent. Company Agents inform the EA about their shipping plans at the beginning of every season while the EA feeds back the amount of fish a company has caught and the remaining level of fish in the fishing banks. Communication between the AA and the companies refers to selling fish, placing ships for auctions or registering for trading ships, making bids, as further described in Section 4.

3 Genetic planning

A company must establish its own plan for sending ships to fish banks with the aim of obtaining the maximum profit and, if ecologically oriented, keeping the minimum level for resource regeneration. Planning is difficult as a company has no knowledge about the fishing actions of other companies and thus does not know either how much fish it will actually catch or the level of resources at the end of each fishing season. To develop adequate plans under these conditions, we propose a genetic solution in which a Company Agent evolves both its fishing plans and the "presumed" plans of the other companies.

The company agents are developing their plans by using the model of cooperative co-adapted species. In this model, each subcomponent evolves separately by a genetic algorithm, but the evolution is influenced by the existence of the other subcomponents in the ecosystem [8]. In our case, a species corresponds to a company's plan; each individual of a company population represents a possible plan for sending the ships owned by the company to fishing banks during all the K fishing seasons. The population of a company's plan is evolved separately but individuals in a population are evaluated in conjunction with the best individuals in all other companies. The fitness function of an individual is computed as the ratio of estimated profit over the "ideal" maximum profit. If a company is ecologically oriented and an individual plan destroys fishing banks the fitness is decreased with $20 * x\%$ for x depleted resources. The estimated profit of a plan is the profit a company would obtain if it follows that individual plan while the other companies are following the plans specified by their best individuals. The maximum profit is the profit a company would obtain if it sends all its ships to the most profitable resource during all seasons and it is able to fish that fishing bank to maximum capacity. This maximum profit is an "ideal" measure that cannot be attended usually even if the company is alone in the system; it is used just as an indication of what a company should aim for. The best individual of another company is the one that brings the highest estimated profit to that company, and it preserves the resources if the company is ecologically oriented. If the company profile is not known then the company is presumed to be ecologically oriented.

We have run the genetic algorithm with three representations for the plans of a company; in what follows we detail two of them. In the first representation (Figure 1.a), an individual is formed of K groups of NS_i subgroups of bits. Each of the K groups corresponds to one season, each of the NS_i subgroups corresponds to one ship of company i, and the subgroup codifies the fishing bank where a particular ship is sent during a season. A heuristic repair function is used during evolution to keep the bank values in the required interval. The heuristic function uses a probability distribution that depends on the resource levels to repair non-existent fishing bank values. The second representation is somehow similar.

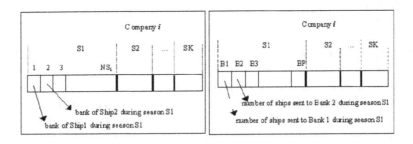

Fig. 1. First (a) and third (b) genetic representation of plans

The third representation uses K groups of bits in which one group corresponds to a fishing season and for each such group there are P subgroups. A subgroup of bits represents the number of ships to be sent during one season to a particular fish bank (Figure 1.b). A heuristic repair function is used during evolution to keep the sum of ships sent to different resources constant, namely equal to NS_i. The heuristic repair function is designed as follows. Be S the season, $b = B_1,.,B_P$ and $\text{Sum}(\text{Val}(gr_1),...,\text{Val}(gr_P))$ the sum of the values stored in the group of bits gr_b, $b = 1...p$, each value representing how many ships will be sent to the bank b during season S. If for a season S, the sum $\text{Sum}(\text{Val}(gr_1),...,\text{Val}(gr_P)) > NS_i$, then the repair function will first decrease the number of ships sent to depleted resources if such cases exist or will decrease the number of ships sent to less populated fishing banks. If $\text{Sum}(\text{Val}(gr_1),...,\text{Val}(gr_P)) < NS_i$ then the repair function will favor the increase of values corresponding to heavily populated fish banks. This representation can also accommodate plans for ship trading.

Proportional selection was used based on the scaled fitness of the individual and on the stochastic remainder technique. We used a two-point crossover rate of 0.6, a probability of mutation in every individual of 0.0005, a population size of 100, and 400 generations.

Figure 2 shows the average profit evolution of 4 companies for the three genetic representations. To evaluate the performance of our approach, we have considered the following cases: (a) there is no banning on fishing from a central authority and all companies are ecologically oriented; (b) no banning, half of the companies are ecologically oriented, half are profit oriented, and the companies have known profiles; (c) no banning, half companies are ecologically oriented but they do not know the profiles of the other companies; (d) there is a central authority which imposes banning on endangered resources and half of the companies are ecologically oriented while the other half are profit-oriented. In all cases, we have considered the fitness values of evolved best plans, the evolution of resources level (fish banks), the individual accumulated profit of every company and the average profit of all companies over K seasons.

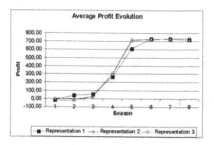

Fig. 2. Average profit for the three GA representations

The results obtained proved that in all cases, except (c), the fishing banks have a healthy evolution and the profit of the companies, after some initial fluctuations, reaches a rather constant level. In case (c), because the default assumption is that all companies are ecologically oriented and half of them are actually interested only in profit, some resources are depleted. However, this is not the case if the profiles of the companies are known, as the ecologically oriented companies are building their plans so as to preserve the resources in the presence of "careless" companies. The results also showed that in case (d), i.e., banning, the ecologically oriented companies are doing better in the long run than the profit oriented ones as they evolve their plans taking into account the preservation of resources. Figure 3 shows the results obtained for case (a) with 4 companies, 4 fish banks and $K = 8$ seasons. Figure 4 presents the results for case (d) with the same number of companies, fish banks and seasons.

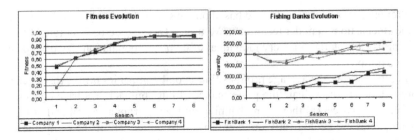

Fig. 3. Results of plan execution for case (a)

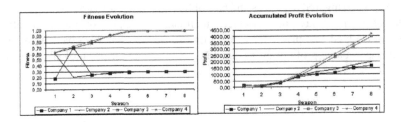

Fig. 4. Results of plan execution for case (d)

4 Planning for ship trading

There are three aspects to be tackled in case of selling and buying ships: how should a company agent decide when to sell or when to buy ships so as to maximize profit, how to organize the auction, and what is the price a company should pay/ask for a ship. We assume that buying and selling ships may take place at the end of every season and that a Company Agent may sell or buy only one ship at a time. A company may buy or sell ships either in a trading auction with the other companies or from a shipyard, which is responsible for manufacturing new ships and destroying old ones. Buying or selling ships from other companies in trading auctions is more convenient than buying from the shipyard.

The process of buying and selling ships may be conveniently modeled using the third genetic representation of a company's plan (Figure 1.b). A Company Agent evolves the fishing plans as described before but uses a different heuristic repair function to account for ship trading. The repair value is updated after each generation of the GA as follows. Consider r_S being the value the repair function must ensure for a season $S(S = 2, K)$ and r_{prev} the corresponding value for season $S - 1$. Initially, r_{prev} and r_S are equal to NS_i, the initial number of ships of the company. Be $r = \text{Sum}(\text{Val}(gr_1), \ldots, \text{Val}(gr_P))$ the current value obtained for season S and banks $b = B_1, \ldots, B_P$ after a genetic update has taken place. If $r \in [r_{prev} - 1, r_{prev} + 1]$ then r_S is updated to r. If $r < r_{prev} - 1$ then r_S is updated to $r_{prev} - 1$, otherwise r_S is updated to $r_{prev} + 1$. The repair function then "repairs" the values $\text{Val}(gr_1), \ldots, \text{Val}(gr_P)$ so that their sum is equal to r_S. When the GA run is over, the CA will have a plan with (possibly) a different number of ships for seasons $2..K$, the number of ships in one season being different with at most one from the number of ships in the previous season. Therefore, the agent will evolve a plan which includes where to send its ships but also when to buy, sell, or keep constant its number of ships.

We have run the GA under this model for the four cases $(a \div d)$ described in the previous section and compared the results with the results of the GAs where the companies were not allowed to trade ships but the resources either allowed for extra fishing capacities or were too weakly populated to support all ships in the system. In all cases, the ecologically oriented companies developed good plans for buying and selling ships, namely buy ships when the resources were flourishing and sell ships when the resources were about to reach the minimum ecological level. On the contrary, the profit-oriented companies had the tendency to buy more ships or, alternately, not to sell ships but attempt to exhaust the resources. The results obtained when running the GA for 4 companies, 4 seasons, case (a), trading with the shipyard, and flourishing resources are presented in Figure 5.

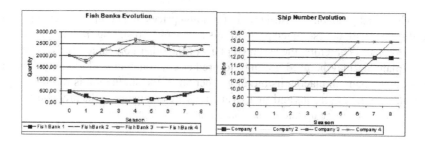

Fig. 5. Results of plan execution when trading ships

The Auction Agent is managing the buying and selling ships to or from the ship-yard or in an auction. The AA keeps a table of indicative prices, *Indic_Price*, for each type of ships in the system. Ships are sold to the shipyard at a price of $0.75 *$ *Indic_Price* or bought from the shipyard (new ships) at a price of $2.25 * Indic\_Price$ Ship prices in an auction may have one of the following values: $\{1, 1.25, 1.5, 1.75, 2\} *$ *Indic_Price*. Thus, the companies have the possibility to trade ships with other companies and obtain a higher price for a sold ship or buy a ship at a lower price that the one of the shipyard. The Auction Agent runs a Dutch auction [9] in which the seller continuously lowers the price until one of the bidders takes the item at the current price. We have selected a Dutch auction because it is efficient in real time and there is no dominant strategy for it.

The bidding strategy of buying companies is inspired from the one proposed [12] as strategy of 1-level agents in a market environment in which agents analyze the past behavior of other agents and try to predict their buying price preferences. A company c wanting to buy a ship determines its bidding price $p*$ using the formula in Eq.1, where $P_{win}(p)$ is the probability of winning the auction with the bid p and, $EstP$ is the estimated profit. , with

$$p* = \arg\max_{p \in P}(EstP - p) * P_{win}(p), \text{ with } P_{win}(p) = \prod_{i \in Comp} P^i_{higher}(p), \quad (1)$$

$P^i_{higher}(p)$ is the probability of bidding higher with bid p than company i and is computed as shown in Eq.2.

$$P^i_{higher} = \begin{cases} \sum_{p'} P(p, p', i), & \text{if comp } i \text{ bids} \\ 1 & \text{otherwise} \end{cases} \quad (2)$$

$P(p, p', i)$ is the probability of winning the auction with bid p if company i bids p' and is defined as shown in Eq.3, where $n(i, p')$ is the probability that company i will bid p'.

$$P(p, p', i) = \begin{cases} 0 & \text{if } p' \geq p \\ n(i, p') & \text{if } p' < p \end{cases} \quad (3)$$

The initial probability distribution of bidding prices of a company was drawn in relationship with the estimated profit of this company as resulting from its presumed evolved plans. Therefore, the assumed plans of the other companies evolved by genetic planning may be used not only to predict environment evolution but also to guide a company in placing its bid in an auction. If the simulation is run for a significant number of seasons and enough auctions are taking place, the probabilities of bidding prices will be updated based on actual accumulated data.

5 Related approaches

Some multi-agent approaches to solve this problem have been proposed in the literature, e.g., [11]. One category uses global distributed planning, in which one agent is elected the planner and is responsible for developing the plans of all companies. These approaches are limited as they evolve a centralized solution and suppose agents are willing to accept global planning. Another category of approaches tries to allocate resources to agents, either by monopolies in which the rights for a resource is given to an agent, or by privatization, in which the resource is sold to an agent. Both cases have drawbacks, as resources can be ruined by competition for monopolies or because not all the resources can be privatized. A third category comprises purely distributed solutions based on self-interested agents that agree on a mutual coercion to prevent the collapse of resources and negotiate the access rights to these resources [1] or that include a game manager which is keen to preserve the resources and negotiate the level of fishing with the companies [5].

There is also some significant work in the domain of using GAs for multi-agent strategy evolution. In cooperative coevolution [8], several populations are evolved independently but the fitness of an individual in one population depends on the merits of the individuals in other populations. In competitive co-evolution [10], individuals from different populations compete for fitness in a shared environment. In [2] a GA solution to the Tragedy of Commons is used to generate an optimized agent society that can avoid social dilemmas, each individual representing the entire agent society; in this case, the GA selects the society that avoids the social dilemma instead of trying to solve the dilemma. In [7] the authors model a social dilemma problem by assuming that each population member corresponds to an agent and investigate some evaluation schemes that are able to avoid the social tragedy. However, they are proposing these schemes for an abstract, simplified version of the tragedy of commons in which there is only one resource with a critical load jointly accessed by the agents.

Compared to those GA approaches, we have taken the approach of [8], we have adapted it to self-interested agents that have their own goals but, in the same time, the common interest of preserving the resources. Besides, we have developed a detailed model of the planning process for a particular application, the FishBanks game, and investigated how different aspects of planning, including planning for different resource exploitation facilities, may be modeled in a GA manner. Preliminary results of our research were reported in [3] but were limited to one genetic representation and there was no mechanism for ship trading.

6 Conclusions

We have presented a multi-agent system with self-interested agents aimed to solve the Tragedy of Commons problem, namely the rational exploitation of natural renewable resources. The agents use genetic algorithms to independently develop their plans and to model the unpredictable world in which they live, achieving thus a sort of implicit cooperation to avoid the overexploitation of resources. The GA solutions we developed account for both fishing plans and trading plans. As opposed to other approaches to the Tragedy of Commons problem, in which the planning process is performed partially or entirely by a centralized authority, the multi-agent solution we propose is inherently distributed.

We have proposed several genetic representations of plans and we have studied them under different assumptions of company agent's behavior and environment features, and we have validated the assumptions under which our models provide a solution to the problem. We have shown that, if all companies are ecologically oriented or if there are some ecologically oriented companies knowing the profile of the others, the agents are able to preserve the resources even in case of lack of knowledge of other companies' plans. We have also validated our model for the case in which companies are allowed to buy and sell ships and we have shown that the GA planning is able to adequately model this process.

Our future work is oriented towards several aspects. First, the Environment Agent may act as one of the planning agents in the system and develop its own prediction of the environment evolution. The EA will check for stagnation of the ecosystem and, following the model proposed in [8], will genetically evolve new species. If new species are evolved then the EA will call new companies to enter the system, aiming thus to increase overall performance. A second direction of research is oriented towards using GA to predict bid prices in auctions and to correlate these evolved prices with the probability model of the other agents presented in Section 4. Third, we want to endow our agents with a GA which will be able to learn other companies' profiles as the Tragedy of Commons problem is best solved in our system when the profiles of the companies are known.

References

[1] Antona, M., F. Bousquet, C. Le Page, J. Weber, A. Karsenty, and P. Guizol. Economic theory and renewable resource management. Lecture Notes in Artificial Intelligence 1534: p.61-78, MABS 1998, editors: J. Sichman, R. Conte and N. Gilbert.

[2] Arora, N. and S. Sen. Resolving social dilemmas using genetic algorithms: Initial results. In Proceedings of the 7th International Conference on Genetic Algorithms, Morgan Kaufman Publishers, 1997, p.689-695.

[3] Florea, A., E. Kalisz, and C. Carabelea. Genetic prediction of a multi-agent environment evolution. In American Institute of Physics Conference Proceedings No.573 - Computing Anticipatory Systems, 2001, p.217-224.

[4] Hardin, G. The tragedy of commons. Science, 162 (1968), p.1243-1248.

[5] Kozlak, J., Demazeau, Y. and Bousquet, F. Multi-agent system to model the Fish Banks game process. In Proceedings of the 1st International Workshop of Central and Eastern Europe on Multi-Agent Systems, June 1-4, St. Petersburg, 1999, p.154-162.

[6] Meadows, D.L., Fiddaman, T. and Shannon, D. Fish banks, LTD. - Game Administrator's Manual. Laboratory for Interactive Learning, Institute for Policy and Social Science Research Hood House, University of New Hampshire, Durham, USA, 1993.

[7] Mundhe, M. and S. Sen. Evolving agent societies that avoid social dilemmas. In Proceedings of GECCO-2000, p.809-816.

[8] Potter, M.A. and De Jong, K.A. Cooperative coevolution: An architecture for evolving coadapted subcomponents. Evolutionary Computation 8(1): MIT Press, 2000, p.1-29.

[9] Rosenschein, J.S. and G. Zlotkin. Rules of Encounter, MIT Press, 1994.

[10] Rosin, C.D. and R.K. Belew. Methods for competitive co-evolution: Finding opponents worth beating. In Proceedings of the 6th International Conference on Genetic Algorithms, Morgan Kaufman Publishers, 1995, p.373-382.

[11] Turner, R.M. The tragedy of the commons and distributed AI systems. In Working Papers of the 12th International Workshop on Distributed Artificial Intelligence, 1993, p.379-390.

[12] Vidal, J.M. and E.H. Durfee. The impact of nested agent models in an information economy. In Proceedings of the 2nd International Conference on Multiagent Systems, AAAI Press, 1996, p.377-384.

Discrete Variable Structure Control Design based on Lamarckian Evolution

Leandro dos Santos Coelho[1] and Renato A. Krohling[2]

[1] Pontifícia Universidade Católica do Paraná, Laboratório de Automação e Sistemas, Rua Imaculada Conceição, 1155, CEP 80215-030, Curitiba, PR, Brazil `lscoelho@rla01.pucpr.br`

[2] Universidade Federal do Espírito Santo, Departamento de Engenharia Elétrica, Campus de Goiabeiras, C.P. 01-9001, CEP 29060-970, Vitória, ES, Brazil `renato@ele.ufes.br`

Summary. Variable structure control (VSC) technique is an important approach to deal with industrial control problems and is an area of very great interest. This paper presents a self-tuning discrete-time VSC design where the optimization task of controller parameters is realized by a new Lamarckian evolution approach based on evolutionary programming combined with a direct search method. Simulation results on a continuous stirred tank reactor, that presents open-loop unstable dynamic and nonlinear behaviors, are shown.

1 Introduction

During the last decade, control scientists and experts have been thinking and given ideas in order to explore future directions of advanced control theory and its applications. An advanced methodology for control design is the structure variable control (VSC) with sliding mode.

This paper presents a self-tuning [1] discrete-time variable structure control design [2, 3, 4] using an optimization method based on Lamarckian evolution [5] or memetic algorithm concepts [6]. The new approach of Lamarckian evolution is realized through of an evolutionary programming (global optimization) combined with a direct search Nelder-Mead simplex method (local optimization). Simulation results are carried out in a stirred tank reactor that presents open-loop unstable dynamic and nonlinearities.

2 Variable structure control systems

This section presents discrete type variable structure control (quasi-sliding mode control) design proposed by Furuta et al. [2], Furuta [3], and Lee and Oh [4] and its application to adaptive control with estimated parameters by a recursive least-squares (RLS) algorithm [1].

The design of Furuta is employed in second order processes with disturbances and unmodeled dynamics. The mathematical model of process is represented by equation of ARMA (Auto-Regressive Moving Average) type:

$$A(q^{-1})y(t) = q^{-d}B(q^{-1})u(t) + \xi(t) \tag{1}$$

with $A(q^{-1}) = 1 + a_1 q^{-1} + ... + a_{na}q^{-na}$ and $B(q^{-1}) = b_0 + b_1 q^{-1} + ... + b_{nb}q^{-nb}$ where $A(q^{-1})$ and $B(q^{-1})$ are polynomials in the delay operator q^{-1}, d is time delay, $y(t)$ and $u(t)$ are output and control variables, respectively, and $\xi(t)$ represents an uncorrelated random sequence. The RLS algorithm is utilized to update estimates of parameters $\hat{\theta} = \{\hat{a}_1, \hat{a}_2, \hat{b}_0, \hat{b}_1\}$ of the ARMA model of second order. The definition of the sliding hypersurfaces is given by:

$$s(t+1) = e(t+1) + k_1 e(t) + k_2 e(t-1) = 0 \tag{2}$$

and the error is given by equation

$$e(t) = y(t) - y_r(t) \tag{3}$$

where k_1 and k_2 are determined so that the error is stable on the hypersurface. The control signal is chosen to be in form of

$$\begin{bmatrix} e(t) \\ e(t+1) \end{bmatrix} = \begin{bmatrix} 0 & 1 \\ -k_2 & -k_1 \end{bmatrix} \begin{bmatrix} e(t-1) \\ e(t) \end{bmatrix} + \begin{bmatrix} 0 \\ 1 \end{bmatrix} v'(t) \tag{4}$$

$$v'(t) = e(t) + k_1 e(t-1) + k_2 e(t-2) + \begin{bmatrix} f_1 & f_2 \end{bmatrix} \begin{bmatrix} e(t-1) \\ e(t) \end{bmatrix} \tag{5}$$

where the component $\begin{bmatrix} f_1 & f_2 \end{bmatrix} \begin{bmatrix} e(t-1) \\ e(t) \end{bmatrix}$ is the switching term of equation, and the component $e(t) + k_1 e(t-1) + k_2 e(t-2)$ driven the state along the sliding hypersurfaces. A relevant definition is a positive definite function given by

$$V(t) = \frac{1}{2}s(t)^2 \tag{6}$$

and the definition of $\Delta s(t+1)$ as the difference

$$\Delta s(t+1) = s(t+1) - s(t) \tag{7}$$

From equation (7) is obtained the equation

$$V(t+1) = V(t) + 2s(t)\Delta s(t+1) + \Delta s(t+1)^2 \tag{8}$$

The control objective will be to make to decrease along the switching hypersurfaces. From equation (8), the following condition is obtained:

$$s(t)\Delta s(t+1) < -\frac{1}{2}[\Delta s(t+1)]^2 \tag{9}$$

Using the equation (7), the control signal have the follow equations:

$$\Delta s(t+1) = [f_1 \ f_2] \begin{bmatrix} e(t-1) \\ e(t) \end{bmatrix} \tag{10}$$

$$
\begin{aligned}
\Delta s(t+1)s(t) &= f_1 e(t-1)s(t) + f_2 e(t)s(t) \\
&< -\delta_1 f_0 - \delta_2 f_0 \\
&< -\frac{1}{2}\left[f_0^2 e(t-1)^2 + 2f_0^2 e(t-1)e(t) + f_0^2 e(t)^2\right]
\end{aligned} \tag{11}
$$

where

$$\delta_i = \frac{1}{2}[f_0 e(t+i-2) + f_0 |e(t-1)||e(t)|], \quad f_0 > 0, \ i = 1,2 \tag{12}$$

$$f_i = \begin{cases} f_0 & \text{if } e(t+i-2)s(t) < -\delta_i \\ 0 & \text{if } e(t+i-2)s(t) < \delta_i \ , \quad f_0 > 0, \ i = 1,2 \\ -f_0 & \text{if } e(t+i-2)s(t) > \delta_i \end{cases} \tag{13}$$

The cost function J that will optimized is

$$J = p\left[y(t+1) - v(t) - v'(t)\right]^2 + r\left[u(t) - u(t-1)\right]^2 \tag{14}$$

The following control law makes the cost function (14) minimal. Using the estimated parameters the above control law becomes:

$$
\begin{aligned}
u(t) = \frac{1}{\widehat{b}_1 + r}\{&y_r(t+1) - \widehat{a}_1 y(t) - \widehat{a}_2 y(t-1) - \left[\widehat{b}_2 - r\right]u(t-1) \\
&- [1 - k_1]e(t) - [k_1 - k_2]e(t-1) - \\
&k_2 e(t-2) - [f_1 \ f_2] \begin{bmatrix} e(t) \\ e(t-1) \end{bmatrix}\}
\end{aligned} \tag{15}
$$

2.1 VSC design based on Lamarckian evolution

The proposed VSC design presents two stages. In the first, the identification of nonlinear process by RLS algorithm is realized. In the second, the optimization of VSC design parameters is considered. The design procedure uses an off-line configuration to get the process model and the model validation is realized on-line with the simulated process. To implement discrete-time VSC based on Lamarckian evolution optimization, three important problems are considered: (i) choice of dynamic sliding surface for the process, (ii) computation of the discrete-time dynamic sliding surface variable, and (iii) self-tuning of the switching control magnitude to reduce chattering.

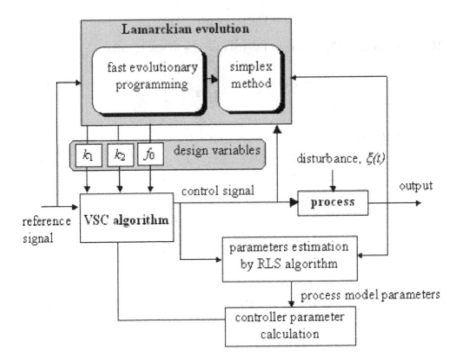

Fig. 1. Optimization procedure of VSC design by Lamarckian evolution.

The proposed optimization procedure of Lamarckian evolution of VSC is presented in figure 1.

2.2 Fast evolutionary programming

Evolutionary programming is a stochastic optimization strategy similar to genetic algorithms, which placed emphasis the behavioral link between an individual and its offspring, rather than seeking to emulate specific genetic operators as observed in nature. The evolutionary programming operates as follows: The initial population is selected at random and is scored with respect to a given cost function. Offspring are created from these parents through random mutations, i.e., each component is usually perturbed by a Gaussian random variable with mean zero and an adaptable standard deviation term. It uses probability transition rules to select generations. Selection is based on a probabilistic tournament where each individual competes with other individuals in a combined population of the old generation and the mutated old generation [7].

The fast evolutionary programming approach uses Cauchy mutations [8] and it is employed in this work. The relationship between the classical evolutionary programming using Gaussian mutation and the fast evolutionary programming using Cauchy mutation is analogous to that between classical simulated annealing and fast simulated annealing. Evolutionary programming uses generation of random values with normal distribution for Gaussian mutation, where

$$p(y)dy = \frac{1}{\sqrt{2\pi}} e^{\frac{-y^2}{2}} dy \tag{16}$$

The mutation operator of evolutionary programming with Cauchy distribution presents a one-dimensional Cauchy density function centered at the origin and defined by

$$f_t = \frac{1}{\pi} \frac{t}{t^2 + x^2}, \quad -\infty < x < \infty \tag{17}$$

where $t > 0$, is the scale parameter. The corresponding distribution function is

$$F_t(x) = \frac{1}{2} + \frac{1}{\pi} \arctan\left(\frac{x}{t}\right) \tag{18}$$

The variance of the Cauchy distribution is infinite. Many studies have indicated the benefits of variance are increased in Monte-Carlo algorithms.

2.3 Simplex method

Direct search method was first proposed in the 1950s and continued been assessed at a steady-rate during the 1960s. The most famous simplex-based direct search method is the Nelder and Mead method [9].

One of basic concepts of the simplex optimization method is that from the current simplex set of points (solutions) a new point is constructed. The Nelder and Mead method is based on creating a sequence of changing simplices, but deliberately modified so that the simplex 'adapts itself to the local landscape'. Operations of this method are: reflection, expansion, and contraction. Details and improvements of classical simplex method can be founded in Wright [10].

2.4 Hybrid method of Lamarckian evolution

One of the disadvantages of evolutionary programming in solving some high dimensional optimization problem is its slow convergence to a good near optimum. The Lamarckian evolution approach proposed in this paper consists of a fast evolutionary programming combined with a simplex method for improved convergence rates and better performance for parameters tuning of VSC.

Evolutionary programming produces the offspring of the next generation through the evaluation function to determine its fitness. This procedure can benefit from the advantages of Lamarckian theory. By letting some of the organism's "experiences" be passed along to future organisms. Following a Lamarckian approach, first would try inject some "smarts" into the offspring organism before returning it be evaluated. A traditional hill-climbing routine could use the offspring organism as a starting

point and perform quick, localized optimization. The hill-climbing procedure is realized by simplex method.

The adopted criteria for Lamarckian evolution is the realization of local search
after of fitness evaluation of fast evolutionary programming in each generation of w%
best fitness values of population members. The Lamarckian optimization procedure
is presented in figure 2.

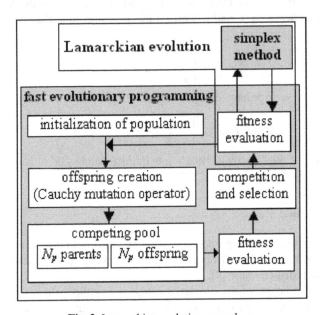

Fig. 2. Lamarckian evolution procedure.

3 Simulation results

3.1 Description of case study

The case study of a continuous stirred tank reactor (CSTR) control is presented in
this work. The dynamic equations of CSTR represent the nonlinear process are given
by:

$$\frac{dx_1}{dt} = -x_1 + D_a (1 - x_1) e^{\frac{x_2}{1 + \frac{x_2}{\phi}}} \tag{19}$$

$$\frac{dx_2}{dt} = -(1 - \beta)x_2 + BD_a (1 - x_1) e^{\frac{x_2}{1 + \frac{x_2}{\phi}}} + \beta u \tag{20}$$

$$y = x_2 \tag{21}$$

where x_1 and x_2 represent the reagents concentration (dimensionless) e reactor temperature, respectively. The control input, u, is the temperature of the cooling jacket surrounding the reactor. The physical constants are D_a, φ, B and β which represent the Damköhler number, the activation energy, reaction heat and the heat transfer coefficient, respectively. The nominal parameters of the system are $D_a = 0.072$, $\varphi = 20$, $B = 8$ and $\beta = 0.3$. In this case the process exhibit the open-loop unstable behavior [11].

The simulation of dynamic behavior of reactor is realized by conversion of the equation (19), (20) and (21) for the discrete equations system, by the utilization of Euler method [12]. Consequently, the resultants equations are:

$$x_1(t+1) = x_1(t) + T_s \left[-x_1(t) + D_a(1 - x_1(t)) e^{\frac{x_2(t)}{1 + \frac{x_2(t)}{\varphi}}} \right] \tag{22}$$

$$x_2(t+1) = x_2(t) + T_s \left[-(1-\beta)x_2(t) + BD_a(1 - x_1(t)) e^{\frac{x_2(t)}{1 + \frac{x_2(t)}{\varphi}}} + \beta u(t) \right] \tag{23}$$

$$y(t) = x_2(t) \tag{24}$$

where T_s denotes the sampling time (adopted) and t is the t-th control step. The relation between the system output, y, and control input, u, can be obtained by substitution of the equation (23) in the equation (24) with $t = t+1$:

$$y(t+1) = x_2(t) + T_s \left[-(1-\beta)x_2(t) + BD_a(1 - x_1(t)) e^{\frac{x_2(t)}{1 + \frac{x_2(t)}{\varphi}}} + \beta u(t) \right] \tag{25}$$

where the input signal is the control signal, $u(t)$, and $y(t)$ is the process output.

Simulation results of VSC design

The effectiveness of proposed self-tuning VSC based on Lamarckian evolution optimization is demonstrated through controlling an open-loop unstable nonlinear CSTR.

- Prediction model: The prediction model uses $na = nb = 2$. The parameters of polynomials $A(q^{-1})$ and $B(q^{-1})$ are estimated using RLS algorithm. The estimated parameters are initialized in simulation with $\hat{\theta} = \{\hat{a}_1, \hat{a}_2, \hat{b}_0, \hat{b}_1\} = \{0.2, 0.2, 0.2, 0.2\}$. The diagonal of the initial covariance matrix is set to 1000.
- Reference trajectory: The desired reference signal is given by $y_r(t) = 1.0$ for the samples 1-30, $y_r(t) = 5.0$ for the samples 31-60, and $y_r(t) = 2.0$ for the samples 61-90.

- Objective function: The objective function of VSC is given by equation (14) and the control law is set to the equation (15).
- Optimization procedure: The Lamarckian evolution methodology is used in optimization of k_1, k_2, and f_0 of VSC. In the sequel it illustrates the main features of the Lamarckian evolution employed:

(i) *Fitness function*: The fitness function to be maximized is represented by equation, $fitness = \frac{k_s}{1+J_{ga}}$, where the scale coefficient is $k_s = 15$. The performance index (same equation (14)) is given by $J = p\left[y(t+1) - v(t) - v'(t)\right]^2 + r\left[u(t) - u(t-1)\right]^2$ with $p = 1$.

(ii) *Constraints for control signal*: $-10 \le u(t) \le 10$.

(iii) *Solution structure*: The optimization variables k_1, k_2, and f_0 are coded by as real variables. The adopted population size is set to 50 individuals.

(iv) *Design parameters of Lamarckian optimization*: The selection of fast evolutionary programming uses 30 individuals for competition. The optimization procedure of simplex method is set for 30% of best individuals of population of fast evolutionary programming in each generation.

(v) *Stopping criterion*: The adopted termination criterion is the number of generations is equal to 100. In this case, the number of experiments (runs) for each design is set to 10. Table 1 summarizes the performance and design parameters of VSC optimized by Lamarckian evolution. Figure 3 shows the performance of best VSC design for 10 runs using Lamarckian evolution.

Table 1: Simulation results for VSC.

r	k_1	k_2	f_0	J_{ga}
0.01	0.5548	1.2250	1.8479	56.7661
0.10	3.3792	0.1615	1.4010	47.0415
0.20	4.8182	0.6283	2.2300	49.5692
0.30	1.6419	0.4585	2.4665	48.4104
0.40	2.3040	0.4868	2.9319	47.9655
0.50	7.1811	2.3324	2.6677	55.7353
0.60	13.2056	5.4040	3.4017	52.2271
0.70	13.3710	2.2271	4.2988	51.4571
0.80	14.8549	2.8938	4.6854	50.9188
0.90	14.4830	2.5882	4.6179	55.2627
1.0	20.0674	3.4316	5.7853	55.3929

The VSC designs for J_{ga} with r set to 0.50, 0.60, and 0.90 present an adequate performance in control of the process. In these designs, the control design presented behaved well with a small overshoot without a deterioration of the quality of the control signal.

However, the performance of controllers was affected by nonlinearity of CSTR process mainly for reference signal $yr = 5.0$ (samples 31-60). Furthermore, the controller designs (for $r = 0.50$, 0.60, and 0.90) obtained fast response, reasonable control activity, and good setpoint tracking ability. The best result was for $r = 0.90$ and it is presented in figures 3 and 4. The good performance indicated by the VSC confirms the usefulness and robustness of the proposed method for practical applications.

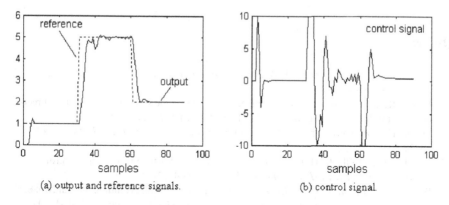

(a) output and reference signals. (b) control signal

Fig. 3. Best results of VSC design based on Lamarckian evolution for $r = 0.9$.

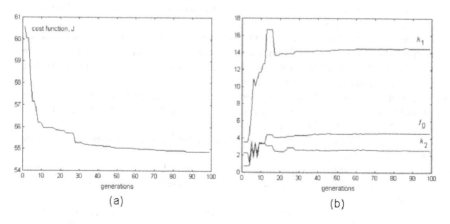

(a) (b)

Fig. 4. (a) Cost function convergence of best individual of population for $r = 0.9$, (b) Parameters convergence of best individual of population for $r = 0.9$.

4 Conclusions

This paper presented the development of a Lamarckian evolution method to the VSC design. In the paper, the effectiveness of the proposed control schemes were shown in simulations of a CSTR process. The utilization of Lamarckian approach avoids the tedious manual trial-and-error procedure and it presents robustness in tuning of VSC design parameters.

The aim of future works is to investigate the use of Lamarckian evolution combined with other computational intelligence methodologies, such as fuzzy systems and neural networks for multivariable processes optimization.

References

1. Wellstead, P. E. and Zarrop, M. B. (1991) *Self-tuning systems: control and signal processing*, Chichester: Wiley
2. Furuta, K., Kosuge, K., and Kobayashi, K.(1989) VSS-type self-tuning control of direct-drive motor, In *Proceedings of IECON*, Philadelphia, PA, pp. 281-286
3. Furuta, K. (1993) VSS type self-tuning control *IEEE Trans. Industrial Electronics* 40(1), pp. 37-44
4. Lee, P. -M. and Oh, J. -H. (1998) Improvements on VSS-type self-tuning control for a tracking controller *IEEE Trans. Industrial Electronics* 45(2): 319-325
5. Whitley, D., Gordon, V. S., and Mathias, K. (1994) Lamarckian evolution, the Baldwin effect and function optimization. In *Parallel Problem Solving from Nature*, PPSN-II, Davidor, Y., Schwefel, H. P., and Manner, R. (eds.), Springer-Verlag, pp. 6-15
6. Moscato, P. and Norman, M. G. (1992) A 'memetic' approach for the traveling salesman problem — implementation of a computational ecology for combinatorial optimisation on message-passing systems. In *International Conference on Parallel Comp. and Transputer Applications*, IOS Press
7. Coelho, L. S., Coelho, A. A. R., and Krohling, R. A. (2002) Parameters tuning of multivariable controllers based on memetic algorithm: fundamentals and application. In *17th IEEE International Symposium on Intelligent Control*, ISIC'02, Vancouver, British Columbia, Canada (accepted for publication)
8. Yao, X. and Liu, Y. (1996) Fast evolutionary programming. In *Proceedings of the 5th Annual Conference on Evolutionary Programming*, Fogel L. J., Angeline P. J., and Bäck, T. (eds), San Diego, CA, MIT Press, pp. 451-460
9. Nelder, J. A. and Mead, R.(1965) A simplex method for function minimization. *Computer Journal* 7: 308-313
10. Wright, M. H. (1996) Direct search method: once scorned, now respectable. In *Proceedings of the Dundee Biennial Conference in Numerical Analysis*, Addison Wesley, UK, pp. 191-208
11. Chen, C. -T. and Peng, S. -T. (1997) A nonlinear control strategy based on using a shape tunable neural controller *Journal of Chemical Engineering of Japan* 30(4): 637-646
12. Åström K. J. and Wittenmark, B. (1984) *Computer controlled system: theory and design*, Prentice-Hall, NJ

GA-P Based Search of Structures and Parameters of Dynamical Process Models

López, A.M.[1], López, H.[1], and Sánchez, L.[2]

Oviedo University, Spain
[1] Electrical Engineering Department. C. Universitario de Viesques. Edificio 2, 33204, Gijón, Asturias. e-mail: [antonio,hilario]@isa.uniovi.es
[2] Computer Science Department. C. Universitario de Viesques. Edificio 1, 33203, Gijón, Asturias. e-mail: luciano@lsi.uniovi.es

Summary. The most effective approaches for evolutionary identifying dynamical processes depend on iterative trial-error searches in a hierarchical fashion: a new structure is proposed first; then, its set of parameters is numerically determined, and the process is repeated until a model accurate enough is found.

Canonical Genetic Programming has been used to automate this search; but its output can be difficult to interpret. Because of this reason, the use of hierarchical learning methods, that combine GP search of structures with deterministic optimization algorithms, has been proposed. We will show in this paper that the output of such methods can be further improved with non hierarchical algorithms. In particular, we will show that the use of GA-P improves the interpretability of the models and does a better model search than previous approaches.

Keywords: GA-P algorithms, Genetic Programming, System Identification, Hierarchical Models .

1 Introduction

Most of the evolutionary methods for system identification from sampled data focus in nonlinear state space-based models. For this kind of models, the objective of the learning process is the production of a set of difference equations defining the dynamics of the process. Unfortunately, for practical purposes, a set of equations that relates all state variables between them is hard to manage in all but small sized problems. Modular representations are usually preferred, because they allow to determine groups of variables affected by specific parameters.

Genetic Programming has been applied to learn such modular models. One of the first examples was given in [9], where a structured Genetic Algorithm, in a tree based representation, is used. The set of functions that was proposed contained only two-input quadratic functions, which are not the building blocks that control engineers expect to find in structured models. Some implementations nearer to usual practice

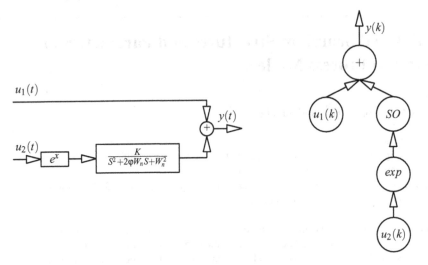

Fig. 1. Block diagram representation of a system (left) and its tree based representation (right). "SO" stands for "Second Order" and "*exp*" for "exponential function"

can be found in [2, 4, 5, 6, 10, 19, 23] and other, less common approaches to model the dynamics of a system, are described in [7, 17, 24]. Most of these schemes introduce dynamic considerations by means of extended terminal sets, that include either input and input-output delayed variables.

One of the most complete methods is described in SMOG [15, 16]. The problem is addressed there as a search of a diagram block based representation of a model of the process in a tree codification (see Fig. 1). The function set used includes continuous time blocks defined in the domain S, making the dynamical considerations intrinsic to the search. Recently, a similar approach has been used for the induction of process controllers in [11].

Under the considered approach (see Fig. 2), hierarchical evolutionary algorithms are applied: canonical GP is used for the evolution of model structures and combined with deterministic numerical optimization methods (Hooke and Jeeves algorithm) for parameter tuning. An iterative search of structure and parameters is done: each model considered is parametrically tuned by means of Hooke-Jeeves algorithm as a previous step to fitness evaluation. Genetic operators defined for evolution affects only the structure of the models.

We will show in this paper that, according to our experimentation, better results can be obtained if a new representation and a new set of genetic operators are used. The representation that is proposed in this paper is adapted from an idea first proposed in [8], and shares characteristics with GA and GP algorithms, being able to search in parallel in both structure and parameter spaces.

The focus will be put not only at the capabilities of the solutions to reproduce the sampled data used for training or validation. They will be also structurally com-

```
1. Initialize random population of models.
2. Tune parameters of models (Hooke-Jeeves algorithm).
3. Calculate fitness.
4. Selection of models and application of genetic operators.
5. Go to 2).
```

Fig. 2. SMOG evolution. Canonical GP is used for structural search and Hooke-Jeeves method is used for parameter tuning

pared with a known model for the target system. This way, they can be analyzed as explaining methods of the underlying relationships in the data.

2 Structure of the Paper

The outline of this paper is as follows: in Sect. 3, the scope of application of this method is introduced. In Sect. 4, the parallel search of parameters and structure done by the GA-P algorithm is described. Then (Sect. 5) an experimental validation of our proposal is done, modeling both a synthetic and a real process and comparing the results with those obtained with previous works. The paper finishes (Sect. 6) with the concluding remarks and future work.

3 Scope of Application

Our interest is focused over a class of physical systems involving common non-linear features, to which conventional methods are hardly applicable. Being a GP based modeling approach used, the definition of the functional set will define the scope of application of the algorithm. The GP will evolve a set of diagram block representations of the process. A diagram block is, in turn, a series, parallel or feedback association of subsystems. Series association is intrinsic in GP. Parallel association will be allowed by means of arithmetic operators, such as + and -, and feedback representation will be allowed by means of an special operator [1] described next.

Regarding the catalog of subsystems, we used only memoryless version of the common non-linear features of physical systems, such as dead zones or saturations [14, 20]. All the dynamic behavior is delegated to linear elements: we include in the function set a reduced group of linear models (first and second order dumped linear subsystems, unitary delay and static gain) such that it is possible to get higher order systems by means of series association.

4 Proposed Algorithm

Most of the approaches to learn models of dynamical processes are based on a hierarchical search. Nevertheless, there is an inherent drawback with the hierarchical

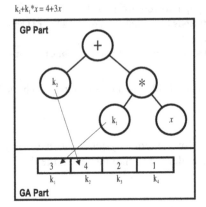

Fig. 3. Representation of a generic individual in GA-P algorithms. Individuals have two parts: a tree based representation and a chain of numerical parameters

learning of models: the searches of the numerical parameters best suited for the structures being produced by GP are, themselves, multi modal problems [1]. Therefore, deterministic methods fall frequently into local minimum points and, as a consequence of this, a good structure can be assigned a low score in the search process. Despite this problem, hierarchical approaches are able to find good models because GP can produce several times the same structure with different initial values for the numerical parameters. Thus, the deterministic algorithm will eventually find the global minimum. But under this context the GP is not only being used to search different structures but also to search different numerical starting points, a problem in which GP is known not to perform too well.

In previous studies [1], we have tried the replacement of Hooke and Jeeves method with a real coded genetic algorithm, obtaining good numerical results. Anyway, such a hierarchical approach is a highly consuming task, because many resources are wasted in the identification of structurally invalid systems. An strategy that does not need the GA to converge before examining a new structure, and that does not discard too soon structures that may be valid, is needed. The GA-P algorithm was selected for the search.

GA-P [8] is an hybrid between genetic algorithms and genetic programming, that was first used in symbolic regression problems. Individuals in GA-P have two parts: a tree based representation and a chain of numerical parameters. Different from canonical GP, the terminal nodes of the tree never store numbers but linguistic identifiers that are pointers to the chain of numbers (see Fig. 3).

The behavior of the GA-P algorithm is mainly due to its crossover operator. Later in this section it will be described in detail how we adapted it to the problem at hand; let us say for the time being that either the tree parts or the chains of parameters may be selected and crossed, thus the GA search of the parameters and the GP search of the structures are being done in parallel. This way, individuals structurally fitted will have more possibilities to undergo an intensive parameter optimization, while those structurally unfitted will tend to disappear. A niche strategy [21] is used in the evolutionary process, preventing the search to fast fall into local minimum points.

Representation

Structure and parameters parts of the representation are defined as follows:

- Structural component. Tree based representation makes it impossible to model a wide set of systems, such as those involving nested or not unitary feedbacks. The reason is that a block diagram is not a tree when it includes feedback, but a directed graph. The proposed representation (see Fig. 4), mixes a link nodes list with ideas from [16] and [22]. A special feedback node is used. Both input and the feedback branches originate in it. The terminal nodes of the feedback branch (marked as "**") are recessive. This way, standard structural modification operators can be applied at any point in the individual to evolve structures.

  It also contains a third link to another node from which the feedback signal will be taken, converting the representation in a graph. This pointed node must be contained in the path between the feedback node and the output node of the system. Otherwise, feedback node looses its significance. This consideration must be present in the creation and modification of individuals as a consequence of structural genetic operators. When an individual does not accomplish this condition after an structural modification, invalid feedback nodes are reinitialized.

  Algebraic loops are neglected by means of the implicit inclusion of a unit delay in the feedback branch. To prevent series associations of delays, dynamic blocks used respond instantly. But, known the fact that physical systems never respond instantly to an excitation, a unit delay is also implicitly linked to the output of the model.

- Parameters component. It contains a vector of values with the parameters of the model to be evolved by the GA component of the algorithm. It is used a real value codification based on [3].

Genetic Operators

Two sets of operators are applied in the evolutionary process:

- Structural Genetic Operators. Subtree crossover [13] and internal crossover [12] are used. Subtree, node and a special operator for feedback mutation operators are also used. This set of operators only affect the structural component of the individuals involved, not the parametric one.

  All of the structural operators act over tree based representations. Therefore, feedback links are inhibited during the process.

- Parameter Genetic Operators. Two structurally identical individuals are selected from the population for each application of this set of operators. They only affect their parameter component, not the structural one. Real based genome crossover operator is defined for the parameters of the model as a random movement of a

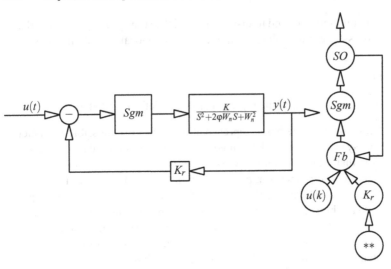

Fig. 4. Block diagram representation of a feedback system (left) and its genetic representation (right.) "SO" stands for "Second Order" and "*Sgm*" for "sigmoid function". Also, "******" stands for a recessive terminal

vector in the direction of the other. After crossover, a mutation, a direct search or both can be applied to the resulting offsprings depending on predefined probabilities. Mutation is defined as a crossover with a randomly generated individual. Direct search is performed by means of Nelder & Mead algorithm [18] run for a few iterations.

5 Numerical Results

To validate our approach, as a first test, an empirical control system of a first order process with a proportional saturated controller and a sensor without dynamics (see Fig. 5(a)) was modelled by means of the defined GA-P strategy. It was also compared with a hierarchical process. Both approaches were stopped after certain number of evaluations of the objective function.

Experiments were repeated 10 times each. Table 1 contains validation errors for each experiment. Observe that GA-P improves slightly the results, but the differences are not significant. The gain with GA-P is in the identified structure (see Figs. 5(b) and 5(c), where the best models obtained by both approaches are shown.) Observe that only little deviations are present in the parameters values, a problem which could be easily solved by the application of more intensive optimization procedure over that structure. In this case, GA-P found exactly the structure of the target model, explaining very well the data relationships. In contrast, the hierarchical method was trapped in a local minimum of the structure. It is only capable of fitting the sampled data.

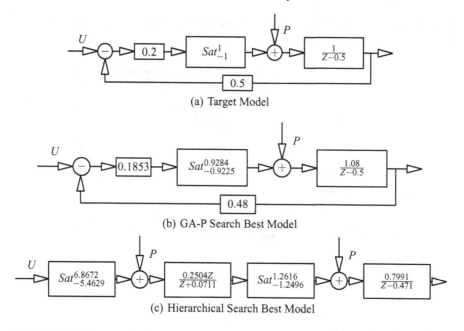

(a) Target Model

(b) GA-P Search Best Model

(c) Hierarchical Search Best Model

Fig. 5. Modelling a synthetic example. Upper part: target model. Central and lower parts: structures of the learned models. "Sat_a^b" stands for "saturation" block with limits in a, b

As a final test, a real system was modelled by means of the proposed scheme. A DC motor was selected, in order to have information enough to contrast the GA-P solution with a known model for the process (usually a first or second order dumped linear subsystem with a non-linear dead zone component).

Experimental conditions were the same as in the preceding section. Table 1 contains the numerical validation errors for each experiment. From it, it can be concluded that the best result was found at experiment 10, shown in Fig. 6. Solution is close to a known model for the system: the search scheme is capable of capturing the most significant relationships in the data. This figure also includes a comparison between the motor and the model responses using a squared input signal. Observe that the behavior is correctly reproduced and the noise is smoothed as expected.

6 Concluding Remarks and Future Work

The identification of nonlinear systems from sampled data is a multimodal problem either in structure and parameter spaces. We have shown that "state of the art" hierarchical learning algorithms can be trapped in these minimum points and be unable to find the right structure in certain cases. We have solved this problem by introducing a parallel evolutive search of parameters and structure that does not waste time optimizing parameters for invalid structures neither discards structures too early.

Table 1. GA-P (left) and hierarchical (center) modelling errors for the synthetic problem. Right: modelling errors for the direct current motor

Experiment	Error		Experiment	Error		Experiment	Error
1	0.00017		1	0.00206		1	0.9196
2	0.0004		2	0.00301		2	0.7755
3	0.00005		3	0.00129		3	0.7354
4	0.00004		4	0.00184		4	0.8433
5	0.00019		5	0.00287		5	0.9223
6	0.00005		6	0.00112		6	0.9259
7	0.00005		7	0.00111		7	1.1809
8	0.00006		8	0.00107		8	1.0134
9	0.00029		9	0.00147		9	1.0976
10	0.00007		10	0.00263		**10**	**0.6933**
Average	0.00014		Average	0.00185			

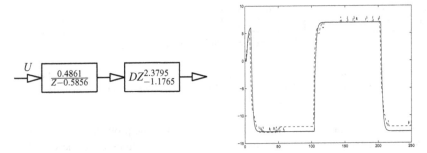

Fig. 6. Modelling of a direct current motor. Left: best model found ("DZ_a^b" stands for "dead zone" block with limits in a, b). Right: Comparison of model (continuous line) and system (plotted line) responses

While being able to process more complex problems than its predecessors, this learning algorithm is not complete. In practical situations we need to be able to incorporate expert knowledge to the system, either in the form of structural restrictions or by means of closed submodels with known expression around which a joint model should be evolved. In a near future, we plan to incorporate a measure of structural quality to the fitness function and use multicriteria evolutionary algorithms to obtain a family of solutions with balanced precision and complexity from which the control engineer can choose.

References

1. H. Lopez A.M. Lopez and L. Sanchez. Graph based gp applied to dynamical systems modeling. *Connectionist Models of Neurons, Learning Processes and Artificial Intelli-*

gence, pages 725–732, 2001.

2. V.M. Babovic, B. Brozkova, M. Mata, V. Baca, and S. Vanecek. System identification and modelling of vltava river system. In *Proceedings of the 2nd DHI Software User Conference*, 1991.

3. L. Davis. *Handbook of Genetic Algorithms*. Van Nostrand Reinhold, New York, NY, 1991.

4. S. Dzeroski, L. Todorovski, and I. Petrovski. Dynamical system identification with machine learning. In *Proceedings of the Workshop on Genetic Programming: From Theory to Real-World Applications*, pages 50–63, Tahoe City, California, USA, 1995.

5. A.I. Esparcia and K.C. Sharman. Some applications of genetic programming in digital signal processing. In *Late Breaking Papers at the Genetic Programming 1996 Conference*, pages 24–31, Stanford University, CA, USA, 1996.

6. G.J. Gray, D.J. Murray-Smith, Y. Li, and K.C. Sharman. Nonlinear model structure identification using genetic programming. In *Late Breaking Papers at the Genetic Programming 1996 Conference*, pages 32–37, Stanford University, CA, USA, 1996.

7. H. Hiden, M. Willis, B. McKay, and G. Montague. Non-linear and direction dependent dynamic modelling using genetic programming. In *Genetic Programming 1997: Proceedings of the Second Annual Conference*, pages 168–173, Stanford University, CA, USA, 1997.

8. L.M. Howard and D.J. D'Angelo. The GA–P: A genetic algorithm and genetic programming hybrid. *IEEE Expert*, 10(3):11–15, June 1995.

9. H. Iba, T. Karita, H. Garis, and T. Sato. System identification using structured genetic algorithms. In *Proceedings of the 5th International Conference on Genetic Algorithms, ICGA-93*, pages 279–286, University of Illinois at Urbana-Champaign, 1993.

10. M.A. Keane, J.R. Koza, and J.P. Rice. Finding an impulse response function using genetic programming. In *Proceedings of the 1993 American Control Conference*, volume III, pages 2345–2350, Evanston, IL, USA, 1993.

11. M.A. Keane, J. Yu, and J.R. Koza. Automatic synthesis of both topology and tuning of a common parameterized controller for two families of plants using genetic programming. In *Proceedings of the Genetic and Evolutionary Computation Conference (GECCO-2000)*, pages 496–504, Las Vegas, Nevada, USA, 2000.

12. K.E. Kinnear. Alternatives in automatic function definition: A comparison of performance. In *Advances in Genetic Programming*, chapter 6, pages 119–141. MIT Press, 1994.

13. J.R. Koza. *Genetic Programming: On the Programming of Computers by Means of Natural Selection*. MIT Press, Cambridge, MA, USA, 1992.

14. P.H. Lewis and C. Yang. *Basic control systems engineering*. Prentice Hall, Englewood Cliffs, New Jersey, 1997.

15. P. Marenbach. Using prior knowledge and obtaining process insight in data based modelling of bioprocesses. *System Analysis Modelling Simulation*, 31:39–59, 1998.

16. P. Marenbach, K.D.. Betterhausen, and S. Freyer. Signal path oriented approach for generation of dynamic process models. In *Genetic Programming 1996: Proceedings of the First Annual Conference*, pages 327–332, Stanford University, CA, USA, 1996.

17. B. McKay, B. Lennox, M.J. Willis, G.W. Barton, and G.A. Montague. Extruder modelling: A comparison of two paradigms. Technical report, Chemical Engineering, Newcastle University, UK, 1996.

18. R. Mead and J.A. Nelder. A simplex method for function minimization. *Computer Journal*, 7(4):308–313, 1965.

19. H. Oakley. Two scientific applications of genetic programming: Stack filters and non-linear equation fitting to chaotic data. In *Advances in Genetic Programming*, chapter 17, pages 369–389. MIT Press, 1994.

20. K. Ogata. *Modern Control Engineering*. Prentice-Hall, Englewood Cliffs, New Jersey, 1996.

21. L.A. Sanchez and J.A. Corrales. Niching scheme for steady state ga-p and its application to fuzzy rule based classifiers induction. *MATHWARE*, 7(2-3):337–350, 2000.

22. K.C. Sharman and A.I. Esparcia. Genetic evolution of symbolic signal models. In *Proceedings of the Second International Conference on Natural Algorithms in Signal Processing, NASP '93*, Essex University, UK, 1993.

23. A.H. Watson and I.C. Parmee. Identification of fluid systems using genetic programming. In *Proceedings of the Second Online Workshop on Evolutionary Computation (WEC2)*, number 2, pages 45–48, 1996.

24. M. Willis, H. Hiden, M. Hinchliffe, B. McKay, and G.W. Barton. Systems modelling using genetic programming. *Computers in Chemical Engineering*, 21:S1161–1166, 1997. Supplemental.

Author Index

Index